Menr

MW01493415

Young Center Books in Anabaptist and Pietist Studies

STEVEN M. NOLT, SERIES EDITOR

MENNONITE FARMERS

A Global History of Place and Sustainability

Royden Loewen

Johns Hopkins University Press

Baltimore

© 2021 Johns Hopkins University Press
All rights reserved. Published 2021
Printed in the United States of America on acid-free paper
2   4   6   8   9   7   5   3   1

Johns Hopkins University Press
2715 North Charles Street
Baltimore, Maryland 21218-4363
www.press.jhu.edu

Library of Congress Cataloging-in-Publication Data

Names: Loewen, Royden, 1954– author.
Title: Mennonite farmers : a global history of place
and sustainability / Royden Loewen.
Description: Baltimore : Johns Hopkins University Press, 2021. |
Series: Young Center books in Anabaptist and Pietist studies | Includes
bibliographical references and index.
Identifiers: LCCN 2020052636 | ISBN 9781421442037 (paperback) |
ISBN 9781421442044 (ebook)
Subjects: LCSH: Agriculture—Religious aspects—Mennonites. |
Communities—Religious aspects—Mennonites. | Mennonite farmers—History. |
Mennonites—History. | Sustainable agriculture.
Classification: LCC BX8115 .R69 2021 | DDC 289.7—dc23
LC record available at https://lccn.loc.gov/2020052636

A catalog record for this book is available from the British Library.

Special discounts are available for bulk purchases of this book.
For more information, please contact Special Sales at specialsales@jh.edu.

CONTENTS

Acknowledgments vii

INTRODUCTION 1

1
Sect and Settler in the North: Plowing Friesland, Iowa,
Manitoba, and Siberia 14

2
Peasant and Piety in the South: Planting Java, Matabeleland,
and Bolivia's Oriente 45

3
Something New under the Mennonite Sun: A Century of
Agricultural Change 75

4
Making Peace on Earth: An Agricultural Faith of
the Everyday 115

5
Women on the Land: Gender and Growing Food in
Patriarchal Lands 145

6
Farm Subjects and State Biopower: Seven Degrees
of Separation 176

7
Vernaculars of Climate Change: Southern Concern,
Northern Complacency 208

8

Mennonite Farmers in "World Scale" History: Encountering
the Wider Earth 232

CONCLUSION 265

Appendixes
A. Methodology 271
B. Seven Points on Earth Interview Questions 275

Notes 279
Bibliography 301
Index 325

My list of acknowledgements for help in the making of this book is long, and the notes below are incomplete. This book, building on the Seven Points on Earth research program, launched in 2013, benefited from the support of a great number of people on five continents. At the top of the list are the 159 farmers—80 from the Global South and 79 from the Global North, 95 men and 64 women—who graciously gave of their time to talk to members of the research team; this generosity often was accompanied by meals and the sharing of personal records and led to lasting friendships. (For a full list of these farmers, see the "Transcribed Interviews" section of the bibliography.)

A close second is the Seven Points on Earth research team, seven hardworking and creative graduate and postgraduate students, all of whom have long since advanced their own careers and publication track records: John Eicher, formerly of Iowa City, presently of Altoona, Pennsylvania; Hans Peter Fast, of Zeist, Netherlands; Susie Fisher, of Gretna, Manitoba; Aileen Friesen, of Winnipeg; Danang Kristiawan, of Jepara, Indonesia; Belinda Ncube, of Bulawayo, Zimbabwe; and Ben Nobbs-Thiessen, of Winnipeg. All brought to the project their particular gifts—basic curiosity, linguistic expertise, empathy to the local, knowledge of the global, utter commitment to completing difficult tasks. Their remarkable program of research is outlined in appendix A. I also acknowledge the valuable work of Steve Dueck, Kerry Fast, Sara Jantzen, Anne Kok, and Piet Visser, who undertook much-needed ancillary research.

A wide array of leaders in academia facilitated the research, writing, and initial knowledge mobilization related to the Seven Points on Earth project. Staff at a variety of archives—including the Brethren in Christ Archives in Mechanicsburg, Pennsylvania, the Bolivian National Archives and Library in Sucre and Legislative Archives and Library in La Paz, the Doopsgezinde Library and Archives in Amsterdam, the Manitoba Archives and the Mennonite Heritage

Archives in Winnipeg, the National Archives of Zimbabwe in Bulawayo, the Iowa Mennonite Museum and Archives in Kalona, the State Archives in Omsk, Russia, and the State Historical Society in Iowa City—provided treasured assistance to the research team. I wish to acknowledge David Ibbetson, president of Clare Hall at Cambridge University and Clare Hall staff, who generously welcomed me as a visiting fellow in both 2013 and 2016, allowing me a concentrated period of time to produce a first draft of the manuscript. I am also grateful to the Canadian Historical Association, the US Agricultural History Society, the US Social Sciences and History Association, the Canadian Mennonite University, the Centre for Transnational Mennonite Studies (University of Winnipeg), the Canada Research Chair in Agricultural History program at the University of Guelph, the Climate Change Workshop at Cambridge University, and the Mennonite World Conference, which facilitated presentations of academic papers based on this research.

In addition, I am most grateful to the many people who assisted in creating conduits into the very heart of the seven communities. In Friesland, Professor Piet Visser introduced Hans Peter Fast and me to the pastors Flora Visser, Simon van der Linden, and Peter Lindeboom and provided me with crucial historical materials. I also acknowledge the "Dutchies" from Vrije Universiteit and elsewhere—Piet Visser, Carel Roessingh, Cor Trompetter, Anna Voolstra, Yme Kuiper, Martine Vonk, and Fernando Enns—for assistance at crucial moments. The artist Yvonne Willims and the farmers Doeke Odinga, Menno de Vries, Margriet Faber, and Jacob van den Hoek generously hosted me in their churches, in their homes, and on their farms.

In Iowa, John Eicher was essential to accessing the community. I thank him and his wife, Julia Wicker, for hosting Mary Ann and me at their home in 2014, as well as the various Mennonite churches of Washington County, who connected John to local farmers. I also thank Franklin Yoder for counsel and Marilyn Preheim, who hosted me in her home. John was especially assisted by Mary Bennett, of the State Historical Society of Iowa.

In Manitoba, Paul Bergman, Joe Braun, and Sean Friesen provided special support and contacts for Susie's fieldwork. I want to thank the experts of the West Reserve—Conrad Stoesz, Hans Werner, Lawrence Klippenstein, Jake E. Peters, Doreen Klassen, Adolf Ens, Eleanor Chornobouy, John Friesen, and Ted Friesen—who have provided their insights over the years. I am deeply grateful to Elmer and Hilda Hildebrandt, in Neubergthal, and Joe Braun, of Altona, for inviting me to share my research with local history buffs. Kevin

Nickel, of Rosenfeld, and Terry Mierau, of Neubergthal, hosted the entire Seven Points on Earth team on their farms.

In Java, I thank Adi Dharma for hosting me in Semarang in 2011 and Utami for guiding me to Margorejo. I am profoundly grateful to Heru Puji Utami and Marina Margaretha Kristiawan, of Jepara, and Pastor Suharto and Bu Suharto, of Margorejo, for hosting me in their homes in 2014, as well as to the farmers Kasmito Kliwon, Krismiati, Retno, Sumarmi, Sutarmi, and Yoga Saptoyo for welcoming me into their homes. I also thank Lawrence Yoder, of Harrisonburg, Virginia, for graciously sharing his unpublished English translation of the Gereja Injili di Tanah Jawa (GITJ) history and Alle Hoekma, of Amsterdam, for steering Danang along a fruitful research path.

In Matabeleland, my colleague Eliakim Sibanda was key. I thank him for introducing me to Connie Nkomo, of Bulawayo, as she and her sons graciously hosted Mary Ann and me in 2003. I'm also grateful to Pastor Sabelo and Sichelesile Ndlovu and their family, as well as to Belinda and Siphatho Ncube and their lovely children, who hosted me in their homes in 2016; special thanks to Zie Nyoni, our driver. In turn, Ben and Lorraine Myers, at Messiah Village, and Alden Braul and Aida Vidal Vega, of Winnipeg, hosted Belinda on her various research trips. The sitting bishop Sinda Ngulube and Mama Susan Ngulube, former bishop Danisa Ndlovu and Mama Treziah Ndlovu, Pastor Mbongeni Ncube and Mama Julia Ncube, Deacon Denny Ndlovu, and John Masuku provided invaluable support for Belinda. Denny and John, as well as Josephine Siziba, Alfred Gumpo, Eldah Gumpo, Eldah Siziba, and Timothy Sibanda, welcomed me to their farms during my visit in 2016. Daryl Climenhaga, of Providence University, Manitoba, generously loaned sources related to Matopo Mission.

In Siberia, Tatiana Smirnova, Olga Shmakina, Paul Toews, and Marina and Walter Unger were fellow collaborators in the "Germans in Siberia: History and Culture" conference in June 2010, an event that connected me to Apollonovka. This conference also introduced me to the historian Peter (Piotr) Epp, of Isul'kul; he and Elena Epp hosted and aided Aileen in her research. David and Anna Epp, Ivan and Anna Dirksen, and Andrei and Valya Pauls also provided Aileen warm support. I also want to thank the Ivan and Elizveta Rosenbach family, of Apollonovka, for hosting me in their home in 2010.

In Bolivia, the Mennonite Central Committee personnel based at Centro Menno, in Santa Cruz, assisted both Ben and me, as well as Kerry Fast, in our fieldwork. Dick Braun and Ramont Harder Schrock took me to Riva Palacio in 2004 and 2012, respectively, introduced me to a number of farmers, and facilitated visits with the schoolteacher Abram Thiessen and Bishop Johann Hamm. The farmers Wilhelm Buhler and Peter Hamm offered particular assistance to Ben during his research. Kennert Giesbrecht, editor of the Canada-based *Mennonitische Post*, was most helpful in connecting our team with Riva Palacio farmers.

At the University of Winnipeg, I wish to acknowledge the many supportive administrators. Andrea Dyck, Sandy Tollman, and Jennifer Cleary provided crucial administrative assistance for the project, while the senior administrators Neil Besner, James Currie, and Glen Moulaisson offered deeply appreciated support for the Mennonite Studies program during these years. Fellow rural, family, world, and oral historians Emma Alexander, Peter Denton, Ryan Eyford, Alexander Freund, James Hanley, Serena Keshavjee, Mark Meuwese, Alexander von Plato, Eliakim Sibanda, Mussie Tesfagiorgis, Janis Thiessen, Sharon Wall, Hans Werner, and other colleagues from the Department of History endured my ceaseless talking about this project. I want to thank my first Mennonites and Environmental History class, in 2016, who produced some of the first research materials for this project and served as an irreplaceable early sounding board: Caltin Dyck, Thomas Epp, Matthew Franchuk, Sabrina Janke, Sara Jantzen, Jen Klassen, Jennifer Logan, Rebecca Penner, Kelly Ross, Christoper Sundby, Joe Warkentin, and Maysen Zelinksky.

I acknowledge the work of the Winnipeg-based filmmaker Paul Plett, of Ode Production, who helped put this story on film in the publicly accessible feature documentary *Seven Points on Earth*, released in 2017. The farmers Jacob van der Hoek, of Friesland, Dave Yoder, of Iowa, Jeremy and Megan Hildebrand, of Manitoba, Hadi Pitoyo, of Java, Hetta Dube, of Matabeleland, Peter Epp, of Siberia, and Peter Hamm, of Riva Palacio, Bolivia, offered insights into faith and farming in their filmed interviews.

I thank the two anonymous readers of the manuscript for their very helpful comments, as well as Steven Nolt, the series editor, and Greg Britton, editorial director at Johns Hopkins University Press, and David Carr, at University of Manitoba Press, for overseeing the binational publication of this book. I am grateful for copyeditor Joanne Allen's deft hand and Cynthia Nolt's careful guidance on photographs. I salute my learned coffee friends—Harold Dueck, Gerald Friesen, Charles Loewen, and David Smucker-Rempel—who listened and offered counsel.

Finally, I wish to acknowledge Mary Ann, my editor, companion, and fellow traveler; our children, Bec, Meg, and Sasha, who loved *Seven Points on Earth*; and our grandchildren, Remy and Kay, who will.

Mennonite Farmers

Introduction

The Mennonite farmers studied in this book are not a homogeneous community of culturally quaint agriculturalists. Rather, they are a disparate group cultivating the soil in seven rural communities around the world, in places profoundly different from one another. Located in seven countries—Bolivia, Canada, Indonesia, Netherlands, Russia, the United States, and Zimbabwe—on five continents, these places are set in starkly distinctive climatic zones, maritime, semiarid, and tropical, among others. Similarly, they represent a variety of cultural constructions; indeed, the names of the farmers alone—Margriet, Gerald, Sumarmi, Joe, Zandile, Vladimir, Abram—reveal particular ethnicities despite common *Mennonite* identities.[1] Together, the climatic conditions and the cultural artifacts have resulted in specific approaches to agriculture. They reflect Wendell Berry's idea of "local culture" in which farming is shaped variously by historical memory, a moral imagination, community loyalty, and the natural environment. Ironically, these local worlds have also been shaped by amorphous global forces.[2] As such they constitute a global history of the Mennonites literally from the ground up; they also offer a detailed look into the disparate everyday challenges of food production around the world.

The story of these Mennonite farmers, and indeed of farmers in every country, makes for an intensely local narrative. I share Joachim Radkau's feeling "that the secrets of history are hidden above all in the microcosms and therefore elude the habitual globetrotters."[3] The microcosms in this book take us not only to seven countries but into the heart of seven communities within them. In order of their founding they include a constellation of Mennonite farms in Friesland, in northern Netherlands; Washington County and neighboring counties in Iowa; the Rural Municipality of Rhineland in Manitoba; the village of Margorejo in Java; Matopo Mission in Matabeleland; Apollonovka in Siberia; and Riva Palacio colony in the Santa Cruz department in Bolivia. Yet, despite the minutiae that compose the stories of these places, I also share Radkau's sense that to understand farmers' everyday life is to fathom something fundamental about global agriculture itself.[4] To my mind, this quotidian rural world reveals the heart of that history in the twentieth century, that is, the arrival of modern agriculture and its remaking of community and environment. Indeed, the far-flung places studied in this book experienced the broad process of modernization—that social response to industrialization introducing standardization, formality, and rationalism—as the foundation of community and economic endeavor.[5] A study of *seven* places in no way accounts for all the variables of agricultural history, but perhaps the common association of the number seven with wholeness or totality, for example, "the sum of the spiritual . . . and the material," signals the overall aim of this book.[6]

Certainly, each of these seven places underwent some form of agricultural transformation during the twentieth century. At that century's beginning, all these farms produced a wide mix of commodities in the name of self-sufficiency, and they reared animals for food, draft power, and land fertilization. At the end, most engaged in some form of specialization, used fossil-fuel-powered machines, and relied on technology writ large—mechanical, chemical, or genetic. But as this book demonstrates, the seven communities experienced these factors in vastly diverse ways. These modern technologies became available at different times in the various countries. Structures of power and credit, as well as national food policies, differed as well, with some governments considerably more hostile than friendly to local concern. Social upheaval—war, civil war, revolution—histories of colonialism, systemic racism, and multiple other factors all augured against homogeneity.

This diversity is apparent even among Mennonite farmers linked by common religious faith. Disparateness colors the varied stories of Margriet, Calvin, Sumarmi, Joe, Zandile, Vladimir, and Abram and many others told in this book. They are historical narratives of ordinary farmers in relationship with particular natural environments in distinctive local communities.

From the Global North we learn of Margriet Faber, of Tzum, Netherlands, who for decades now has risen each morning to direct her dairy cows to the milk parlor from green pastures on a farm she acquired from her father, who taught her how to predict the weather by surveying the horizon of the flat, treeless plain of Friesland. We listen to the soybean farmer Gerald Yoder, of Washington County, Iowa, explaining why he planted GMO soybeans on the rolling, fertile plains of the southeastern corner of the state, thankful that the plow culture of his South German ancestors has ended and the soil now is safe from erosion. We see how Joe Braun turned a pocket of the flat, clay soils of the Rural Municipality of Rhineland, in southern Manitoba, into an organic farm, contesting the economies of scales around him and, in a way, replicating the cultivation practices of his ancestors, settlers from Russia, in the 1870s. We learn how Vladimir Friesen, of Apollonovka, Siberia, disdained the dictates of the Soviet-era state farm, the massive *sovkhoz* encompassing his Mennonite grandparents' once pristine wheat farm bordered by birch forests.

From the Global South, in Margorejo, Java, we hear from Sumarmi, who has overseen rice planting in the warm waters of her land, all the while recalling how her grandparents procured their allotment on a Mennonite mission farm in the shadow of Mount Muria in fields carved out from the jungle next to the Java Sea. We follow the story of Zandile, of Matopo, who has hoed the spring-fed soil made fertile by dung hauled in by donkey-drawn wagon, within the semiarid savannah of Matabeleland, land that Cecil Rhodes once deeded to white US missionaries. And we meet Abram Enns, of Riva Palacio, Bolivia, who tells how he, a horse-and-buggy traditionalist, hired bulldozers to clear-cut a farm in the eastern lowland forests to grow soybeans, an action that led to dust-bowl conditions until he rediscovered the dairy production his immigrant parents had known in Mexico, and before them, his grandparents in Canada.

These stories of Mennonite farmers interacting with nature in their particular places become especially meaningful when they are compared with one another. Indeed, in comparison they provide crucial insights into the nature and stages of twentieth-century global agricultural history. They offer the grist by which to understand J. R. McNeill's broad assertion in his authoritative *Something New under the Sun* that with the coming of the twentieth-century new technologies and economic structures made humans "capable of total transformation of any and all ecosystems," a process "enmeshing farmers" somehow, everywhere.[7] Comparative analysis allows for the unpacking of this complex process. As Philippa Levine argues, comparative history compels us to work "through multiplicities rather than with a single variable" and thus "push past overly simple and frequently binary readings of power to more complex and rewarding analyses."[8] And this analysis can be achieved through a commitment to empirical study. Indeed, at the foundation of such scholarly consideration is the case study, set in specific time periods and geographically bordered locales. Thus, while the environmental historian J. Donald

Hughes acknowledges the need to see natural phenomena that "extend throughout the biosphere, transcending every national frontier," he insists that it is through specific case studies, "well known to the author," that we can understand humans' relationship with their particular ecosystem.[9] Certainly, such studies require scholars to be flexible, recognizing selected communities' distinctive and contingent trajectories. Another leading environmental historian, L. G. Simmons, speaks of empirically based "world scale" approaches to environmental history that "consider and weave together many types of information."[10] They take into account overlapping degrees of orality and literacy and variations in types of government and personal records, sources produced with a wide range of subjectivity.[11] But in the end the case study in comparative perspective sheds light on the enmeshing of the local with the global.

The global story of the Mennonite farmers in this book comprises one set of such case studies.[12] Once a small European religious reform movement emphasizing the everyday ethics of simplicity and nonviolence, the Mennonites have become a global people with a remarkable record of environmental engagement in highly disparate rural settings. Perhaps they began as Swiss and Dutch sixteenth-century urban radicals, the so-called Anabaptist left wing of the Reformation, but this movement was also propelled by farmers. Many were restive and apocalyptic agriculturalists drawn to a nonviolent form of

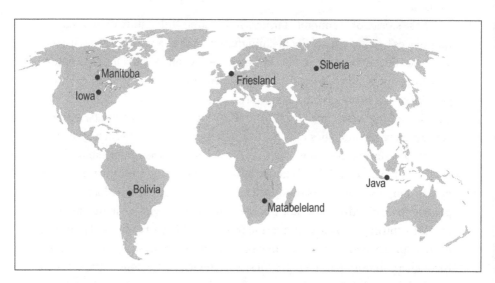

The jurisdictions encompassing the seven Mennonite communities of this study.
Map by Paul Plett.

The Seven Mennonite Farm Places at a Glance
(in order of their founding)

	Netherlands	United States	Canada	Indonesia	Zimbabwe	Russia	Bolivia
Jurisdiction	Friesland	Iowa	Manitoba	Java	Matabeleland	Siberia	Santa Cruz
Place	Friesland	Washington County	Rural Municipality of Rhineland	Margorejo	Matopos	Apollonovka / Waldheim	Riva Palacio Colony
Nearby city	Leewarden	Iowa City	Winnipeg	Pati	Bulawayo	Omsk	Santa Cruz
Founding year	1530	1847	1875	1882	1898	1911	1967
Church(es)	Doopsgezind (ADS)	(Old) Mennonite; Mennonite Church USA; Old Order Amish; Beachy Amish; Amish-Mennonite	Bergthaler; Sommefelder; Mennonite Church Canada; Evangelical Mennonite Mission Conference (EMMC)	Gereja Injili di Tanah Jawa Mennonite (GITJ)	Brethren in Christ Church (BICC)	Mennonite Brethren (MB); Evangelical Baptist, Siberia	Old Colony Mennonite (Alt-kolonier or Reinlaender)
Farm focus	Potatoes, dairy	Corn, hogs, soybeans	Wheat, canola, sunflowers	Rice, casava, watermelons	Cattle, vegetables	Wheat, dairy, hogs	Soybeans, maize, dairy, cattle
Currency	Euro	US dollar	Canadian dollar	Rupiah	Zimbabwe dollar	Ruble	Bolivar
Land unit	Hectare	Acre (2.47 ha)	Hectare / acre	Bau (0.7 ha)	Hectare	Dessiatina (1.1 ha)	Hectare
Annual precipitation	785 mm (31 in.)	890 mm (35 in.)	500 mm (20 in.)	2,900 mm (115 in.)	400 mm (16 in.)	380 mm (15 in.)	1,000 mm (39 in.)
Average temperature January	+9°C	−6.5°C	−18°C	+25.4°C	+26°C	−18°C	+26°C
Average temperature July	+22°C	+24°C	+20°C	+25.4°C	+17°C	+20°C	+20°C
Climate	Marine	Continental	Continental	Tropical	Tropical	Continental	Tropical
Soils	Clay/sandy/peat	Loess/silty/clay	Black chernozem/lacustrine	Silty-alluvial/sandy	Sandy/clay	Black chernozem	Alluvial/clay/sandy
Interview language	Dutch/Frisian	English	English	Javanese	Ndebele/English	Russian/Low German	Spanish/Low German

Anabaptism after they failed to overthrow feudal lords during the widespread Peasants' War in the mid-1520s. And even those Anabaptists who were pacifist from the beginning, such as Menno Simons, the Dutch priest whose name became associated with the largest of the surviving Anabaptist groups, the Mennonites, sometimes hailed from farm families. Then too, the teachings of these peaceful Anabaptists—a rejection of all violence, the full separation of church and state, loyalty to the local congregation, and insistence on a simple life and humility[13]—often marked the very cultural requirements for a truly sustainable rural life.

The Mennonites' rural engagement certainly grew in diaspora as persecution sent large numbers of Swiss and Dutch Anabaptists across national borders. In places well away from their birthplaces, in far-flung rural districts, they accepted religious peace in exchange for tilling difficult lands. This rural diaspora placed the Swiss Mennonites on southern German soil, the Dutch Mennonites on lands in northern Poland and southern Russia, and eventually both groups in the United States and Canada, as well as in locations throughout the Americas. The diaspora gave birth to the idea that the religious teachings on peace and simplicity were best lived out in agriculture, away from the cities and centers of power, close to nature and within the local.[14] This idea became infused in the work of Mennonite missionaries in late-nineteenth-century colonial Africa and Asia. And the idea held as the Mennonites, in various locales, evolved into or became closely associated with separate rural-based, Anabaptist-oriented subdenominations, including the Amish in the United States, the Brethren in Christ in Africa, and specific Mennonite Brethren–derived Baptist congregations in Russia. In each of these circumscribed church bodies, and in each of the seven places studied in this book, twentieth-century farmers recognized that working in nature could secure the goal of agrarian simplicity and close-knit communities of faith.

Indeed, their common roots as Mennonites have linked these seven communities. Each has had some sort of history with teachings of the "Anabaptist vision," which in turn have affected agricultural practice, at least indirectly. Those tenets—variously listed as simplicity, humility, communitarian closeness, and nonviolence—shaped the religious discourse studied in this book. And so did the common Mennonite emphasis on everyday ethics, an insistence that the true Christian was a *Nachfolger Christi*, one who "followed Christ in life," rather than one who possessed an abstract doctrinal "faith in Christ."[15] But the point here is that given the remarkable diversity within the global Anabaptist community, as well as variables of local history and environmental specifics, differences among these communities eclipsed commonalities. The sectarian nature of the Mennonites, that is, their emphasis on the local congregation and their incipient anticlericalism, ensured diversity. Each of the communities differed, both internally and from the others, on what this religious teaching meant for everyday life, resulting in a remarkably diverse church body. Indeed, as the anthropologist Philip Fountain points out, the global Mennonite community today is a "sometimes rather precariously and tenuously, interconnected movement . . . notable for diversity" and for the fact that its members' "identification with being Mennonite 'stretches' or 'bridges' across a series of divides."[16]

This diversity, however, in no way reduced the importance of religion in everyday life. As in most of the world's agrarian places in the past, religion informed not only these farmers' local worlds but also their daily work on the land in some way. In general terms, religious teaching permitted and even encouraged the cultivation of the soil. It gave meaning to the biosphere and, in Michael Redclift's words, linked "personal identity" to the environment and offered "normative value" to human ingenuity on the land.[17] In everyday terms, it promised the cultivators of the soil cultural legitimacy, even spiritual meaning. As holders of a Judeo-Christian worldview, Mennonite farmers everywhere have retold the Old Testament accounts of creation, citing the seemingly contradictory ideas of humans having "dominion" over the earth and "taking care of it."[18] And they have understood this bifurcated teaching in light of other cultural variables, including locally based folklore and nature-based mythology. Their faith has opened debate on appropriate levels of technology, the meaning of profit and success, and the definition of sustainable agriculture. It has intersected with inherited knowledge on how farmland should be kept environmentally healthy for the next generation. It has addressed how farmers should overcome social barriers between landowners and the landless, affected ideas on gender equality with regard to equitable landownership, and reinforced ethnic social boundaries that defined the rural community. It has conditioned the level of acceptance or rejection of national food security goals and the global marketplace.

If religion was locally dynamic, its intersection with intractable local environments made it even more so. True, their very interaction with nature itself, as food producers, has meant that the Mennonite farmers of these seven communities shared a common story. They have all struggled to grow food for human consumption on the earth's most treasured of resources, the soil. The daily imperative of working the land has been the same: with an eye to the weather, they have risen each morning and intervened in nature, using water and sunlight to enable the soil to turn its nutrients and teaming microorganisms to the task of producing a common human necessity, calories and nutrients—food. As farmers, too, they have worked mostly outdoors, their lives swayed by the vagaries of climate and the concerns of the biotic: disease, scourge and infestation. All have searched for nutrients to maintain or increase the soil's fertility, and all have taken steps to save the soil, whether from the sea, erosion, compaction, salinization, or loss of organic carbon. And in the very act of cultivating the soil, all have made their claim of ownership or occupation of it seem natural and uncontested. If translators were available, farmers from these seven locations could visit all day with one another about their common concerns, hopes, and histories.

And yet, stark environmental differences separated these farmers from one another. One obvious difference was the mix of commodities produced in each of the seven locales. In simple terms, Friesland has produced potatoes and butter; Siberia and Manitoba, wheat and oilseeds; Iowa, pigs and corn; Java, rice and cassava; Matabeleland, cattle and viscos; and Bolivia, soybeans and cheese. Then, too, the seven communities have experienced various political and national contexts, oftentimes shaped by events of immense consequence. The community in Friesland has been most directly affected by world war and by continental market regulation. Iowa, Manitoba, and Siberia are the colonial products of the mass relocation of European settlers to the world's grasslands in the nineteenth century. Java and Matabeleland were indelibly shaped by European imperialism in the nineteenth and twentieth centuries. Bolivia is the twentieth-century product of postcolonial agricultural expansion into equatorial bushlands. In each of the seven places the local has been part of the global account of increasing impositions on nature.

The stories of these places have also been distinct from one another because they relate to particular ecozones and ecotones. Four of the seven communities, for example, lie in the earth's temperate zone, with distinctive growing seasons. Three are located in the tropical zone, with precipitation rates that range from the marine to the semiarid. The seven places are also built on a variety of soils. They include the flat, subterranean brown, loamy clay belts in the vicinity of Witmarsum, Sneek, Joure, Holswerd, Makkum, and other places in Friesland. They also include grasslands: the gentle rolling plain of rich, glacialized loess soils of Washington County and its environs in southeastern Iowa; the open prairie of black, lacustrine clay soils in the Rural Municipality of Rhineland in southern Manitoba; the birch tree–bordered black chernozem clays of Apollonovka and neighboring villages in southern Siberia. And they include a variety of landscapes within the world's tropical zone: the yellow and red silt-based clays of Margorejo, near Mount Muria, on the northern coast of Java; the spring-fed, loamy garden soils amid semiarid pasture land of Matopo, in Matabeleland; and the sandy-clay, alluvial soils of cleared bushland of Riva Palacio in the Bolivian Oriente. These soils, found in various regions of the wider earth, support a range of fauna and wildlife common within those regions. But they also shaped distinctive local narratives.

The paradoxical argument I make is that this localized Mennonite interaction with the environment—in its varied social, cultural, and ethical dimensions[19]—always takes place within a global context. But this global link presents itself in a variety of ways. First, it is merely an account, in the environmental historian Donald Worster's words, of "the long, intricate experience that ordinary people have had with nature . . . the world over."[20] Worster's own career hinged on his 1979 broad swipe against reckless human action resulting in the Kansas Dust Bowl, but he quickly moved beyond this American story by arguing that there was nothing exceptional to the plowing of semiarid plains in one part of the world; human greed has threatened sustainability everywhere. Second, it's a simple matter of understanding that the local and the global are linked in some way. In his *Unending Frontier*, John F. Richards argues that the consequences of brazen human intervention in nature may be "best understood in a holistic global perspective," but they are "best seen in the details at the local level."[21] Ultimately the one can't be known without also grasping the other. Third, it is an account of the power of the global over the local, sometimes seen as overt courtship of national governments by multinational corporations. Vandana Shiva's *Making Peace with the Earth*, for example, issues the clarion call to stop the destructive impact of global, industrial agriculture on the local, especially with the lure of cheap food, and particularly in the Global South. This powerful global agricultural impulse has also expressed itself as hegemony, inscribed in the very cultural values that have shaped the meaning of success and respect.[22] And finally, the global is seen in the agency that ordinary farmers the world over seek to exercise in the face of this power imbalance. In a play on words, the US historian Paul S. Sutter sets out an agenda for an "agrarian turn" in an article titled "The World with Us." He applauds agricultural history's new concern about food security and a sustainable environment, but dismisses any idea of

a "unitary nature" and instead sees "many discrete human and nonhuman actors," each seeking "a modicum of control over their lives" within a range of
"agroecological instabilities" both in the United States and beyond.[23]

This story of the "many" points of instability, defying an intractable "unitary," is not only at the root of what I try to accomplish in this book; it consti-
tutes its central challenge. Gaining even a basic knowledge of seven places is
difficult at best, and reaching Hughes's bar of case studies that are "well
known" to the historian is, of course, impossible to achieve for seven communities; scholars can easily spend a lifetime studying just one community. Still,
six of the seven places portrayed here, all but Iowa, were places I had visited
before this project began. In each place I had met with farmers, attended
church, walked in the fields, overnighted, and built some form of social network with regional scholars or local lay historians. Importantly, they were
also places in which a graduate or postgraduate student with appropriate linguistic skills and education in history or the social sciences could be identified and hired. All the researchers—John Eicher, Hans Peter Fast, Susie Fisher,
Aileen Friesen, Danang Kristiawan, Belinda Ncube, and Ben Nobbs-Thiessen—
were either native to the regions encompassing these communities or versant in national languages as a result of extensive graduate field research.
These researchers, as I explain in appendix A, allowed me to capitalize on my
academic position as Chair in Mennonite Studies at the University of Winnipeg, with its global links, as well as on my subvocation as a Mennonite farmer
on a "half section" of grain land. It was as both an academic and a farmer that
I sought entry into the heart of these seven communities.

Most important, though, was the skill and integrity of the entire research
team. Each of the team members built a rapport with local farmers, resulting
in a sharing of texts, both literary and oral, the very archive undergirding this
study. Certainly, the second major challenge of this project was to obtain the
"many types of information" that Simmons had called for, primary sources
that could reveal "empirically" the inner workings of the farms in these places.
From earlier work on Mennonite settlements in Manitoba, the American Midwest, the Russian Empire, and Bolivia I had a sense of the range of documents—
newspapers, state records, and personal diaries—that would be available for
four of the seven places. But Friesland, Matabeleland, and Java were especially strange to me; fortuitously, local research in these places yielded a variety of memoirs, rich local histories, mission reports, letter collections, and
even sermons and poetry. Although uneven in quality and quantity, these documents nevertheless provided answers to common, pressing questions.

An important companion to written texts, as noted above, was oral history, with its common aim of listening to local farmers answer questions about their histories with the land, taking into account experiences shaped by gender, class, and generation. These oral accounts linked the texts of these places. Indeed, oral history, conducted by the seven researchers in the seven places, proved crucial to overcoming the unevenness of text-based historical sources. This approach allowed farmers to share what they had heard the land say to them, facilitating what Debbie Lee and Kathryn Newfont refer to as moments when "the land speaks." It allowed the farmers to share what they imagined had occurred on the land in the past, moments of both desecration and restoration.[24] Reflecting the ideas of Alessandro Portelli and other oral history experts, this method revealed the "truth" of an encounter, for example, not just the exact details of a particular tilling of the soil but, more importantly, the meaning of that tilling over time.[25] Perhaps we came to the farmers with a uniform set of questions, put to them by outsiders, aware that the very questions raised could elicit specific answers, but as interviewers we made every effort to convey to the interviewees that no particular response was expected. And as will be evident, while canned answers cannot be avoided, the disparateness of stories shared by neighbors is testament to the open-ended nature of this study's interview instrument. Still, oral history entails a particularly close, even subjective interaction with the very creators of the text, a relationship I implicitly acknowledge in this book by referring to the interviewees by their first names.

Significantly, an ethics protocol is a central feature of oral history. This task was the foremost concern of the first of three "Seven Points on Earth" team workshops, one held in Amsterdam's historic Allgemeine Doopsgezinde Societaet in 2013. At this workshop our seven-person team reviewed the broad contours of Mennonite history, pondered the challenge of environmental history, and considered the intricacies of oral history methodology.[26] On the basis of this exchange, the team outlined the interview questions, reproduced in appendix B. Those questions focused on changing farm methods, inherited forms of agricultural knowledge, religion and the soil, gender and land, natural disasters and climate change, state policy and transnational connections.

These questions also came to serve as the outline for this book's chapters. As chapters 1 and 2 show, all farmers described in this book have been linked to modernizing, contentious global flows of agricultural knowledge. Emanating from seemingly inexorable centers of imperial power, this knowledge has traveled overseas from Europe to the Americas, along imperial routes into

Africa and Asia, and into the most recent settler societies of Northern Asia and South America. In chapter 3 the flows become temporal rather than spatial, with a focus on how farmers both resisted and applauded the tumultuous changes in global agriculture over time. Chapter 4 grapples with the soul of the community, revealed in religious teaching on "peace on earth," articulated in seven different ways, by one farmer in each of the places studied. In chapter 5 we consider the cultural variable of gender, focusing on how women in five of the seven places negotiated the nexus of the patriarchal farm household in different ways, in colonized and decolonialized settlings (Java and Matabeleland) and in white settler communities (Iowa, Siberia, and Bolivia).

The last three chapters ask questions that take us beyond the boundaries of each community. Chapter 6 introduces the state and its nationalist policies of agriculture and traces a range of relationships, from the seemingly most complementary in Manitoba to the most strained in Siberia. Chapter 7 differentiates farmers in yet another way, with reference to their ideas on climate change; again we visit but five places, those showing an overt concern with the climate's history in the economically vulnerable Global South and those revealing a certain complacency toward it in the rich, highly developed Global North. Finally, chapter 8 seeks to understand degrees of transnationalism, or global ties, in the lives of farmers and their approaches to the land; it contrasts the ironic ways in which the traditionalist Old Colony Bolivian Mennonites and the highly integrated Iowa Mennonites negotiated that wider world but then accounts for the variations of this relationship in the remaining five places. As noted above, these seven themes—colonial origins, technological change, religious belief, gender relations, climate change, government policy, and the global turn—are themselves "seven points on earth." It is my hope that readers will come to know these farmers from these seven vantage points as fellow workers in the fields, creating a common, sustainable agriculture.

This book seeks to understand the history of human relationships with the earth by focusing narrowly, over vast stretches of the earth, to find general answers to a set of pressing question: how can we steward the earth, enjoy our roles as cohabitants of it, and continue eating in health-giving ways? In all societies the act of food consumption is encased in cultural meaning: in each of these seven Mennonite communities, prayers of thanksgiving are offered before meals, food consumption is measured, and food sharing is venerated, thus sacralizing the act of eating. But farmers in each of the seven communities also tell and retell stories of human interaction with the soil that produce the food humans consume. Oftentimes those stories ascribe a certain sacred value to soil, featuring prayers at planting or harvest, a kneeling in the earth, or imagine forbears doing so. These narratives also reveal a common concern with the health of the soil even if farmers' actions have been inadequate. In the past, farmers have planted trees to stop erosion, legumes to regenerate fertility, and specific crops to address environmental challenges related to drainage, weeds, pathogens, or drought. In short, farmers have worked hard to keep the earth healthy for each new generation. The local dimensions of these seven disparate places illuminate a common challenge, the imperative to let the earth continue nurturing life.

The world community needs to know more about this historical challenge. In history, religion has indelibly shaped the way farmers approached the land, both for good and for bad. Where religion legitimized colonial structures, focused on dominion, or even encouraged an eschatologically oriented carelessness, it failed the environment. But in its critique of unchecked greed, its veneration of local community, and its linkage of nature to the divine, religion has offered a moral compass. When the Mennonite farmers of the seven communities farmed a place, they both tilled and tended to it as a place, holistically—socially, culturally, and ethically. They never just farmed *in* a place; they farmed the place. And where they did so thoughtfully and wisely, they also aimed to farm sustainably. The act of farming a particular place defined who they were, and in turn it shaped the nature of that place. Their vocation was both a tilling and a tending, a cultivation of and a caring for, to invoke the creation story of Genesis. This struggle is at the heart of the seven histories in this book. Collectively they tell a global story of an intricate relationship with the land.

CHAPTER 1

Sect and Settler in the North

Plowing Friesland, Iowa, Manitoba, and Siberia

In 1617 a group of Dutch Mennonite entrepreneurs and their families moved eastward from the province of Friesland into the Kalkwijk-Lula peatland district of the neighboring province of Groningen. Here they turned a profit by mining peat and selling the transformed landscape as farmland. These entrepreneurs, members of the conservative Old Flemish Mennonite Church, were known for their plain clothes, foot-washing services, and strict church discipline. But what made this group especially distinctive was that in Groningen they turned from trading to farming and in so doing sought the religious values of simplicity and humility. Then, almost a hundred years later, in 1711, this group of plain agriculturalists joined with Mennonite refugees from Switzerland to form a new, rural congregation. The union reinforced the link between agriculture and simplicity. Perhaps the Dutch had been wealthy enough to support the Swiss fugitives through the Amsterdam-based Mennonite Committee for Foreign Needs, but together the Dutch and the Swiss committed themselves to living as a close-knit community with farming as a preferred way of life. Of particular note, a 1744 document outlining the basis of this union, included a confession that in the past congregants had become "involved, with too much lust and love in buying and selling," presumably of peat fuel and land.[1] Rural simplicity had been rediscovered and given religious meaning.

A second, more material aspect of this story is also significant. It is that these united farmers eventually found their economic success by shifting from dairy farming to potato production. A 1745 letter from one of the Swiss farmers, Samuel Peter, spoke of a cattle epidemic, of "poor beasts . . . dying in front of our eyes," nineteen animals at his farm alone.

The key to sustainable agriculture in Groningen would be the lowly potato, brought north to the Netherlands by Swiss Anabaptist refugees, who knew it as "bread for the poor."[2] Indeed, the Dutch agricultural historian Jan Bieleman credits "Mennonite refugees" for having first introduced the potato to Groningen's "Peat Colonies," from where it was then adopted by farmers in neighboring Friesland in the form of "monocrop potato production."[3] And it was a descendant of these Swiss–South German refugees, Geert Veenhuizen, who in 1888 developed the famous Eigenheimer breeding potato, which would also find its way onto farms in Friesland. These neighboring provinces—Groningen and Friesland—would become the European node for commercial potato production.[4] Thus it was that a beleaguered group of Anabaptist migrants became the conduit for one feature of the great Columbian Exchange; the potato, brought to Europe from the Americas, had made its way into the northeastern corner of western Europe, allowing that region to turn what had been a basis of subsistence into an export commodity. Ironically, the humble "bread for the poor" would become the very foundation of a commercialized agriculture that challenged rural simplicity in Friesland.

This meeting of two groups of Mennonites—the Dutch Old Flemish and the Swiss–South German Anabaptists—points to three grand themes of the history of agriculture. First, an early modern expansion of agriculture, claiming soils from an existing natural condition, whether it was peat, forest, swamp, sea, or grassland. Second, modernization, in particular the arrival of monocrop cultivation based on plant breeding and commodity commercialization. As Alfred W. Crosby writes in his *Ecological Imperialism*, a "biological expansion" led to a "scattering" of crops, weeds, and animals through "similar latitudes . . . in temperate zones," aided by the lack of pathogens and the abundance of precipitation in places "rich in photosynthetic potential."[5] This expansion occurred within Europe, but even more importantly in the history of agriculture, it occurred as European settlers migrated overseas. The Dutch and the Swiss in Groningen had traveled only a relatively short distance from home, but European-descendant Mennonite farmers in the Global North would create their stories of settlement and agriculture in far-flung places in North America and Northern Asia, sites where they hoped to build local, agrarian communities of faith. Within this diaspora lay a third broad theme, one reimagining the land itself. This thought process legitimized

a "great land rush," to quote the title of John Weaver's global history of settler societies. At the root of it all was the Enlightenment-based idea that by
putting their plows to the soil, farmers added "value" to the land and revealed
a teleology making such settler activity seem natural and inevitable.[6] Here
too was a transformation backed by power. In his *Unending Frontier*, John F.
Richards points to the central event in which "technologically superior pioneer settlers invaded remote lands," displacing Indigenous peoples or, at the
very least, shaking local societies to the core.[7]

This story of northern origins and colonialism also underpins the study of
the Mennonite farmer in the Global North. Indeed, farmers at each of the four
northern places in this global story—Friesland, Iowa, Manitoba, and Siberia—
were part of this wider, European-oriented narrative. Friesland, of course,
was the province of a colonial power and grew food for that power for centuries. Each of the communities in Iowa, Manitoba, and Siberia were white settler societies facilitated by expansive colonial powers, American, British,
and Russian. Despite their disparate locations in Europe, North America, and
Northern Asia, these four Mennonite farm communities shared a common
story of agricultural progress, reshaping their natural environments, sending
their commodities to distant market places, and accepting increasingly complex technologies to manage those environments.

And yet this grand story of a flow of power and knowledge from western
Europe and beyond always returns to the local, to the farmers and settlers
themselves. In small places, farmers drew their tools through the soil, employing tested genealogies of agricultural knowledge, constructing narratives of climate and weather, and negotiating rural social boundaries, all the
while informed by folklore, inherited cosmologies, vernacular belongings, and
local codes of ethics. It's a story of the way they made sense of agriculture. In
this quest, religion mattered, shaping the farmers' approach to the lands they
cultivated, just as activities on those very lands shaped aspects of their religious life. When the Mennonite farmers intervened in the ecological world,
they did so both as citizens of a wider earth, growing food for the marketplace,
and as Mennonites, variously responding to religious teachings emphasizing
charity, simplicity, and nonviolence, each with implications for how land was
farmed. Within the wider world they were willing participants in colonial
projects as settlers; at the local level they aimed to be the "quiet in the land,"
tilling and tending the soil in peace, close to nature, in harmony with local
community.

Friesland: The Irony of the Call to Simplicity

This culture of linking religion and land, however, was much more important for some Mennonite communities than for others. Ironically, it was least strong in Friesland, the birthplace of Menno Simons, the namesake of the Mennonites. Certainly Menno's own religious pilgrimage as an Anabaptist, literally a rebaptizer, is a story of nonviolence; most notably he strove to turn apocalyptic and socially radical Anabaptists into peaceful followers of Christ, smoothing their relationship with the state and preaching a modest, pious, humble way of life. But the link between simplicity and rurality in Friesland was tenuous at best.[8] For example, in his subversive message of separation from the wider world of power and ecclesial glitz Menno lauded all humble artisan classes, of which the peasant was only one. And given the progressive nature of Dutch agriculture, farming could easily signal encroaching capitalism, including rising wealth and ostentation.

Indeed, the story of agricultural modernization is at the heart of the history of rural Friesland. Agriculture was highly productive and forward-looking in Friesland, which was in close proximity to some of Europe's fastest-growing cities. Moreover, the natural environment of Friesland lent itself to a commercialized and innovative agriculture. Temperatures in this part of Europe have varied over the centuries, with the latter part of the sixteenth century and the first half of the seventeenth being a period of severe winters within the so-called Little Ice Age.[9] But in relative terms over time both the summer and winter months have been moderate, with an average temperature of plus 3°C in January and plus 17°C in July, and there has been a defined rainy season in both winter and spring.[10] In addition, farmers benefited from Friesland's flat, naturally treeless plain, including its easily cultivated rich soils recovered from the shores of the North Sea and the inland Ijsselmeer. Certainly, Mennonite farmers here tilled different kinds of soils, including the coveted marine clay districts in northwestern Friesland and the somewhat less fertile lands of sand and peat in the southeast. Even the marine clay belt of the western half of the province was not homogeneous; near the sea, behind natural *kwelder* levees, the land was sandy and quite high, ideal for potatoes, while farther inland clay pan soil nourished meadows and a rich dairy culture.[11]

This diverse agricultural landscape, argues Jan Bieleman in his sweeping *Five Centuries of Farming*, has fundamentally shaped Friesland's history since 1500, culturally, politically, and economically. Because the drainage of water in Friesland as elsewhere in the lowlands was the foremost concern of farmers

Friesland province in the Netherlands and selected rural communities home to
Doopsgezind (Mennonite) farmers, including Witmarsum, the 1496 birthplace of
Menno Simons. Map by Weldon Hiebert.

over the centuries, a vibrant and cooperative civic culture developed by
the sixteenth century. At its social core, argues Bieleman, the local associations required to build and maintain dikes resulted in a region "without feudal structures" and served as the genesis of a "powerful, united free peas-
antry." This social fact also propelled Friesland's economy, which in turn
enabled the claiming of ever more land from the sea. By 1506 significant

sources of new lands had become available as the old Middelzee, historically cutting Friesland into two parts, was closed, creating large new, fertile polders, and by 1545 a serious reclamation of the sea had begun along the Frisian coast in general.[12] Coastal farmers especially gave a nod to the commercialization that the rich soils of their communities allowed.

Other factors propelled Frisian agricultural progress. Bieleman's history describes a population that grew almost without interruption from the sixteenth century well into the twentieth. Friesland's town-based population, already 25 percent of the province's total by 1500, more than doubled by 1650, with Leewarden alone having a population of 16,500 by the 1660s.[13] Thus, demand for a range of foodstuffs increased: grains, meat, and especially butter, whose production increased sevenfold between 1525 and 1650. In addition, this lowland, crisscrossed by an intricate network of canals, allowed for the easy transportation of farm products to urban markets, local and international. Even as they maintained their dairies, Frisian farmers plowed up meadows to produce more urban-destined commodities, especially legumes such as beans and peas. By 1650 farmers in Friesland were noted for their commitment to increasing nitrogen in the soil, applying cattle manure to the land but also contracting for easily spread "night soil" of numerous towns. In the eighteenth century they also began planting red clover for its nitrogen-fixing properties, intercropping possibilities, and soil-tilth improvement. By 1800, as both commodity prices and the size of cattle herds increased, Friesland became known throughout Europe for its "high standards" in animal care and hygiene, with tie-up stall barns and strict separation of feed and manure. But two commodities especially made Friesland famous: its butter, providing London with a quarter of its need; and its potato, with production increasing fivefold between 1750 and 1800, making it northwestern Friesland's major crop.[14] Meanwhile flax held its own, rapeseed and sugar beets became alternative crops, and pig fattening dominated agriculture on the less fertile sandy soils. Only wheat production fell, as long-distance transportation systems brought in cheap grains from North America and the Baltics. Signs of agricultural innovation propelling commercial activity abounded.

Mennonite history in Friesland reflects the story of Frisian agriculture. True, most early Anabaptists here were urban artisans, residents of Haarlingen, Leewarden, Sneek, and other towns, but at least one prominent historian, Harold S. Bender, has suggested that Friesland was home to many who were in fact "rural."[15] Other historians have found in agrarian, utopian groups, such as the fifteenth-century communalist Brethren of the Common Life, following the teachings on simplicity by Thomas à Kempis, the theological antecedents of Dutch Anabaptism.[16] Social historians linked to the polygenesis school of Anabaptist origins, on the other hand, have argued that everyday environmental conditions, such as repeated harvest failures and food shortages, led desperate Frisians to an early fascination with violent, apocalyptic ideas advanced by the most radical of Anabaptists.[17] Authoritative biographers have also placed Menno Simons in an agriculture milieu, one asserting that he was "born to a Dutch peasant family living in the village of Witmarsum," another that he was raised by parents who "were very probably dairy farmers."[18] Certainly Menno's close associates included at least some farmers: Tjaert Reyndertz of Harlingen, for example, arrested for harboring Menno, declared a heretic, and executed in 1539, was a peasant.[19]

The progressive nature of Dutch agriculture, of course, meant that these rural connections were no guarantor of Menno Simon's emphasis on simplicity and humility. Indeed, for Menno, agriculture was hardly a preferred vocation. True, he could somewhat romantically tie the natural environment to spiritual virtue, at one point paraphrasing the Song of Solomon to encourage his followers: "Verily the unfruitful, cold winter is swallowed up, and the fruiting, delightful May has come in the land. The lovely, fair flowers spring forth everywhere; the turtledove drops his liquid note."[20] But in mercilessly castigating the arrogant and selfish, he also implicated those who were closest to nature, that is, greed-filled landowners who "add farm to farm" and in the process "ignore the cause of the poor and needy."[21] At another juncture, Menno criticized all men, including peasants, as sexually harassing women, naming "lords, princes, priests, monks, noble or ignoble, citizens or peasants" as those who stalk women the way "the dog pursues the hare." Anabaptist farmers in Friesland were simply not idealized as bastions of virtue.[22]

The linkage of land and humility in Friesland would shape the lives for but a minority of Mennonite farmers. Farmers on the islands of Terschelling and Vlieland, just off the coast in the North Sea, did follow the ascetic teachings of Jan Jacobs from 1580 to 1612. The Janjacobsgezinden worshipped in plain meetinghouses and avoided contact with "worldly" people,[23] including members of the neighboring, liberal Mennonite Waterlander affiliate, who worked as Greenland whalers, North Sea fishers, and Baltic sailors. Ironically, writes the art historian Nina Schroeder, seventeenth-century Mennonites embraced the Dutch countryside more often for reasons of wealth and luxury than as a way of securing a plain and simple life. As an example, Jacob Isaacksz's 1665 painting *Mennistenhemels* (Mennonite heavens), while featuring a rustic,

House and windmill on the path to Menno Simons memorial at Witmarsum in Friesland, Netherlands, 1967. The monument has become an international tourist attraction. Photo by Frank H. Epp, Mennonite Archives of Ontario, Waterloo.

country abode in Holland "with bleaching fields in the foreground" and the city of Harlem in the distant horizon, was owned by a well-to-do urban Mennonite merely heralding his rural retreat. Indeed, the Dutch Golden Age, after the 1660s, marked a time of heady investment and often country-based opulence for Mennonites, including those from Friesland.[24]

Over the centuries, this vibrant economy ensured that no identifiable rural Mennonite communities developed within the province. The economic historian Cor Trompetter has observed that Frisian Mennonites who purchased farmland tended to be urban merchants seeking investment opportunities other than stocks and bonds. And when they chose to operate their farmland, they did so as overseers, leasing the land to peasants or hiring farm-workers rather than becoming cultivators themselves. Trompetter argues that while Friesland may have "had very good soils, indeed some of the best in the country," fertile land did nothing to help form "specific Mennonite agrarian communities" in Friesland.[25] Although census records, such as the 1849 record for the municipality of Utingeradeel (in and around Akkrum), in the center of the province, listed almost 2,000 Mennonite residents, or 48 percent of the population, it was an unusual place, and its concentration of Mennonites was a matter of happenstance, not design.[26] Indeed, only seven other municipalities in Friesland listed at least 600 Mennonite residents, and in no place did they come close to the concentration at Utingeradeel. Significantly, this same census designates only 5.8 percent of Friesland as Mennonite, with 62 percent Reformed Frisians and 27 percent Catholic Frisians.[27] The Mennonites who did farm were scattered throughout Friesland, and most of them easily blended into a rapidly modernizing agriculture.

The economic vibrancy of Friesland also resulted in vast class difference between landowning and land-renting, or peasant, Dutch Mennonites. Piet Visser, for example, argues, on the basis of censuses, that after 1810 about 30 percent of Frisian Mennonites were farmers, of which the vast majority, 75–80 percent, were landowners.[28] Often linked to urban wealth and exhibiting opulent lifestyles, they held themselves apart from the rural poor. Although both classes might well attend the same Mennonite church, the poor occupied the back pews, meaning that they were also well away from the central heater in wintertime. The church addressed this poverty not by financing the purchase of land for poor farmers but with personal appeals for outright charity earmarked for the most destitute and based on a moralistic ideology of spiritual egalitarianism. To an extent, geography reinforced this class division. Visser reasserts that by the early nineteenth century most Frisian congregations could be found either on the richest of the fertile coastal land north of Harlingen, where Mennonite farmers focused on potatoes, sugar beets, and corn, or in south-central Friesland, the Greidhoek (Meadow Area), where dairy farming was prominent.[29] Fewer congregations were located on sandy and swampy areas of the east, De Wouden (The Woods), and those that were comprised mostly poorer tenant farmers.

While farming was not idealized, tilling the soil was spiritualized nonetheless. The 1685 inside cover of the Dutch minister Thielman van Braght's edited volume *The Martyrs Mirror*, the thousand-page compilation of martyrs stories from the sixteenth century, features an etching of a rustic peasant, spade in hand, digging in the soil. Above him are the Latin words *fac et spera*, "work and hope." Visser has noted that the depiction is not especially significant, as it was a common imprint in books published in the Netherlands. Both the literary scholar Julia Kasdorf and the historian David Weaver-Zercher add to this argument by suggesting that in any case the etching references a spiritual truism, of struggle and eternal life, and is not a celebration of a special place for the farmer in Dutch Mennonite life.[30] Examining Frisian agriculture at the local level, as a focus on Mennonite farmers allows, reinforces the overall grand theme of a progressive agriculture in Friesland. Any struggle to keep Anabaptist ideals of humility and simplicity alive addressed both the urban *and* the rural districts of Friesland. Moreover, given the geographically scattered nature of the Mennonites here and this agricultural zone's relative richness, any historical Anabaptist teachings on simplicity became distant ideals, more easily spiritualized than lived out in daily life.

Iowa: Servants of Faith on the Edge of the Frontier

The story of agriculture among Dutch Mennonites contrasts starkly to that of the Mennonite farmers in the diaspora of the Global North's grasslands. The second place in this study, Washington County and environs in southeastern Iowa, was, in comparison with Friesland, a highly cohesive Mennonite community and decidedly agricultural in nature. Here long-settled Swiss-descendant Mennonite and Amish farm families from central Pennsylvania, as well as recent immigrants directly from Europe, leapfrogged over Mennonite and Amish settlements in Ohio, Indiana, and Illinois to settle on the very "edge" of the so-called American frontier. Iowa's natural environment fit the cultural aim of building a close-knit, rural community remarkably well. Its low rolling hills, known for fertile soils, temperate climate, and well-watered seasons, seemed ideal for this venture.

The stories of Mennonite and Amish farmers who settled in Iowa are also intertwined with the broad themes of US agricultural history. But here, the starting point is the dispossession of Indigenous people, the Native Americans, from their lands. Cornelia F. Mutel's *The Emerald Horizon: The History of Nature in Iowa* describes a rich grassland, 28 million acres of soil evolving from the action of wind and water on glacial till, that was home not only to native hunting societies but also to farming communities.[31] By 1000 CE this grassland also featured semisedentary Indigenous villages that practiced "intensive horticultures" of corn, as well as sunflowers, squash, dwarf barley, and other small grains. Then, when the white settlers arrived in the early

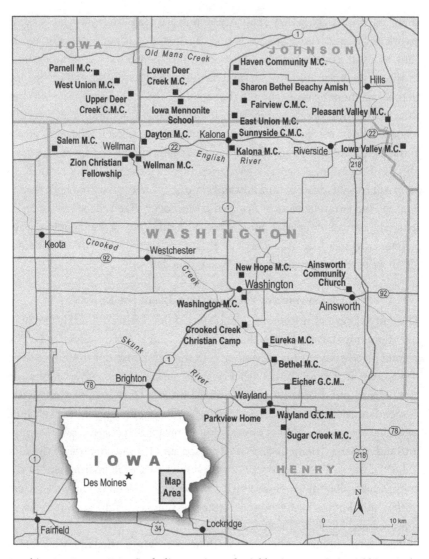

Washington County, Iowa (including sections of neighboring counties), and historical Mennonite churches in communities dating from 1847. Map by Weldon Hiebert.

nineteenth century, writes Mutel, the region underwent "the most rapid and complete ecological transformation in Earth's history."[32] It also witnessed the significant defeat of Black Hawk and his Sauk nation in 1832, a generation after the Louisiana Purchase, in which the legendary chief turned down a treaty with the United States, declaring that land simply could not be sold. As

John W. Hall argues, "The Black Hawk War . . . although militarily negligible, represented a turning point in the history of the Old Northwest," making it apparent that the "Great Father," the president of the United States, "ultimately intended that the Indians surrender the chase in favor of intense agriculture."[33] In southeastern Iowa itself, the Meskwaki nation, allied to the Sauk, was forced from its lands as a result of Black Hawk's defeat. Still, many of its members continued to live surreptitiously along the Iowa and Cedar Rivers and in 1856 purchased privately owned land to make their place in central Iowa permanent.[34] The Mennonite settlers would never know the Meskwaki as the powerful ally they had been to the Sauk.

Rather, they would know Iowa as the edge of the American "frontier," a natural landscape transformed by European-based institutions. The idea of the frontier made famous with Frederick Jackson Turner's classic 1893 thesis that "cheap land of the frontier" resulted in a culture that "broke the bonds of old custom"[35] has, of course, been critiqued by late-twentieth-century historians. Fredrick Luebke, Kathleen Neils Conzen, Jon Gjerde, and many others have observed that European settlers sought a form of cultural "maintenance" and even "ethnic reinvention" with a revered emphasis on the generational succession of farmland.[36] Nevertheless, the focus on the settlers' complicity in an environmental overhaul, on "how best to exploit available resources,"[37] echoed elements of the original frontier idea. Another debate, one highlighting the metropolis, did shift the focus away from the frontier. In his 1992 *Nature's Metropolis*, William Cronan describes a "hierarchy of city, town and country" in which urban entities like Chicago shaped the economy of the rural sections, in particular building a "vast pork hinterland." Iowa farmers answered Chicago's wheat markets and then after 1860 also those for its corn and hogs; the city's lumberyards and barbed-wire factories, in turn, enabled the enclosures for the rooting pigs.[38] The Mennonites' success in building their close-knit, rural enclave required the city on the eastern horizon.

Despite these features of rural life they shared with other ethnoreligous groups, the Mennonites experienced this "frontier" in their own way. As Swiss-descendant Anabaptists—members of both the moderate Old Mennonite Church and the more traditionalist Amish—this community had sectarian roots in rural Europe. True, the first known rebaptism in Europe occurred in the city of Zurich in January 1525, but this Anabaptist act quickly extended into the countryside, where it was welcomed by restive peasants, often associated with the bloody peasant revolts of the mid-1520s. These rural rebels saw in rebaptism a promise of freedom from repressive feudalism and state church. The historian Werner Packull describes one Swiss peasant, Hans Klinger, finding relief in Anabaptism from "sour work," that is, a life of sweat and blood to support the showy "three tall horses" drawing his feudal lord's stately carriage.[39] And the historian C. Arnold Snyder in describing the heresy and 1530 drowning of Margret Hottinger highlights her father, the radical and literate farmer Jacob Hottinger, imprisoned for comparing a defecating cow to the holy Mass.[40] By 1540, according to the Mennonite historian Samuel Geiser, most Swiss Anabaptists were rural peasants given to absolute pacifism and social quiescence. Both the violent rural phase and the confrontational urban phase of Anabaptism were already bygone religious expressions.[41]

As important to Iowa settlers as these stories of Anabaptist beginnings were stories of rural life in the diaspora. This narrative recalls the seventeenth-century Mennonites who fled persecution in Switzerland and traded their agricultural expertise for religious freedom in the German-speaking lands of the adjacent Palatinate and Alsace Loraine. Based on an agricultural "reputation which made them eagerly sought after as renters and managers of the larger estates," writes Harold S. Bender, these Mennonites found their freedom, albeit as "strangers and pilgrims on the earth." But even as mere land renters, they focused on land improvement and eventual landownership. Swiss–South German Mennonites and Amish expressed these values when in the early eighteenth century many migrated to North America, and to central Pennsylvania in particular, in search of land and lasting religious freedom. Here they benefitted from William Penn's treaty with the Indigenous Conestoga, as well as from the Susquehannock'a decimation from infectious disease, in effect handing the colony's interior to the settlers.[42] Many of these farmers purchased large tracts of land in what became Lancaster County and then, just a few decades later, also in Maryland and Virginia.[43] By the early 1840s even these sources of farmland were inadequate for their farming ambitions, and Mennonites began scouting for opportunities in the continental interior, at the far edge of the American frontier as they understood it.

This migration story, with a century-long sojourn in Pennsylvania, was but one of two settlement narratives relevant to Iowa Mennonites. The second account focuses on the 1830s and 1840s migration of Swiss-descendant Mennonites and Amish more directly from the German-speaking Palatinate and Alsace Loraine to Iowa, often with only short stays in Pennsylvania and Ohio. This migration also comprised seasoned farmers and included large numbers of the Amish from Alsace Loraine. Since 1693 these rural Anabaptists had followed the strict teaching of Jakob Ammann and embraced agriculture and rural life as guarantors of simplicity and humility. In time they also became known for what the historian Steven Reschly dubs "a coherent system of agriculture," geared to both self-sufficiency and market participation and focused on keeping soils fertile through the use of manure and legumes.[44] The migration also included Mennonite farmers from the Palatinate, who had benefited from agricultural innovation in their region.[45] By 1750 the Mennonite farmer David Moellinger, dubbed "the father of agriculture in the Palatinate," had developed ways for turning higher ground, above the flood-prone Rhine Valley floodplain, into fertile soil by planting clover, spreading pulverized limestone, and employing a regimented plan for crop rotation. Palatine Mennonites welcomed these modern farm methods, but in the 1830s and 1840s they rejected another form of modernity, universal military conscription, a major precipitating factor in their migration to the United States, according to the historian Steven Nolt. Ironically, one form of modernization sent them to America, while another enabled their cultivation of the country's immense interior grassland.[46]

These two migration narratives—one of long-settled Mennonites, mostly Pennsylvanians, the other of newcomer Mennonites and Amish from the Palatinate and Alsace Loraine—reflected the wider American mythology of the frontier. Mennonite and Amish settlers also spoke of "the promise of land and freedom in America." This theme infuses the story of the "first Mennonite to settle in Iowa," John Carl Krehbiel, from the Palatinate, who arrived in Lee County in 1839.[47] It is also the theme of the first Anabaptist settlers in nearby Washington County, who hailed from Pennsylvania. As the Amish memoirist Samuel Guengerich tells it in a 1924 hagiography, the first Amish scouts arrived in the region in 1840, having traveled five hundred miles by riverboat and on foot. Here, they concluded, they had "splendid chances of securing future homes on cheap but good land."[48] By February 1847, one of those scouts, Daniel Guengeruch, and his family had settled in Washington County on land just north of present-day Kalona. The process required Guengeruch, as it did all American frontiersmen, to enter a deed at the state land office, which he did in Iowa City for the sum of thirty dollars. Before the year was over, the Guengeruches had built a log cabin, purchased another parcel of eighty acres for two dollars an acre, fenced ten more acres, and cleared another eight. As Samuel Guengerich puts it, the family was well on its way to establishing "a comfortable home and farm of 150 acres."[49] The Guengeruch family had met the bar of the successful settler family.

The Mennonite and Amish settlers of southeastern Iowa not only built their homesteads but also established distinctive farm communities that would embody the Anabaptist teachings on simplicity and self-sufficiency. This cultural aim is encapsulated in a second text, the diary young Samuel Guengerich kept in 1866, the year he and his wife, Barbara, moved onto their own farm. It reflects his life as a junior householder, supplementing farm earnings with earnings from carpentry and teaching.[50] It also reveals the social boundaries of his world, Amish "side streams" and American "main channels."[51] Guengerich was keenly aware of the two worlds: on Sunday, July 1, he attended an Amish church service in a private home and heard a "striking sermon," but on July 4 he complained that "the people had quite a high time to celebrate the day of Independence." Guengerich also lived within two worlds, the farm community and the urban market, although the latter signaled a set of interactions clearly meant to bolster the self-sufficiency of the farm. On January 4, the day after he wrote that "we butchered our hogs for market," he sold them in Iowa City for "$10.15 per cwt" and then made several purchases, including two food items, "pepper and allspice," the basis for a farm household self-sufficient in food. On April 11 he sold 12.5 bushels of wheat in the city, and on the way home he purchased "4 apple trees, 3 cherry trees, 10 grape [plantings]," again with the aim of household self-sufficiency. Even the way he wrote about the weather seems bifurcated, reflecting the simple, toiling farmer, on the one hand, and the poetic citizen, on the other. On January 11 he complained of so much "sleet . . . that a person can hardly get about," while on April 29 he gushed romantically that "nature begins to clothe the hills a[nd] valleys in its verdant garments," and on August 20 he recorded "a middling fine day . . . in the evening, excellent, pleasant!"[52] The land of Washington County would provide the Guengerich couple with a chance of living in a close-knit community, albeit one from which the wider world could be safely engaged.

Iowa's environment also enriched the Mennonite farm families materially. Settlers would tell of trade with members of the Meskwaki nation, who resided along Deer Creek as late as 1910; but more often they recalled them as beggars, with Meskwaki women traveling by pony-drawn wagons from one farm to the next asking for lard or flour.[53] Indeed, settler records suggest a vast social chasm between the prospering Amish and Mennonite farmers and their "Indian neighbors."[54] The 1876 inventory of the late farmer John Reber, of Johnson County, for whom Samuel Guengerich served as executor, is an account of a well-stocked house and a remarkably elaborate set of farm equipment.[55] John's unnamed widow is granted property worth $737 (about 40 acres); rudimentary farm implements—several hayracks, a horse-drawn bobsled, a wheelbarrow, a stirring plow, a corn plow with seeder, a harrow—as well as horses, cows, calves, hogs, piglets, and shoats. Indicating the widow's relative financial strength, she "resumed" the ownership of two workhorses, six milk cows, one bull, and three young cattle. In addition to items reserved for the widow's sustenance and future farm needs, the "General Assets" of the farm included the technologically sophisticated implements of a progressive American farm made good: a Kirby reaper, two Wier corn plows and seeders, a Buckeye grain drill, and a part interest in a Masston threshing machine and a Keystone corn planter, for a total equipment and tool value of $335. Significantly, the estate settlement took well over a year and involved filings with the circuit court of Johnson County. Clearly, the Reber farm was no simple yeoman operation.

By the opening decade of the twentieth century, a second generation of Mennonite farmers had transformed Washington County into a place more fully integrated into Iowa's mainstream culture. According to the local historian Kate Yoder Lind, church tensions had fully separated the Old Order Amish from the more progressive Amish Mennonites and Old Mennonites, the latter gathering in church buildings, meeting for winter Bible conferences, and even sending missionaries to India by 1899. This increasingly wider world of the acculturated Mennonites was also seen in more immediate, localized conduits to the outside, including free rural mail delivery at Kalona after 1898 and a mutual telephone company in 1901. And both the religious and social changes affected the way Mennonites related to the environment. New technologies, including refrigerator railcars, which exported fresh meat eastward, for instance, propelled the commercialization of Mennonite farms. Yoder Lind writes of new concerns with production output, leading the farmers, for example, to replace the "motley group of red, roan, brindle and spotted cows" with purebred stock. A more scientific mind-set now reshaped other aspects of agriculture: the old "plain one-bed" gardens were replaced with standardized plots of multiple straight rows; in the fields, farmers more intentionally rotated crops and made "plans to put tight fencing around the remaining acres of their land so that every field would be hog-tight."[56]

Perhaps here were all the signs of the modernizing farm, yet Yoder Lind sees deeply rooted gendered roles as complimenting these changes: the boys knew "when the soil was ready for spring work . . . when the clover was ready to be cut," while the girls needed to be "able to plant, care for and harvest a garden-full of vegetables . . . manage a flock of poultry . . . milk a cow [and] husk corn."[57] Iowa Mennonites valued a genealogy of knowledge, affirming successive generations' precise sense of belonging on the farm. And in turn they valued their respective farm's place within the identifiable local Mennonite community. Despite their engagement with the wider word, the local rural place served as the guarantor of a life of relative Anabaptist simplicity and bulwark against cultural threat from an ever-present wider society.

Manitoba: Right by Reserve and Land "Improvement"

The story of the Mennonite settler in Manitoba—the third geographic place of this study in order of its founding—has an even more defined ethnoreligious character. Here the environment, measured by vast sections of unbroken grassland and defined by the challenge of adapting to an unfamiliar climate, shaped a particular rural society. In what became the Rural Municipality of Rhineland, that is, the eastern half of the Mennonites' West Reserve[58] (thus named as it lay to the west of the mighty Red River and west of the smaller Mennonite East Reserve), Dutch-descendant immigrants from Russia laid out a cohesive community that was exclusively Mennonite. Indeed, in the 1870s they chose the newly minted, wintry province of Manitoba, in western Canada, over temperate Kansas and Nebraska, for these very reasons: Canada was willing to offer the Mennonites not only a federal guarantee of military exemption but also exclusive Mennonite agricultural land blocks.

In certain broad strokes, the Manitoba Mennonite community resembled settlements on the US grasslands. They were part of a continent-wide story of settlers invading the ancient homeland of Indigenous people. Indeed, the Mennonites arrived in Manitoba just four years after the Metis people's resistance, led by native son Louis Riel, was squelched in 1870. And they came just three years after the 1871 signing of Treaty 1 with the Anishinaabe Cree, which initiated a lengthy relocation of the Indigenous inhabitants to places that included Roseau River Reserve, adjacent to the Mennonites' own West Reserve. The Mennonites would benefit from what James Daschuk refers to as the federal government program of "clearing the plain," the often forcible extinguishing of Indigenous land rights, ensured by an inhumane manipulation of food aid.[59] Thus, when Mennonite delegates from Russia arrived in 1873 to scout the land, they were party to a scene of Indigenous defeat. As white colonists the Mennonites readily commented on what was considered a passing culture. The historian Lawrence Klippenstein's study of the delegates' diaries concludes that the Mennonite scouts' opinions of the Metis and Indigenous people varied widely. They spoke of the Metis as a "civilized class, . . . very obliging and hospitable," with an occasional "farmer [who] knew his business," but they also spoke of them as "lazy farmers of mixed Indian blood." As for the Anishinaabe, they were described at one juncture as

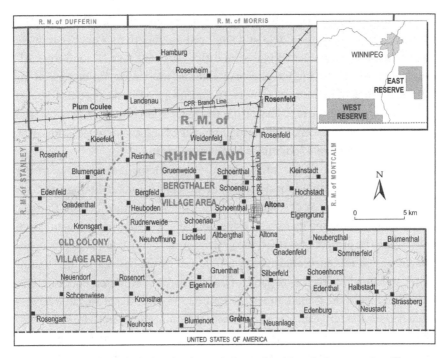

The Rural Municipality of Rhineland, Manitoba, and its historical Mennonite villages, including Neubergthal, founded in 1876. Map by Weldon Hiebert.

"gayly decorated," potentially "vengeful," and not "attractive neighbors" and at another as stubborn rejecters of Christianity, with one delegate quoting an Indigenous elder as declaring, "I will remain Indian; I will believe in the Great Spirit and follow the faith my fathers taught me."[60] No matter the varying descriptions, they represented the racialized gaze of the settler. Ultimately, the Mennonites' primary concern was with the veracity of US railroad propaganda depicting the Indigenous of Canada as vengeful and intent on disrupting white farmers' settlement plans.

Evidently the Mennonite settlers in Manitoba, like those in Iowa, considered any threat from the Indigenous inhabitants benign. According to the geographer John Warkentin, the 1875 survey of the Mennonite West Reserve was undertaken with apparent obliviousness to Indigenous people. And while the Metis had chosen wooded lands close to rivers, all the other white "settlers were ignoring the forty odd miles of open prairie between the Red River and the [Pembina] Escarpment."[61] From the Mennonite perspective, the treeless plain was one for the taking. Thus, in 1875 Jacob Y. Shantz, the Ontario

Mennonite assigned to help the first Mennonites settle the East Reserve a year before, hired two Metis riders and explored the length of what became the West Reserve. Shantz's aim was straightforward: to move beyond the variously rocky or low-lying East Reserve and find more room on better soil for the next round of Mennonite settlers. In the end, he easily secured the request for a second Mennonite reserve from the federal government. By July 1875 Mennonite settlers were registering homesteads on the new West Reserve, and by 1877 a total of 426 families had settled on it.

A second grand theme in the Manitoba settlement story links to the metropolitan-hinterland thesis, which historically is said to distinguish Canada from the United States, with its fixation on the frontier thesis.[62] Canada, in particular, extended its refrain of "peace, order and good government" from Ottawa to the project of settling the prairie west. Treaty 1, for example, negotiated by federal representatives, was followed by the passage of the Dominion Lands Act in 1872, the dispatch of the North-West Mounted Police in 1873, and the start of the Canadian Pacific Railroad in 1881, all setting an irreversible path to western Canadian colonization. Studies by Ryan Eyford on state-backed "white settler reserves" in Manitoba's Interlake region and by Shannon Stunden Bower on government-directed prairie-draining projects that transformed Manitoba's "wet prairie" underscore the metropolitan-hinterland approach to western Canadian settler history.[63] As Gerald Friesen concludes in his by now classic *The Canadian Prairies: A History*, an export-oriented prairie agriculture, accounting for nearly half the world's export market of milling wheat by 1928, was assured because the state organized an "entire society . . . to facilitate this activity."[64]

Given the nature of western Canadian settlement, the Mennonite communities in Manitoba became much more cohesive than those of their counterparts in Iowa. Unlike Washington County in Iowa, the West Reserve in Manitoba was closely monitored by a federal government intent on seeing it succeed. Indeed, Ottawa granted the Manitoba Mennonites a special "hamlet clause" that overrode the homestead system requiring that families reside on their own 160-acre "quarter section" parcel of land. Manitoba Mennonites were allowed to superimpose their own medieval-appearing open-field system with quaint *Strassendörfer*, literally "street villages," onto the legally surveyed, square-mile blocks, the checkerboard land system that extended across the Canadian prairies.[65] Importantly, group settlement of the type the East and West Reserves facilitated meant that each of the three Mennonite denominations that transferred their entire congregations to Manitoba in 1874 and 1875 could observe a certain territoriality. The small, financially well-to-do Kleine Gemeinde (literally, "small church") and the poorer but less religiously strict Bergthaler Mennonites (named after their colony in Russia) carved out the eight-township East Reserve into distinctive sections. On the massive, seventeen-township, or 400,000-acre, West Reserve, the Old Colony or Reinlaender Mennonites, most traditionalist of the three denominations, settled the higher, sandy loam soils of the western half.[66] The eastern half, the focus of our study, a land of mostly low-lying, heavy clay soil, drew the more moderate Bergthaler, who had rejected the soils of the East Reserve. The land blocks reinforced the Mennonites' sense of local community in more ways than one.

Perhaps these settlements seemed to constitute transplanted communities *par excellence*, as Warkentin argues in his cultural geography of the Mennonites of southern Manitoba. Still, the land proved immensely challenging. While Iowa's winters were colder and its summers were hotter than those in southern German lands, the Manitobans found themselves in even more starkly different lands than those of their old homeland in southern Russia. Certainly the Manitoba plain was fertile and inviting. As one early traveler, Alexander Henry, put it in 1800, after crossing what would become the West Reserve, here was "one level plain, without a hill or stone" and with a grass stand that "would be rather long were it not for the buffalo."[67] But the natural advantage of soil fertility was diminished by its heavy clay composition and the region's short growing season.

True, the Bergthaler Mennonite settlers' corporate memory was one of environmental adaptation. They knew that their ancestors, who had farmed subterranean soils in Friesland in the sixteenth century and northern Poland in the seventeenth and eighteenth centuries, had been forced to adjust to the semiarid, open grasslands of New Russia in 1789.[68] A short, two-hundred-kilometer secondary migration within Russia, from Khortitsa to Bergthal Colony in 1836, just north of the Black Sea port of Mariupol, reinforced the specific requirements for an agriculture on semiarid plain: black summerfallow, tree planting, and strategic plowing. But the move to Manitoba in 1874 and 1875 was quite another matter.[69] The temperature in the new homeland, for one, was severe in comparison with that of the old; while Manitoba's average July temperature of plus 20°C was similar to that of their homes in Russia, Manitoba's winter, with an average of minus 20°C in January, seemed bitterly cold, leaving the settlers with not much more than a 100-day frost-free zone. In addition to this stark difference, the eastern half of the West Reserve was made up of difficult-to-cultivate black clay chernozem soil, once the lake bottom of the ancient glacial Lake Agassiz, technically a lacustrine plain.[70] This heavy clay soil also possessed poor internal drainage, a fact exacerbated by the utter flatness of much of the eastern West Reserve. With an elevation drop of only a half to one and a half meters per mile and but a few natural streams, the land was prone to flooding. Only on occasional sections of land did the Bergthaler settlers find good, deep, well-drained loamy soils that graced the western half of the Reserve. Significant adaptation to the new homeland would be required.

Like the Mennonites in Washington County, Iowa, those in Manitoba's West Reserve eventually prospered. Indeed, many Bergthalers who had settled on low-lying or rocky lands of the East Reserve in 1874 relocated to the West between 1878 and 1881 to embrace much-improved grain-marketing opportunities.[71] Gerhard Ens's 1984 history of Rhineland Municipality, the eventual legal jurisdiction encompassing the eastern portions of the West Reserve, describes how "between 1884 and 1900 the amount of cultivated acreage increased every year, doubling in the six years between 1884 and 1890 alone."[72] Wheat yields of up to thirty-five bushels an acre in 1892, coupled with high prices of seventy-five cents per bushel, often resulted in gross annual profits of more than two thousand dollars on a typical farm. In addition to wheat production, as Joshua MacFayden shows in his 2018 work *Flax Americana*, West Reserve farms became known across Canada as eager growers of such specialty crops as flax.[73] The mid-1890s marked hard times, with weed infestations and low prices, but by the turn of the century steam engines were common, as were seed drills, binders, self-feeding threshing machines, and gang plows.[74] Early service centers such as the towns of Gretna on the US border and Altona in the east-central half of the reserve grew with the arrival of regional railroads, introducing Jewish, German, and British Canadian townsfolk, as well as new markets, to the insular West Reserve.[75] As Susie Fisher has recently argued in her innovative study of the Bergthaler villages, Mennonites transplanted their "seeds from the steppe" to the West Reserve, achieving an "affective nature of landscape construction" on a prairie "reimagined by European emigrants."[76] The Mennonites' wheat fields transformed the prairie, as did their gardens, flower plots, and the rows of cottonwoods that soon towered over most villages.

The nature of adapting to the West Reserve's specific environment is recorded in the first settlers' writings. The memoirist Peter A. Elias, who arrived here in 1879, wrote of the first task of finding hay for the family's cattle, then constructing the family's interim, subterranean sod house, the *semlin*,

An autumn hog-butchering bee beside a Mennonite house barn (note the white building attached to the barn on the left) in Neubergthal, Manitoba, likely sometime in the 1920s. Photo by Peter G. Hamm, Mennonite Heritage Archives, Winnipeg.

and a rough animal shelter. Next came the imperative of scoring firewood for Canada's harsh winter, a task made difficult by a huge snowfall in October. As Elias put it stoically, this "deep layer of snow" gave "us plenty of time to think about when spring might come." Even a generation later, Elias's letters to relatives back in Russia still referred to Manitoba's challenging climate. An 1897 letter written in late April, for example, notes that while spring seeding had finally commenced, it was late by Russian standards, as "we had a lot of snow this winter," so much so that at one spot "in our garden there still is a drift two feet deep." And while the wheat planting was completed, it was still too cold and wet for the oats and barley.[77] Yet in a later letter, one from January 1900, Elias crows about a well-established farmstead, complete with an "expensive" quarter section of 160 acres valued at five thousand dollars. Moreover, his farm boasted "fruit trees, plums, cherries, gooseberries, currants," a garden of "watermelons" and a bumper crop, that is, "652 bushels of wheat from 16 *dissiatinas* [17.6 hectares]," and similarly good yields of oats and barley.[78] Despite the harsh landscape, Elias could boast of success in a new environment.

Another set of early documents describing Mennonites' process of adapting to their new environment comes as an "outsiders' gaze." In the decades following 1876 a number of traveling Canadian, British, French, American, and even Russian writers and government officials visited the West Reserve and wrote about their encounters with the Mennonites. All were struck by the sight of reestablished old-world villages and freshly plowed land but noted the difficulty of adapting to the western Canadian grassland. J. W. Down, of the Canadian Department of Immigration, who visited in 1876, saw a people busily erecting sturdy frame wood or log houses, satisfied that they had already seeded almost fifteen hundred acres to wheat for export. He was concerned, though, that the Mennonites had not yet mastered the simple formula for breaking the West Reserve's "beautiful rich black soil": start with two-inch-deep plowing in June or July, when the grass, "full of sap . . . will readily rot and decay," then return for a four-inch plowing in fall, harrow the land just before winter, and wait for the spring thaw to reveal land ready for planting.[79] Other visitors from the 1880s spoke of the challenge of annual flooding. The American author Henry J. Van Dyke, who visited the village of Blumenort in 1880, commented on its neat "line of a dozen low thatched houses," especially impressive as many of farmers had been "forced to move twice on account of the wetness of the land."[80]

Then there was the challenge of weed control in heavy clay land. The Catholic priest Fr. Jean-Théobold Bitsche, for example, who visited in 1883, described simple "well maintained" homes but was critical of Mennonite fields, for "nowhere does one find more weeds than in their crops," resulting from "their poor system of ploughing," presumably their transplanted practice of springtime plowing, which compacted soils and retarded domestic seed germination.[81] Other observers, however, held out hope for successful adaption from these seasoned agriculturalists. W. Frazer Rae, of Scotland, applauded the Mennonites in 1881 for their innate agricultural acumen, for old practices of incorporating wheat straw into the soil and keeping cow manure for fuel rather than, like most settlers, "burning straw in their fields and casting their manure into the river." Moreover, they were committed to adaptation, appreciating "improved ploughs, thrashing machines and harvesters" they saw in Canada "and buying the novel implements of agriculture wherewith to cultivate the soil."[82] The sum of these outsider reports announced a settlement with a singular focus on learning to farm in a new environment.

Adapting to the high-precipitation clay belt and to the short growing season was the overwhelming concern of Manitoba farmers at the local level. John Warkentin notes how difficult it was for the Mennonites from semiarid, sandy Russia to make the transition to Manitoba. For at least the first decade many West Reserve farmers still plowed in the springtime as they had in Russia, meaning that "seeding was delayed and the crop was caught by frost in fall."[83] In time the settlers learned to replace old ways with new: they plowed their heavy clay land in fall and only lightly turned it in spring. Perhaps, as Warkentin asserts, "nowhere else in North America [had] a peasant culture from Europe been so thoroughly re-established."[84] Still, the strength of this communitarian culture was premised on learning the limitations and opportunities of flat, wet, and wintry Manitoba.

Siberia: Settling an Eastern Land

This preoccupation with land and community was remarkably similar to Mennonite concerns on the vast plain of southern Siberia. Here thousands of Mennonite immigrant settlers from New Russia (later Ukraine) began plowing the grassland a generation after western Canada was opened for settlement.[85] And in similar fashion to farming in the North American interior, the Mennonites on Russia's eastern "frontier" relied on the promise of a transcontinental railroad, a nation-state's program of encouraging European settlers despite historical Indigenous landownership, and a national policy of attaining food security by expanding cultivation. Here too, a colonial power advertised for settlers from afar to plow a grassland. This was the big picture behind the settlement of a set of Mennonite villages centered around Waldheim, later redubbed Apollonovka, in Siberia, some one hundred kilometers northwest of Omsk. It explains how the residents of this village district, founded in 1911 by members of the pietistic Mennonite Brethren Church, were attracted to Siberia from New Russia in the west and in short order built a successful farm community.[86]

Indeed, like the stories of Mennonites in Manitoba and Iowa, the Siberia narrative can be told by referencing the broad themes of settlement history: the extinguishing of Indigenous power, a rush for land, a biotic expansion, and the mythmaking of an endless frontier indelibly shaped by transplanted cultures. Alfred Crosby's classic *Ecological Imperialism* may describe Siberia as "the neo-Europe manqué . . . not far from Europe" with plant life similar to that of northern Europe, yet he also notes that Siberia was historically distinct, with its "fierce climate" and vast geography, a pastoral zone with Indigenous Siberians reliant on reindeer. It is surprising, writes Crosby, that Siberia ever became a "white" place culturally similar to the settlers' previous homes in western Russia.[87] As Aileen Friesen has recently argued, a central cultural force in this transformation was simply an effort by "church, state and settler alike . . . to [plant] Orthodoxy in Siberia," a transplantation that succeeded despite the

The Omsk region of Siberia, Russia, and its historical Mennonite villages, including Waldheim, founded in 1911 (later Apollonovka). Map by Weldon Hiebert

requirement to "define" the old religion within the context of colonization.[88] And yet, more heinous was the fact that as in North America, Indigenous peoples' tragic encounter with smallpox cleared the way for white settlers; by 1630, writes Crosby, smallpox had crossed "the Urals from Russia and pass[ed] through the ranks of the Ostyak, Tungus, Yakut, and Samoyed like a scythe through standing grain," sometimes killing three-quarters of a single nation.[89] The disaster diminished the Indigenous presence in Siberia to such an extent that when Russia's peasants began moving eastward in large numbers, 5 million by 1914, they deemed Siberia a vast, mostly empty land of opportunity. By the time the first Mennonites arrived in Omsk around 1900, they spoke of the Indigenous Kirghiz people as a harmless people, engaged mostly "in pastoral pursuits in summer, living in tents in true nomadic style." And they referenced the native Cossacks as a similarly benign people, engaged in farming, albeit in "a very primitive way," at least "not in the way we are used to it in the south."[90]

A second overarching theme in Siberia settlement history is environmental adaptation, specifically the need for settlers to farm within a continental climate on a semiarid plain. Southern Siberia was, after all, part of a vast grassland stretching eight thousand kilometers from New Russia in the west to northern China and Kazakhstan. In *The Plough that Broke the Steppe,* David Moon argues that a central challenge in introducing commercial agriculture to southern Siberia was its low annual precipitation of only 15 inches, or 380 mm per year, and its "hot summers and cold, freezing winters."[91] The promise of Siberia lay in its rich soils; like those of the Ukrainian steppe in the west, they were easily cultivated black chernozems, interlaced with "dark chestnut soils (*temno-kashtonovye pochvy*)," albeit "with some areas of salty soils (*solontsy*)." Just how to harness this dry, precarious soil was the preoccupation of state-backed scientific studies of farming, led by the famous soil specialist Vasilii Dukuchaev. He preached the treachery of destroying native grasses and woodlands and counseled "working with, rather than . . . combating, the steppe environment," especially pressing the case for tree preservation. Moon refers to Mennonites in particular as a people who successfully adapted to this type of low-precipitation agriculture in New Russia. Indeed, soon after their arrival from West Prussia in 1789, they put a millennia-old lowland agriculture behind them and eventually developed "techniques to conserve scarce moisture in the fertile black earth." Certainly, these early Mennonite settlers had invited soil erosion and dust storms by plowing up the native grasses and cutting the sparse woodland. But learning the hard way, argues Moon, Mennonites created dryland farming techniques that "proved useful in assisting the expansion of farming on the steppes further east" in Siberia.[92]

Historians of the Mennonite settlers themselves also agree on their primary concern with land and environment. By the time the Mennonites began settling Siberia, they had of course learned both to farm semiarid plains and work with colonialist governments. John Staples, for example, highlights the fact that the first Mennonite settlers in New Russia were flummoxed about how to make the semiarid lands productive. But he also emphasizes the work of temporal Mennonite leaders like the agricultural reformer Johann Cornies, who, through his government-backed Agricultural Society, enforced a four-crop rotation that included moisture-preserving black summerfallow and an ambitious tree-planting program.[93] In his essay on the Mennonite experience in Siberia, the historian John B. Toews emphasizes the state-issued settler incentives, similar in nature to those of the late 1700s. He describes the Siberian Mennonites' close working relationship with the federal Peresetencheskoe Upravlenie (Resettlement Administration), created in 1896, which served millions of other settlers, especially from 1907 to 1910. He highlights the land parcels of 48–50 desiatinas (53–55 hectares) in Siberia, slightly down from the 64 desiatinas (70 hectares) each family had been allotted in New Russia. In addition, inexpensive lands, low tariffs, and state credit for the purchase of inventory assisted the poor, although Toews asserts that "no leveling of class status was necessary, for the church had always emphasized brotherhood (*Bruderschaft*)."[94] Land in Siberia would allow for the replication of a closed agrarian and sectarian community.

But the Siberian settlers' own writings reflected a rather more significant change, one related to the requirement of adapting yet once again to a new environment. As Hans Werner argues in his 2012 piece "Siberia in the Mennonite Imagination," letters by the settlers in German-language newspapers in Russia or the United States were by nature positive in outlook, veritable booster invitations to come to Siberia despite reports of harsh weather and primitive conditions. They emphasized their own abilities to pioneer, adapt to weather extremes, cultivate Siberia's chernozem soils, and exploit "railroad-based markets."[95] The entrepreneur Peter Wiens, who had arrived in the Omsk region in 1897, dismissed the idea of "Siberia as a land filled with dangerous wild animals" in a letter in the *Odessa Zeitung*.[96] Siberia had all the potential of a "civilized" place, he wrote. Werner observes that most settlers, however, boasted not about cultural potential but about the richness of the land. Sometimes they described the function of Siberia birch forests, intersected by an occasional spruce grove. They were not simply trees but a crucial defense "against the snowstorms and . . . excellent fuel," wrote the farmer J. D. Enns in a 1902 letter in the *Mennonitsche Rundschau*. The matter of fertile soil was the dominant theme in many other letters too. In one 1904 piece in the *Odessa Zeitung* an unnamed writer described the "topsoil . . . as fertile and deep" even though the land was flat and dry, that is, "not well endowed with streams other than the large rivers, the Ob and the Irtysch."[97] Other writers boasted about Siberia's bumper crops. Perhaps the summer of 1903 had been cool and rainy, wrote Peter Wiens, but farmers simply adjusted, bringing their sheaves of wheat under the barn roof and threshing it inside during wintertime, resulting in a massive wheat harvest of 200 puds (ca. 3,300 kilograms) per desiatina (ca. 48 bushels per acre), with individual plants bearing five and six kernels per head.[98] Siberia had become the new homeland of promise; few wanted to return to the temperate old homeland of New Russia, or "return to Egypt," as one Siberia booster put it in biblical language in 1904.[99]

The promise of Siberia also infuses historical records from the residents of Waldheim itself. The short memoir of the farmer Abram J. Dyck describes the village as at once separated from wider Russia and the center of a close-knit, multi-sited Mennonite community. He begins his narrative by describing Waldheim's founding in 1911 about forty-eight kilometers north of the Gorkoja station on the Trans-Siberian Railroad. Indeed, he emphasizes Waldheim's remoteness, surrounded as it was by birch forest. But he also describes its tight communitarian nature. In fact, Waldheim itself was made up primarily of members of four extended families—the Toews, Dueck, Regier, and Thieszen clans—most from the single village of Blumenfeld in Molochna (Molotschna) Colony in New Russia. Furthermore, it was surrounded by three smaller villages, including the village of Wissenfeld (later Pokovyrovka) and two even smaller family-based places, essentially the estates of the Bekker and Wall families, named Bekkerchutor and Berezovka. Of these villages, asserted Dyck, Waldheim was the center, home until the end of the 1920s to the district's school and church as well as to the Mennoband Cheese Factory, a cooperative.[100]

Even more important to its development than its social character was Waldheim's environment. Waldheim's shallow wells produced excellent water, wrote Dyck, and its low-lying land yielded knee-high grass in summer and adequate wintertime hay for the cows, thus providing good milk flow year round. Years with abundant water and naturally fertile soil resulted in bumper crops in 1912 and 1914, giving the village a head start; indeed, Waldheim and the surrounding area quickly gained a reputation for predictably good crops. The most impressive natural element, though, was the birch forests, which had so stirred the settlers from the open plain of New Russia that they named the village Waldheim, "forest home." Indeed, the 900-desiatina (1,000-hectare) tract of land that made up the village district had equal portions of cultivated land and birch forest, interspersed with meadow, with the forest such a dominant part of the landscape that the first village mayor, the *Vorsteher*, Jacob B. Dueck, had difficulty overseeing a survey of it.[101] Despite this challenge, the settlers loved their forest; in fact the village's Mennonite Brethren church choir was said to be the best in the Omsk region, in part because it practiced in the woods, "absorbing the songs of the wild birds."[102] Here was a local rendition of the grand settler narrative.

A second Waldheim memoir, by Catarina Dueck Rahn, only four years old when her family moved to Siberia in 1911, engages with environmental questions related to landownership. Perhaps the environment here was less than idyllic, but Rahn's parents moved simply to obtain "their own land." Indeed, her family had considered investing in a flour mill at Khortitsa Colony in New Russia, but as Rahn puts it, "My mother did not want to hear of it" and persuaded not only her husband but her entire wider Regier family to move to

Siberia. Rahn details the process by which these settlers jointly purchased the land tract from the landowner Tjeljatnikow and then divided the total "in accordance with the quantity each one had purchased." The system somehow granted her parents a farmstead of 150 desiatinas (165 hectares), double the typical size and including arable, meadow, and woodland.[103]

But if land was the hope, the fierce climate, and especially Siberia's winter, was the main challenge, according to Rahn. Because there were eight children in the family, her parents built a large, fort-like dwelling of vertically lodged birch logs, the cracks covered with clay, and a roof of straw. Later, her family admitted that a "sod house would have been much warmer than the house with the thin walls."[104] But even when it was soon replaced by a fine-looking seven-room red-brick house and adjoining barn, the challenges of winter shaped domestic space; the summer room, for example, doubled in winter as the place where heating wood was stacked. The large barn was also constructed with an eye to winter: it housed the cows and horses but also the chickens, which needed a place warmed by the large animals. Rahn's family also housed the farm machinery in the barn and created space to thresh their wheat in winter. As a teenaged girl, she found that winter even challenged her own social space, keeping girls from meeting up with boys. "In the summertime," writes Rahn, "it was easier, as we could go out in nature, the girls [secretly] exiting the village street at one end, the boys at the other, and then meeting one another" in the birch forests, away from the inquisitive eyes of parents or preachers.[105] Unlike Dyck, Rahn emphasized Siberia's winter; like him, she saw the forest as a life-giving element.

Two final memoirists, both born in Waldheim in its early years, place their memories teleologically with reference to the Russian Revolution, which thwarted the dreams of the villagers just six years after settlement. Abram Enns, born in Waldheim in 1912, retells pioneer stories of the village, calling it a primitive place, "a wilderness" that needed taming. The work, he says, was simple and "labor-intensive," marking a time of horses and oxen, scythes and sickles, of two-meter-long threshing flails and only later a horse-drawn threshing stone. And yet, he writes somewhat wistfully, it was a time when all Mennonites, both poor and rich, could farm, some on purchased or rented land in the village, others on their own *chutors*, or estates.[106] Johann Epp, born in Waldheim in 1915, presents a similar narrative of hope and struggle, with a foreboding subtext that anticipates the short-lived dream at Waldheim. His family too had found in "thinly settled" Siberia the answer to the "urgent question—how long will there be farmland for everyone" in New Russia? But the Epp pathway to landownership in Waldheim would be arduous. At first the family tried farming in the large Mennonite settlement on the semiarid Kulunda Steppe in Altai, to the south of Omsk, but tragically Epp's father died of a blood clot in 1915, and hope was lost.[107] Two years later, in the spring of 1917, good fortune was restored when his mother married a man who owned land at Waldheim, one hundred kilometers to the north. As Epp describes it, Waldheim was a virtual paradise, lying in "a magnificent region [with] good, fresh, black earth for cultivation, immense yields of wheat and fresh air, thick berry-rich woods, with flocks of birds, all praising the creation." Here, he writes, everyone could "happily agree, 'on God's earth it is beautiful.'" Certainly life on the farm entailed hard work, but his parents worked the land with confidence, and in short order "the earth brought a hundredfold of fruit." It all seemed pre-ordained, that "the Lord would bring about even more blessings."[108]

This statement turned out to be ironic, for while Waldheim's physical environment was promising, Russia's political situation, and the 1917 Bolshevik Revolution in particular, was about to dash the villagers' dreams of landownership. In the broader narrative, the Waldheim settlers had hoped to tap into a nation-state's agenda to develop its conquered hinterland. At the local level they faced a strange new environment, but one with immense potential. Letter writers in particular spoke of this hope of Siberia, even given Waldheim's remote location and Siberia's short growing season. However, intense suffering lay in store for them, and postrevolution difficulties fundamentally undermined this environmental equation.

Conclusion

The story of agriculture among Mennonite communities in the Global North was one of engaging the physical environment in increasingly intense ways. In Friesland, farmers turned the sea bottom and peat bogs into productive farmland. In Iowa, Manitoba, and Siberia, Mennonite settlers plowed the grasslands of so-called "unending frontiers" and benefited from the rich natural fertility of those soils. At each of the four places farmers honed an agriculture based on localized knowledge. The Mennonites in Friesland built on centuries of experience in farming low-lying coastal lands; those in the diaspora—Iowa, Manitoba, and Siberia—transferred agricultural knowledge but also adapted by observing practices in their new homelands.

Agriculture in each of the four places also relates to a broader story. At each place Mennonite farmers produced crops that found their way to urban markets both near and far by rail. They were the markets of the metropolis, agents of the global marketplace, that is, the commodity purchasers in Amsterdam, Chicago, Winnipeg, and Omsk. Indeed, this story can be told as one of farms located within empires: the centuries-old Friesland fed an imperial center, Amsterdam; the three settler communities, located within powerful empires, were shaped by similar programs to create national agricultural economies based on diminished Indigenous rights, transcontinental railroads, and orderly allocations of federal land. Mennonite farmers benefited from such programs.

The story of farming in each of these four places in the North is also an account of a link between the cultural and the environmental. Indeed, the Anabaptist farmers of both Dutch and Swiss descent often tied their desire for farmland to religious goals. Humility and simplicity ordered early Dutch Anabaptist teachings, but agriculture in progressive Friesland was hardly a guarantor of these teachings. In the diaspora, though, rural life was given religious meaning, albeit with distinctive local dimensions. As the Swiss Mennonites fled into nearby German-speaking Alsace Loraine and the Palatinate, they developed a particular idea of the sojourning farmer, emphasizing simplicity and community, values they transplanted to Pennsylvania and to the American "frontier" state of Iowa. As Dutch Mennonites migrated to northern Poland, then to New Russia, and from there in the 1870s to Manitoba, they grew accustomed to rural life on exclusive land blocks, secured by Mennonite institutions and a growing sense that agriculture could well secure old Anabaptist values. Meanwhile, many Mennonites who remained in Russia found their destinies in an eastern frontier, where old pieties and ethnic exclusivity were reinforced in birch forests and the promise of farmland but then upended by revolution shortly after settlement.

This story of agriculture in the Global North has both global and local dimensions. The more obvious ones are global in nature, stories of agricultural expansion, broadly based knowledge exchanges, urban and imperial markets, and a culture that offered the farmer occupying new sources of agricultural land some form of legitimacy. But it is the local accounts, in their particular religious understandings and climatic adaptations to specific regions, that add substance to the global story. It is a global story of the reach of empire made visible by a local narrative of tacit knowledge.

CHAPTER 2

Peasant and Piety in the South

Planting Java, Matabeleland, and Bolivia's Oriente

In 1854, just three years after he was dispatched to Java by the Dutch Mennonite Missionary Society, Pieter Jansz met the Javanese mystic and recent convert to Christianity Ibrahim Tunggul Wulung.[1] Jansz, the first Mennonite missionary of the modern era, had become discouraged with the Dutch East Indies, and his mission work on an oppressive Dutch plantation was flailing. He found inspiration in Tunggul Wulung's approach to land and religion. Described as a tall, slender ascetic with a powerful, charismatic gaze, Tunggul Wulung was an unlikely associate of the missionary Jansz. For one thing, the mystic was stridently anticolonial. Moreover, he infused his newfound Christianity with strands of other faiths, with elements of Islam but also, more importantly, with Javanese animistic views on nature.[2] Tunggul Wulung even claimed special mystical powers, melded Christian and Javanese cosmologies, and continued using Javanese "spirit-craft" rituals.[3] It was enough to challenge Jansz's western sensibilities and evangelical missionary outlook.

Yet Jansz was attracted to aspects of Tunggul Wulung's Christianity, especially his idea of the rural Christian farm village. It was to be a society founded on old Anabaptist principles of meekness, equality, and simplicity and located in the middle of "primeval forest," well removed from the corrupting influence of colonial masters. Significantly, it also would be based on newly accepted Christian teachings on nature, especially those contesting old Javanese taboos. As Jansz put it in his diary, Tunggul Wulung's plan was to build a "village in the fearsome forest by the sea without fear," based on the theology that creation, although inherently spiritual according to Javanese teaching, also celebrated a transcendent creator in control of nature. Indeed this belief was a central theological tenet, encased in the mantra "God the Father, God the Son, God the Holy Spirit / the Three of them are one / Dangerous places, evil infested woods, all poisons . . . harmless / By God's grace we find safety forevermore."[4] This far-reaching environmental reimagination, hinting at a syncretistic Christian-animistic outlook, lay at the very foundation of Tunggul Wulung's village vision. It would also order the farm village established in 1882 by Pieter Jansz's son, Pieter Anton Jansz, at Margorejo, in the shadow of Mount Muria on the northern shore of Java, near several of Tunggul Wulung's farm villages.

The cultural boundary between the Global North and the Global South is, of course, an imprecise construction. Farming in both places was similar in many respects. When Alfred Crosby writes of agriculture in "similar latitudes in temperate zones," for instance, he has in mind not only the grasslands of the North but also similar places in the South.[5] And they might well have been settler societies in the Argentine pampa, the South African Transvaal, the Australian outback, and the Rhodesian inland plateau, places where white farmers built their neo-Europes. Then too, Western colonial power, which extended its imperial arm into North America and Siberia, also reached into every continent in the Global South.

But in other respects differences between the white settler society of the Global North and colonized land of the Global South were stark. The lowlands and deltas of Java, as well as those of India and China, for example, were not the domain of white farmers, even where colonial masters controlled agriculture. Here labor requirements determined that colonizers allow local farmers to keep their lands or at least to labor on vast plantations, producing the "soft drugs" tea, coffee, and sugar or the industrial products cotton, rubber, and indigo. In other places, such as Rhodesia, deeply rooted local agricultures were maintained, even as a minority of white farmers were granted the best land. And by the time the last of the agricultural frontiers, that is, the jungles and forests of Latin American places, developed in the mid-twentieth century, it was independent, postcolonial nation-states, sometimes acting from revolutionary impulse, that led agricultural modernization programs.

Some of these differences between the white-settler Global North and the colonial and postcolonial Global South are revealed also in the Mennonite agricultural story. As experienced in the South, it was inherently colonial. But it was less often an account of building neo-Europes on grasslands and more often one of farmers in jungle, bushland, and savannah, propelled by post-colonial politics and religion indelibly inscribed in landscape. In both Java and Matabeleland, native agriculturalists converted to Christianity and worked within a colonial framework, while maintaining elements of their nature-based religious teachings. In Bolivia, on the other hand, antimodern white farmers, driven by strict religious teaching on simplicity and communitarian cohesion, entered the last frontiers of farming in the twentieth century at the behest of a postcolonial nation-state.

In none of the three places in the Global South was agriculture directly controlled by a colonial power. Rather, local communities were given latitude to organize agriculture as they saw fit, with farmers able to pursue ideas about land and religion that were not fully aligned with capitalism and modernity. In securing these aims, however, the three Mennonite communities studied here diverged sharply, shaped as they were by locally grounded practice, lore, and even mythology, each responding to local soils and climates.

Java: Mysticism, Colony, and Control

Of the three places in the Global South, the farm village of Margorejo, in Java, was especially given to local agricultural expressions, even though it was set within a colonial power structure.[6] It was a rural place that imperial powers knew could not be settled, but merely controlled and exploited. Indeed, this was the situation in Java generally, the most populated island of the Dutch West Indies, when Pieter Jansz arrived in 1851. Although accompanied by his wife, Wilhelmina, only Pieter came to be heralded as the very first Mennonite to join the modern missionary movement, emulating Catholic, Anglican, and Methodist missionaries' work within the European colonial structure in Africa and Asia. The missionary vision may have faced opposition in some Mennonite quarters for being bolstered by colonial power and undermining ideas about simplicity, but the Dutch Missionary Society saw an opportunity in the strong arm of Dutch colonialism to spread the gospel.

When Jansz arrived in Java to spread Christianity among a predominantly Muslim population, he had no firm plan of action. But eventually the idea of a rural village, allowed by the colonial government and based on Javanese rice culture, took form. Margorejo was the largest and most successful of these villages located on the north coast of Java and founded by Pieter Anton Jansz in 1882. It was nicely located between the gentle, shallow Java Sea and the eastern foot of the dormant volcanic mountain Muria, rising sixteen hundred meters above sea level.[7] In the simplest of terms, Margorejo was a farm village

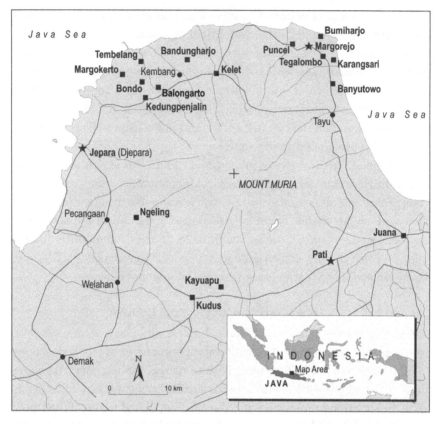

The Mount Muria area in northern Java, Indonesia, and its historical Mennonite villages, including Margorejo, established in 1882. Map by Weldon Hiebert.

comprising three geographic zones: mountainous foothills covered by dense teak forests; dry, higher soils; and low-lying wet fields, easily flooded for rice production.[8] In more complex terms, the mountainside soils varied from deep, yellowish-red tropical Latosols to clay-based Grumusols infused with molten lava, while the low-lying "bottom land" consisted of silty alluvial soils, and soils closer to the coast were somewhat higher, sandier Regosols.[9]

What mattered most was that because of Java's consistently warm, wet weather, these rich soils produced crops year round. Indeed, the two annual hot months, April, with an average temperature of plus 26.5°C, and October, with an average of plus 26.4°C, differ little from the cool months of January and July, each with an average of plus 25.4°C. Of immense significance for agriculture, Java receives a remarkable 2,800 mm (115 inches) of rain per year, mostly during the rainy period from November to March, with a monthly average of more than 300 mm (12 inches). Important too, this region receives good precipitation even in the dry season of July and August, with an average of 170 mm (6.5 inches) each month.[10] The missionaries and the local farmers knew that this region could experience severe flooding and indeed much more serious climatic eruptions, such as earthquakes, tsunamis, and volcanic eruptions. But they also knew its remarkable fecundity as the abiding characteristic of this environment.[11]

Margorejo was also affected by the intersection of Dutch colonialism and Java's ancient tropical rice culture. In this respect it was like much of the rest of Java. Clifford Geertz's now classic 1963 *Agricultural Involution* may have focused on cycles of poverty in Java, but it also provided insight into everyday life in the Javanese village. At the root of this society was the fact that "all aspects of culture [were] fully interdependent," with climate a central variable, meaning that there was "something vaguely . . . tropical about the Javanese [culture]." Tropical fertility ensured that while Java accounted for only 9 percent of Indonesia's total landmass, by the mid-twentieth century it sustained 66 percent of the country's population, in large part because 70 percent of its landmass was dedicated to agriculture, half of it consisting of smallholder-controlled wet rice land. This ecological equation was infused with deeply rooted traditional knowledge: "The micro-ecology of the flooded paddy fields," writes Geertz, came from two sources: the "specialized and technical," which included learned know-how about seed selection, water management, weed control, and fertility; and the "commonsensical, resting on a vast, unexamined accumulation of proverbial, rice-roots wisdom." The latter, argues Geertz, was based on a complex, long-standing agrarian culture of knowledge whose intricacies could never be fully known to the outside observer. And few aspects of rice production changed over time. Indeed, an 1870 agrarian land law ensured the maintenance of old ways: the Dutch masters would leave most lands in Javanese hands and allowed Java's culture of gender-equal, partible inheritance to remain even though it undermined land consolidation and efficient agriculture. Indeed, as Geertz puts it, a "dense web of finely spun work rights and work responsibilities" was required to bring order to a matrix of land fragments.[12] The result was that from 1871 to 1940 the average size of the Javanese family farm remained somewhat stable, at just under one hectare. An old social order prevailed even under Dutch rule.

This intersection of indigenous rice culture and Dutch colonial land policy was apparent in the history of Margorejo. A comprehensive history of the ethnically Javanese Mennonite church in Indonesia, the Gereja Injili di Tanah

Jawa (GITJ), by Sigit Heru Sukoco, a Javanese Christian and member of the GITJ Historical Commission, and Lawrence Yoder, a US national who was a longtime resident of Java, tells the story of Margorejo in all its dimensions, including the environmental. Based on Dutch-language mission reports submitted to Pieter Jansz's mission board in Amsterdam and on oral history, Sukoco and Yoder paint a picture of Jansz as increasingly seeing the farm village as a tool in mission work. A first step in this story was Jansz's choice to work among local Javanese farmers rather than among urban Chinese-Javanese residents despite his own urban background in the Netherlands as the son of a book peddler. Second was his commitment to education, even if it meant establishing his first school on a massive plantation dedicated to government-approved export crops—vanilla, sugar cane, and indigo—using forced labor. But third was his increasing repulsion toward the exploitive nature of the colonial system of which he himself was a part; as he put it, the system simply undermined "normal and good human ethics." Baptizing but five converts in four years, Jansz complained in an 1854 report that colonialism was deadening the spiritual: "the countryside . . . is beautiful and green," but "no one is . . . seeking the water of life."[13] And he blamed the Dutch colonial masters in part for what he considered widespread spiritual malaise.

It was thus that Jansz adopted the mystic Ibriham Tunggul Wulung's vision of a Christian farm village, at once generating a helpful social milieu for the spread of Christianity and also protecting the basic rights of recent converts. In 1856 Jansz created a fledgling version of such a village near the city of Jepara, on the northwest coast of Muria Peninsula. Tunggul Wulung, however, headed to the more isolated eastern side of the peninsula, where in 1861 he established his first Christian village. While Jansz stuck with the village close to Jepara, he both was leery of and admired Tunggul Wulung's work. In an 1870 report to the mission board, Jansz complained of his inability to appeal to Javanese farmers: he reported that he lacked knowledge of the Javanese culture, the ability to speak in riddles and cross religious lines, and even the long beard required of a guru. And yet, guardedly inspired by Tunggul Wulung's persona and attracted to his idea of a Christian farm village, Jansz moved forward to a new phase in his mission work.

In 1874 Jansz wrote and published a short book in Dutch, *Landontginning en Evangelisasie op Java* (Land reclamation and evangelicalism in Java), describing in detail how such a farm village might function from a Dutch perspective. It would, of course, be a Christian village aiming for the highest of moral standards, but it would also be profitable, as residents would receive marketing direction and learn to mitigate risk, taking "profits to pay for losses in a bad season." In addition, agriculture would be complemented by medical facilities and schools. Finally, such a farm village would be independent of the Dutch colonial government, and to that end, Jansz's plan prohibited any "government spies" in the village.[14]

Jansz's farm-village plan stalled, however, when it was presented to a parsimonious mission board in Amsterdam. Eventually it was his son, Pieter Anton Jansz, born in 1853, shortly after his parents arrived in Java, who saw the idea come to full fruition. As a young adult, Pieter Anton traveled to the Netherlands for his education. But within two years of his return to Java, in 1880, he wrote to the Amsterdam mission board shaming it for having less courage than "owners of capital who are not religious" and warned of ever more land falling "into the hands of greedy industrialists" complicit with the Dutch government. He ended his report with the logic his father had developed over his first twenty years in Java: only the right social milieu, supported by a Christian education, could guarantee a missionary's success. How could the mission board's love for schools be effective, he asked, if such "schools are not located in a Christian environment?"[15] Driven by this passion, Pieter Anton Jansz set out to establish his own mission village some eighty kilometers from his father's quarters in Jepara, near two of Tunggul Wulung's villages on the northeast coast of the peninsula, just north of the town of Tayu. Here he chose a piece of land described as "wooded with smaller trees, and not far from the shore of the Java Sea." In August 1882 he received a seventy-five-year lease for a 114-hectare (288-acre) parcel of land from the Dutch government.[16]

Based on the ideas from his father's 1874 book, Pieter Anton Jansz began with a village charter, one that eventually comprised forty-seven articles.[17] A close examination of the original document indicates that the younger Jansz would leave few things to chance. Articles 1 and 2 required that "only those in good standing would be allowed to reside in Margorejo." And as an early sign of deference to the colonial government, the next three articles declared that the villagers would be law-abiding: "no trade or possession of opium" or "strong drink" would be allowed, nor would the village farmers be able to intrude on the government's coffee and timber monopolies, either by trespassing on government landholdings or by "clandestinely selling or buying coffee or wood." Articles 6–10 defined the general moral nature of the village; it would bar idols, which were deemed "contrary to both Christian and Mohammeden religions,"[18] as well as gambling, polygamy, and "especially adultery." But of utmost significance for a particular approach to the environment, villagers were also forbidden to host traditional, sensuous Tayub dancing parties, featuring female dancers, or Wayang puppet exhibitions, both rooted in Javanese spirituality and performed at planting and harvest times.[19]

A third group of articles effectively created the template of a just and equitable society. Article 11 forbade the charging of "high interest rates," as well as any absentee landownership. Also among these socially oriented articles were those ordering all farmers to contribute to a tithe for purposes of mutual aid and religious practice after their sixth year in charge of a piece of land. Article 15 compelled all children to attend school from the age of seven and all adults to "learn to read" unless old age prohibited it. Article 12 placed an element of enforcement on these various proscriptions, including the note that the penalty of disturbing the peace was expulsion from the community.

If these articles brought a certain egalitarianism to the village, they also underscored the village's power structure. The principal authority would clearly reside in the "renter" of the tract, that is, the white missionary, who was accorded executive power. No marriage could be enacted or dissolved without his permission, church services called by him must be attended, and no one could host an outside guest overnight without his consent. He could also order particular cultivation practices, prohibit farmers from working outside the village, and enforce building regulations as well as standards of cleanliness. The secondary authority was the democratically elected village chief, chosen from the ranks of the male Javanese farmers. While he could not be a minister, he had to be elected from a list provided by the "renter." The village chief's oversight was especially focused on building and maintaining the village's physical infrastructure, including "roads, bridges and water pipes." Clearly, Pieter Anton Jansz hoped to use the farm to reorder Javanese morality, but he also anticipated the creation of a separate Christian society, ironically one quite distinct from the rather integrated Dutch Doopsgezinden of Jansz's homeland in the Netherlands.

The last set of these village regulations, articles 18 to 46, contained a fifth group of requirements, ones that provided farmers with particular rights and also certain responsibilities. The most important regulation here was article 26, which dictated that "all residents when they arrive and as long as supply lasts" were entitled to a plot of farmland, and in fact they could qualify for "more land" if they but "improve" their initial piece. While they were permitted to fully cultivate their land anytime within two years of arrival, article 20 dictated that they must "level the slopes for terracing as soon as possible." In any case, farmers could not obtain more than "five *bau* of land," or 3.5 hectares, for any one household. Indeed, to ensure this limit, farmers were prohibited from selling land to fellow villagers without permission from the "renter." Further, they had to keep their cattle in a common enclosure, with only their horse allowed on their own land, and more specifically, their own farmyard. And if they did leave the village for an extended period of time, they were required to make arrangements for the land to be cared for.

Yet farmers had specific rights. Farmers requisitioned for village work by the "renter" were to be compensated at twenty-five cents a day, while the elected village chief was not permitted to ask for unpaid work or to extract any bribes. Moreover, at least one male per household could be exempted from communal work at any given time. The final articles, article 45 in particular, noted that the land would be held by a male household head but allowed him to bequeath land to his wife and children, with minor children's land to be held in trust, rented at 50 percent of the yield, until they reached the age of majority. This lengthy charter ensured the foundation of a successful agricultural experiment, indeed the creation of a particular Mennonite farm village in the Global South.

The story of establishing the farm in 1882 is once again reliant on Sukoco and Yoder's research. The authors tell of how a mission school was moved from Jepara and how the village was officially named Margorejo, meaning "road to well-being." Within two years the village had attracted ninety-one persons, all described by Jansz as "very poor" and including followers of Tunggul Wulung and some Muslims.[20] By 1887 some 60 hectares (147 acres) had been cleared, both rice and dryland crops planted, and a grove of lumber-bearing kapock trees established. And because mandatory tithing set in only after the sixth year, the church planted its own farmland to produce the needed resources for mutual aid and religious programming. The church, and presumably Jansz himself, worked to create a particular local economy, implementing a seven-day week to replace the mystically linked five-day Javanese week. It made Wednesday and Saturday the market days and reserved Sunday as a day of rest, riding roughshod over the names of Javanese days—Pahing, Legi, Pon, Kliwon, and Wage—each with its own cultural and environmental significance. Both a physical and a cultural reordering was in the making within five years of the village's founding.

Times were difficult during the first years. Mission reports indicate that relatively few people were actually baptized, and there were instances of adultery and even a murder. By 1890 signs pointed to outright religious decline, with more excommunications that year than baptisms, and a lopsided membership of fifty-eight female members to forty males.[21] Yet in the midst of this turbulent time Jansz received a boost: in 1888 Johann Fast, the first Low German–speaking Mennonite missionary from Russia, arrived in Margorejo to become Jansz's associate. Separated from his Dutch ancestry by 350 years, Fast nevertheless worked well with Jansz and before long even married Jansz's youngest sister, Jakoba.[22] Within the next decade Fast oversaw a certain consolidation of the mission, including the construction of the first church building—described as a wooden structure measuring ten by twenty-one meters—and the further expansion of the mission land block to 175 hectares (433 acres). He would also be the forerunner of a line of German-speaking Mennonite missionaries from Russia who served at Margorejo, including Johann Huebert (1893 to ca. 1905), Johann Klaassen (1899 to 1906), and Nickolai Thiessen (1906 to 1937). Despite the obvious cultural chasm between this new set of missionaries and Margorejo, they possessed a historically conditioned suitability for the farm village.[23] As Sukoco and Yoder put it, there were striking "similarities between the [farm] settlements they had come from [in Russia] and the Christian [agricultural] settlement the mission was trying to develop among the Javanese."[24] In both places farmers lived next to one another, chose a temporal leader, generally reserved landownership to religious adherents, and farmed individual plots of land. Most importantly, both village cultures were undergirded by the idea that a close-knit rural society augured well for religious adherence.

Yet the Russian missionaries' greatest boast was that they promoted a new cosmology in Java. Early reports from Nikolai Thiessen, published in the Russian-based *Die Friedenstimme* shortly after his arrival at Margorejo in 1906, certainly noted cultural differences between the missionaries and the Javanese farmers. In his first missive, of April 1906, he announces that he is "in charge of daily activities" in the village but notes that he is learning Javanese "with pain and suffering."[25] In his second letter, just a few months later, in August, he again references the "mysterious and difficult-to-understand" Javanese language, but now he introduces his own message to the village. Ironically, there are numerous references to seeding, harvesting, drought, yield, and warm weather, but mostly as metaphors for spiritual vitality and not alluding to actual weather and farming. Indeed, in this lengthy submission there is only one reference to agriculture—"our colonists have had a good rice harvest this year. . . . The Lord has spared their fields from flood and insect damage"—but it is a telling one. A new perspective on nature, with monotheistic Creator God in charge, was being heralded. The success of Thiessen's efforts is left in some doubt, as he inadvertently reports also on peasant resistance to his leadership. He complains, for example, that his instructions to the farmers seem to fall on deaf ears, as he needs to repeat his orders every day. When he asks them, "Why did you not do so and so?," the peasants answer variously "I didn't know I had to," or "I forgot," or "It didn't occur to me." He sees this lack of order as inherent in a culture where "parents don't have the heart to discipline their children," and he worries for the next generation.[26]

The resistance described by Thiessen seems to have abated in successive years. In 1910, for example, Johann Klaassen, who had served at Margorejo a few years earlier, returned for a visit. In a subsequent report he describes a rich farm settlement just an hour and a half by wagon on "good roads" from the closest rail terminus. The settlement was graced foremost by the beautiful church building of teakwood featuring a prominent sign announcing, "This a prayer house for all people," and accompanied by an imposing missionary house, perched a meter above the ground to allow the humid air to circulate. He also reports in positive terms on the school, filled with lively children shouting out answers to the teacher's questions, and highlights their enthusiastic "I want to!" to his own question whether someday they would like to reside in heaven "with Jesus, the universal friend of children." He also writes of being impressed by the market, filled with wares announcing in English "made in Germany," as well as household tools for "everything a woman needs" and local products: fresh fruit, vegetable, eggs, chickens, and fish.[27]

But it is Klaassen's report published in early 1914 from Pati, not far from Margorejo, that best describes the lasting contribution of the missionaries in planting a Judeo-Christian idea about environment in Java. Having recently returned to Java from a furlough in Europe, Klaassen describes a Thursday night prayer meeting dedicated to the planting time, similar to one held for the harvest time. As he puts it, neither of these meetings was "strange for the Javanese, for in their pagan Mohammedan religion they also have their particular time in which they pray . . . and before work in the rice field begins . . . they bring offerings of flowers, rice, incense, etc." Klaassen describes their traditional beliefs in a "village spirit" that lives in "large trees or the forest" watching over the "house, the garden and field." He writes that it is not difficult to attract the farmers to the Christian prayer meeting, where they can bow to a much superior divinity, "God, the creator of the heavens and earth, and ask him for help." Klaassen reports on one particular prayer meeting as especially memorable, for it occurred at a time of severe anxiety over a drought. The farmers' earnest appeals to God for rain were heard, he writes, for "that very night, it began to rain, and rained for a long time . . . pleasing the farmers." He was especially moved that sometime later "the Mohammedans were overheard saying among themselves, 'The Christians prayed, and for that reason God sent the rain.'"[28] A new culture of the environment had been established in northeastern Java.

Records from the early years of the new century indicate that Pieter Anton Jansz's vision had come to fruition. By 1900 Margorejo numbered more than a thousand persons, rice paddies covered half the village territory, the school was flourishing, and a three-ward polyclinic was in operation.[29] Margorejo was also a solidly rural Javanese farm village; indeed, a teacher's college constructed in 1902 worked against creating a class of "Black Dutch," that is, prideful Javanese, so that Dutch was not offered as a language of instruction. But significantly, because the students all hailed from farming backgrounds, neither were any Western-oriented agricultural courses taught.[30] Significantly, there was a new cultural understanding of just how the environment worked. This reorientation would shape the very history of Margorejo over the course of the century.

Matabeleland: Land and the Cheap Missionary

The establishment of a "Mennonite," technically a Brethren in Christ, farm community shaped by a specific local culture, colonial discourse, and a distinctive physical environment is also the story of the second site in the Global South. Matopo Mission, established in 1898 in Matabeleland, in Southern Rhodesia, the rogue extension of the British Empire, was first and foremost a Ndebele place. Matopo was, after all, the site in which a powerful subnation of the Zulu Empire, the Bantu-speaking Ndebele, had mushroomed into existence with a population of twenty thousand by 1823.[31] Their leader, King Mzilikazi, was an interloper from Natal who had led his people over the Drakensberg Mountains and then onto the Transvaal, where he had battled Boer farmers before heading north to the Matopo Hills. Here among the giant rocks of Matopo, Mzilikazi and later his son Lobengula subsumed local Kalanga people and settled down to an orderly agricultural way of life.[32] They created a structured culture based on cattle and gardens that rewarded agricultural success. They lived in their *kraals*, a derivative of a Dutch-Boer word for "farm," heeded the counsel of the *abadala* (elder), and prized their *izinkomo* (cattle) and *amabele* (maise fields), which were worked by the *abafazi* (women).[33] These ways, indicative of intricate, cultural underpinnings, were tested in the late 1890s as the Ndebele stood up to a new enemy, Cecil Rhodes's

British South African Company, and waged a pitted war against it. The Ndebele suffered a bitter defeat in 1893 and again in 1896, but the environmentally
rooted culture of the community remained intact, ready for quite a different
battle another day.

This was the social condition into which US missionaries from the Brethren in Christ Church (BICC), closely allied to the Mennonites, implanted
themselves in 1898. They were preceded in the region by the famed Robert
Moffat, of the London Mission Society, who had begun his work seventy years
earlier, but the BICC missionaries were the first to use agriculture as a mission strategy. The BICC base was a land grant of 3,000 acres (1,214 hectares)
obtained from none other than Cecil Rhodes by their own leader, the BICC
missionary Jesse M. Engle, who hailed from a farm in Pennsylvania. Engle's
untimely death in 1900 allowed a gifted, single female missionary, Hannah
Francis Davidson, formerly a college teacher in Kansas, to shape Matopo Mission over the next six years.[34] Neither Engle nor Davidson was trained in
missionary activity, but both knew agriculture, and they set out at once to
build a Christian farm village. By 1900 a fledging community had been estab-

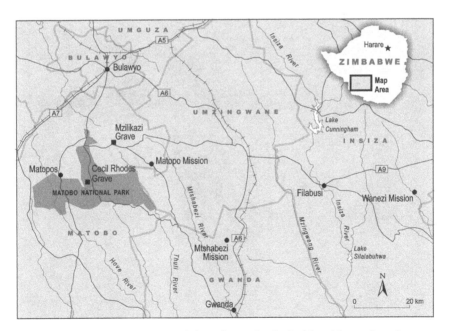

Matabeleland in southwestern Zimbabwe (formerly Rhodesia) and its Brethren in
Christ Church missions outposts, including Matopo Mission, founded in 1898.
Map by Weldon Hiebert.

lished at Matopo Hills. And even though the missionaries sought to impose their own ideas about agriculture upon the locals, Ndebele converts tended their corn gardens and herded their cattle according to ancient ways of understanding nature.

It was a particular agriculture, closely reflecting Matopo's distinct environment. Its climate, mostly a dry savannah biozone, supported the growth of low-profile bushland and open grasslands, while the gardens were spring-fed on rich, loamy soil.[35] In more fulsome terms, the soils have been described as "generally coarse and sandy with pockets of clay soil" along with "local patches of organic soils and peat."[36] The most striking physical feature of this countryside was massive granite batholiths that covered at least half of the local area, often bordering the patchwork of gardens.[37] They were especially imposing, spectacular stacks of house-sized round stones, balancing one on the other. Despite their wasteland appearance, the batholiths were actually life giving, for at their base water seeped from under the rocks into ponds, thus enabling the watering of catenas of rich soils.

Grandmother Danile Ndlou with grandchildren and great-grandchildren at the family *kraal* near Matopo Mission in Matabeleland, 2015. Photo by Belinda Ncube, Bulawayo.

These diverse surroundings held the foundation of a sustainable agriculture. Indeed, they had encouraged settlement from prehistoric times, when Iron Age farmers grew their crops near the natural ponds and springs. Agriculture was also shaped by distinct seasons: the January summer temperature averaged a hot plus 26°C, while the July winter was moderately cool at plus 17°C. The greatest challenge here was uneven precipitation, with rainfalls ranging from a high of 136 mm (5.3 inches) in January, down incrementally each month to a scant precipitation of only .61 mm (.02 inches) in July, and then slowly rising again to the January high.[38] Effecting these rates was a history of El Nino–based droughts that could keep the crucial January rains below 100 mm,[39] but these could be intersected by sudden deluges accompanied by occasional tropical cyclones.[40]

Despite these weather aberrations, not only did the Matopo region possess suitable soil and water for corn but the natural vegetation also featured a wide range of trees for fuel, fencing, and roof frames. Around the batholith rocks were the short albizia tree, the evergreen cassia, the combretum bushwillow, the burl pterocarpus, the spiny ziziphus shrub, and the flowering kirkia.[41] The open woodland areas featured the *Burkea africana*, that telltale iconic deciduous, flat-topped tree of up to eight meters in height, as well as the even taller, round-leafed bloodwood, the *Pterocarpus*, while the more open bushland patches supported the termite-resistant, butterfly-leaf mopane trees. The open areas held a variety of grasses, including somewhat precarious perennial and annuals, but also areas of hardy *Hyparrhenia* spear grass, useful for thatching, while the edges of bushland sustained good-quality, taller *Panicum* grasses that could be grazed.[42] The environment also provided a variety of food sources. For example, this ecosystem sustained some four hundred species of birds, numerous small herbivores, including the rat-like dassie and the small klipspringer antelope. But even with its plethora of grasses and trees, Matopo was hardly a farmer's paradise; its ecosystem also supported an occasional leopard and rhinoceros, large numbers of black eagles and raptors, as well as brown hyenas, warthogs, and zebras, all of which could threaten agriculture.[43]

This multifaceted ecosystem was not, of course, the only variable in the life of the Matopo farmer. Agriculture here, as in Java and elsewhere, was the result of a dialectic between the physical environment and culture, that is, local lore, religious hybridity, and the assumptions undergirding colonial political structures. Terence Ranger's *Voices from the Rocks* speaks of Matopo as a region of resistance to colonial rule, both in the 1890s and the 1960s. But he also argues that its resistance was shaped by religion, arising from animistic ideas and old teachings regarding the Ndebele's "rights (and duties) as guardians of the land."[44] Those rights stemmed from the religious mythologies that gave meaning to both the small springs among the granite expanses and the drought-prone bushland and grasslands. "It is hardly surprising," asserts Ranger, "that a Shona myth of 'the creation of water' is situated in the Matopos." He relates the lore of Mudzanapabwe, the messenger of God, who transformed a sterile "hot and dry" country by striking the granite rock with "a red needled" arrow, thus releasing "life in the big rocks." The water that oozed from the rocks became a sign of divine mercy.[45]

Ranger's thesis has recently been advanced by the BICC historian Eliakim Sibanda, who critically contrasts the views on the environment of the Ndebele farmers and the BICC missionaries. He decries how BICC missionaries undermined Ndebele "immediate and direct" cultural connection to the land. And he highlights Ndebele identity as *abantwana benhlabathi*, "children of the soil," who viewed the soil "as a mother with morals," needing to be revered. But as a mother, it could also nurture "her children, protect and mete out justice."[46] To meet their spiritual goals, argues Sibanda, the Ndebele intentionally kept their farms small, believing that humans, plants, and wildlife all had a right to the land.[47]

This view of nature is apparent, as it was in Margorejo, in the only early written texts describing the community, those by the missionaries themselves. For Matopo, the main source is the 1914 memoir of Hannah Francis Davidson, based on her detailed diary. Davidson records the first eight years of the Matopo Mission farm, from 1898 to 1906, the year she left to establish her own mission farm in present-day Zambia. Along the way she also offers her Western view of the environment, often contrasting it to the Ndebele's. Indeed, Davidson begins her memoir with the unequivocally colonial view of immense and "dark" Africa. On the ship steaming down the west coast of Africa, Davidson and her fellow BICC missionaries pondered just where in vast Africa, that "sealed book . . . with its barbarism, its unknown depths, its gross darkness," they should establish their mission. Their urgency drew on a biblical metaphor: "The field is white. The harvest is ready. Who will go forth in the name of the Master?"[48]

For Davidson, however, the environment was much more than a cultural abstraction. Indeed, on her trip from the United States to Rhodesia in 1898 she carefully observed agriculture along the way. Arriving in Liverpool from the United States and traveling by train to London, she notes the well-kept English fields, although she also suggests that the farming methods seemed an-

tiquated "to people fresh from the farms of western America." She made her
second foray into a foreign countryside after arriving in Cape Town and receiving word from Cecil Rhodes that the BICC would be granted the 3,000
acres in Matabeleland, the recently subdued rebel state two thousand kilometers to the northeast. On the four-day train trip from Cape Town to Bulawayo, she again comments on the land with reference to the United States: the
vast Transvaal, looking "like a desert; not a blade of grass to be seen, [and only]
the red sand . . . covered with bushes" seems very foreign, although other
parts "greatly resemble Kansas prairies." Her highest praise of the countryside
is reserved for the Matopo Hills themselves, whose beauty she once again
compares to that of a place in the United States, this time Michigan, her place
of birth: "To what a beautiful place God had led us for His work. There, spread
out before our eyes, was a beautiful rolling, valley of rich, dark earth. . . .
Here . . . are sparkling fountains of beautiful water, crystal clear, oozing from
under the surface of the rocks, and flowing down the valley. . . . Some surpass
the Michigan lakes in transparency."[49] Even before the farm village at Matopo
had been surveyed, Davidson evaluated lands for their agricultural potential,
not so much as a farmer might, but as an outsider with an agenda, that is, a
foreign missionary.

Throughout her memoir Davidson also assesses the Ndebele approach to
the land. Soon after arriving in Bulawayo, her BICC party procured the necessary wares and supplies for the mission farm and also secured an eighteen-
donkey team to pull a huge wagon when they traveled to Matopo. As they
did so, Davidson pondered Ndebele cosmology. She describes the recent Ndebele uprisings and offers her own Western viewpoint: "The disease among
their cattle, the locust, which devoured their crops, and numerous other trou-
bles were all, by their witch doctors, laid at the door of the white man. *Um-
limo* (their god) also affirmed that their King was still alive and was ready to
assist them in gaining their liberty." She dismisses these beliefs and casts the
farmers of Matopo as impoverished subjects whose spiritual and physical
needs only the white missionary can address. Upon their arrival in Matopo,
she describes a general destitution, with defeated farmers still unable to grow
sustainable corn crops or build cattle herds "to fall back upon as they usually
had in time of grain famine."[50] But to her mind, the Ndebele's lack of means
was owing not only to recent war but also to their old subsistence agricultural
ways, which devalued hard work and profit.

At every turn, Davidson employs her own Western viewpoint to understand Matopo's environment. At one juncture she lauds Matopo's dry climate

as "healthy" and "not unpleasant," even though it is too hot during the day and too cold at night. But then she offers a scientific rationale for it with the pronouncement that "Matopo Mission is located about 20-1/2 deg., south latitude and 29 deg., east longitude, 5,000 feet above sea level." She also offers an evaluation of its agricultural potential, writing that "although it is within the tropics, it has a delightful and salubrious climate the entire year," with a well-defined November-to-April rainy season. Her observations on climate are also intertwined in the very strategy of mission work. A year after arriving at Matopo, the missionaries began their work in earnest; in Davidson's words, it was an "aggressive campaign against Satan and his followers among the rocks and strongholds . . . for we felt that the Lord would have us press the battle to the gates."[51] Significantly, the campaign occurred in the dry season, for according to Davidson, the verdant rainy season was not a good time for missionary work. It was the time when locals were either busy gardening or enduring illnesses, with no time to receive the gospel.

Davidson is equally derisive of the mimetic and communitarian nature of the Ndebele's farmwork. She describes their harvest as groups of neighbors taking turns striking the grain with sticks on the large flat granite rocks. But it was more than just work: "decked out with all their ornaments [the threshers were] divided into two sides . . . [and] perform[ed] . . . a mimic battle with the grain lying on the rock between . . . two lines of battle, each one alternately driving the other before it and at the same time beating the grain with their sticks." All the while they sang Ndebele war songs, intersected with special ballet dances. Davidson was at once appalled, fascinated, and impressed. "The whole was exceedingly heathenish," she writes, "but not uninteresting; and as for the grain, a large amount of it was threshed." From her perspective, the environment was most pristine when it corresponded with Christian ritual. She writes of the very first baptism at Matopo and how nature was invoked in the ceremony: "In August, 1899, nine boys and one girl were by Elder Engle led into one of those sparkling streams and dipped three times in the name of the Trinity, and thus put on the Lord by baptism."[52]

Davidson's ideas on the environment are also encapsulated in her description of the history of the mission farm. She juxtaposes the missionary efforts to growing food with those of the Ndebele farmers, writing dismissively of the former and somewhat glowingly of the latter: "Of course at first there were no vegetables to be had, except such as we could, at times, procure from the natives—corn, sweet potatoes, pumpkins, and peanuts. . . . Elder Engle, alive to the value of the soil and the need of wholesome food, at once secured fruit trees ... bought a small plow and with the two donkeys broke land and planted vegetables." Engle's plot produced food that was both worthy of the mission-ary and able to fetch a profit. Davidson writes of the high prices that vegeta-bles, eggs, and butter fetched in Bulawayo during the 1899 Boer War, mean-ing that the mission farm's "little spring wagon, drawn by four donkeys, went to Bulawayo nearly every week for a time, taking in produce ... so that, dur-ing the darkest days of the war, all our needs were supplied." She also marks "progress" at the mission school with regard to a particular approach to farm-ing, naming it as a form of "industry." In one early report she writes of brick-making lessons and adds that "up to this time the industrial work of the boys had been chiefly on the [mission] farm and in the gardens." To her mind, ag-riculture as the Ndebele knew it did not spell progress but impeded it. Indeed, she writes that the biggest threat to running a proper school was native Nde-bele agriculture, which disrupted the children's school attendance. "If there was anything to be done" on their farms, she complains at one point, "such as digging in the gardens, herding, keeping the animals from the gardens, or running errands, the children must stay at home and attend to it."[53]

Davidson writes of her deep apprehension concerning this form of native agriculture. For one, the Ndebele seemed oblivious to the notion of a creator and too concerned about the spirit world imbedded in creation. She was es-pecially worried about signs of pantheism. In her words, the "natives do not concern themselves about a Supreme God," creator of the grassland, bush, and rocks, telling her that "we came here and found them already created, so we did not concern ourselves to inquire who made them." It seemed a foolhardy construction to her, for even though they had dismissed the idea of a creator, they still saw their own religious deities inscribed in nature. She especially decries how the natives saw the *Amadlozi*, the "spirits of the departed," who either caused illness and other misfortunes or assisted them in attaining their every desire. At treacherous points in life, argues Davidson, the Ndebele fool-ishly summon the witch doctor, who divines the source of the problem. Da-vidson is also apprehensive of indigenous herbal medicine and especially of the way their consumption of powdery herbal substances is "mixed with charms and other superstitious ideas," so much so "that it is difficult to tell wherein the real remedy lies."[54]

Matabeleland's environment was strikingly different from Java's. But in Davidson's account of the farmers who converted to Christianity, it was sim-ilar in a fundamental way: it marked a distinctive intersection, even clash, be-tween Western and native religiosity. It was a struggle of cosmology shaped by both the distinct climate of the place and the respective colonial political structures.

Bolivia: Building a Bridge to the Past

The third place in the Global South, Bolivia, serves as a reminder that white settler societies were not only the domain of the North. In a way, this story of Mennonites and agriculture was merely a part of the last push by farmers with roots in the Global North to colonize frontiers in the Global South. This is the story of Riva Palacio Colony, the largest of several immigrant Mennonite settlements founded just south of the city of Santa Cruz de la Sierra, in the lowlands of eastern Bolivia, the so-called Bolivian Oriente, in 1967. But it is a peculiar part of a story of deforestation and settlement, for Riva Palacio also stood in opposition to capitalist agriculture, intentionally antimodern and communitarian. The Mennonite farmers here were members of the Old Colony Mennonite denomination, formed as a separate group in western Canada in 1875 and then transplanted to northern Mexico in 1922 after objecting to Canada's English-language-school legislation as well as capitalism's increasing hold on Mennonite farmers.[55] Within the broad mountain valleys in northern Mexico's Sierra Madre the Old Colonists had built farm communities committed to old agrarian ways. The arrival of modern technology—rubber-tired tractors, electricity, and the car—after the Second World War, coupled with perennial shortages of farmland, however, threatened the foundation of this horse-and-buggy community. The result was the mass migration to Bolivia and the cultivation of its eastern frontier.

In Bolivia, the Old Colony farmers entered an unusual and ironic relationship with a postcolonial state bent on agricultural modernization. Indeed, Bolivia came to recognize that the physical feature it was famous for—its Altiplano, the Andean home to minerals (famously silver and later tin) and high-altitude crops such as quinoa—was no guarantor of either economic stability or national food security. To reach these goals, Bolivia would need to turn the tropical forest of its fertile eastern lowland into agricultural land. It would also need to embrace the so-called Green Revolution, encouraged by US-based foundations, and allow the use of hybrid seeds, insecticides, and synthetic fertilizers. Beginning with the Easter Revolution in 1952 and the revolutionary policies of the Movimiento Nacionalista Revolucionario (MNR) government under Victor Paz Estenssoro, Bolivia committed itself to a policy of national food security and a diversified economy based on these elements of modern agriculture. This policy would remain in effect even during political turmoil,

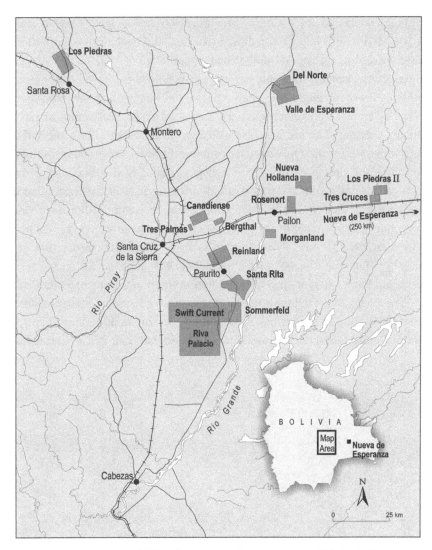

The original Mennonite colonies of Santa Cruz department in Bolivia, including Riva Palacio Colony, established in 1967. Map by Weldon Hiebert.

surviving a 1964 coup that put General René Barrientos Ortuño in power and then enduring his own mysterious death in 1967. And it came with a national and international campaign to encourage farmers to migrate to the eastern lowlands.

As many as 250,0000 Indigenous farmers from the Altiplano responded by 1980, as did some 3,200 farmers from Japan (including Okinawa), and so too did the traditionalist Old Colony Mennonites from Mexico.[56] By 1970 some 4,500 Mennonites had accepted the invitation, and by the 1980s, with thousands of additional arrivals, Bolivia had a population of some 12,000 Mennonites. The reason for their interest was clear: Bolivia possessed a geography suitable for their religiously ordered horse-and-buggy transportation, steel-wheeled tractors, and electricity-less houses.[57] Indeed, these cultural markers, in a setting of ample farmland well away from modern life, promised to keep the traditionalist Old Colonists' communities cohesive, their farms small, and their life simple. A charter of privileges granted by the Bolivian government to a small group of Paraguayan Mennonites in 1955, then extended in 1962, assured the Old Colonist Mennonites military exemption, local self-rule, and church-run schools. Both this provision and the natural environment provided the basis for Riva Palacio agriculture and also for the agriculture of a number of neighboring colonies with roots in Mexico—Santa Rita, Sommerfeld, and Swift Current—just fifty kilometers south of the city of Santa Cruz.

The Old Colonist migrants in Bolivia encountered a major challenge, however. The country's physical environment was fundamentally unlike the humid, continental climate of western Canada that their grandparents had

Two Old Colony Mennonite buggies passing on Brecha 7 roadway, linking two villages at Riva Palacio Colony in Bolivia, at the end of the rainy season, April 2012. Photo by Kennert Giesbrecht, *Mennonitische Post* (Steinbach, MB).

known from 1875 to 1922, and it was also unlike the high-altitude, semiarid plain of northern Mexico that their own parents had called home from 1922 to 1967. By contrast, the Bolivian eastern lowland was covered by dense bushland and experienced extremely hot, humid summers and temperate winters. This region's range of soils, precipitation, and temperatures also was strange. Indeed, so was the vast alluvial plain of these lowlands, capturing rain and runoff from the Andes to the west and sending water northeastward into the vast 100,000-square-kilometer drainage basin of the 1,500-kilometer-long transcontinental Rio Grande. The soils of the far western side of the lowland area, in the vicinity of Riva Palacio, were generally well-drained soils, relatively fertile, albeit wanting in organic matter.[58] But even within the bounds of Riva Palacio the soils could vary greatly. The low ridges featured the most fertile "dark brown, loamy sands," locally known as El Palmar soils, while lower-lying areas possessed less fertile, reddish-brown clays with ironstone concentrations and the least fertile dark grey clays. As a whole, though, because the soils were technically "old soils," with a high degree of clay in the subsoil, they didn't absorb water well, meaning that intensive rainfall could flood the land despite relatively good drainage otherwise. If soils varied in the colony, so could rainfall; the annual average for the region of about 1,000 mm (or 40 inches) per year was distributed unevenly in the colony, which straddled the humid Amazonia of the north and the semiarid Grand Chaco in the south. Moreover, in the months from July to October moisture was markedly deficient, while there was a rainy season in the summer months of January to March. Indeed, parts of the wider Rio Grande plain could well be inundated with perched water for weeks at a time.[59] This feature became especially apparent once the low-lying bush of chaparral shrubland and xerophyte woodland were cleared, removing natural means of water absorption.

This is the land the Bolivian government set aside for settlement, for Bolivians and non-Bolivians alike. The development of these eastern lowlands is described by Ben Nobbs-Thiessen in a *Landscape of Migration*, featuring a complex interaction between three parties—a postcolonial state, settlers in tropical bushland, and Indigenous inhabitants—each with its own aspirations.[60] Nobbs-Thiessen argues that it was not coincidental that when the MNR government began developing the eastern lowlands after Bolivia's 1952 Easter Revolution it welcomed groups such as the Mennonites. Self-sufficient, quiescent groups could further the national agenda of food security. Indeed, the process reflected a fundamental aspect of earlier settlements in Russia and Canada: a federal government opened up land for settlement by dismissing the concerns of Indigenous inhabitants. In the case of Bolivia, the Ayoreo, the Guaraní, and the Guarayo were discounted as potential inhabitants of the Oriente because their Indigenous culture was said by state officials to encourage "low density land use." Indeed, the Bolivian government presented the Oriente as a fertile plain "abandoned" by any potential Indigenous inhabitants, while the response to its campaign to encourage the Andean Aymara and Quechua peoples to resettle in the lowlands was seen to be anemic at best. Mennonite settlers, on the other hand, were quietly welcomed as more certain allies in advancing the national project of replacing "a colonial, *extractive* state, centered on the mining economy of the Andes, with the creation of a modern *cultivated* state based in the east."[61] Clearly, this was a country that did not need the modern citizen, integrated culturally into the nation-state, so much as an agrarian people willing to cultivate the nation's soil and contribute to its economy.

Early diaries describe the process by which the settlement occurred from the viewpoint of Mennonite migrant farmers. Some diaries available for this study were simple travelogues, without much reflection on either religion or the environment.[62] The 1968 diary of the farmer Johann K. Peters, who moved from Campo 38, near Cuauhtémoc, Mexico, to Bolivia, for example, is but a rudimentary record of the financial costs of the migration and travel dates. Peters notes in brief how his group of settlers left home on March 18 by bus, arrived in Panama City on April 4, traveled by ship to Arica, in northern Chile, and then from April 12 to April 16 continued on by bus to Santa Cruz via mountainous La Paz and Cochabamba. It's an outline of a journey that is definitive and unrelenting, with the ocean and mountains serving as impenetrable walls to any temptation to return.

The more detailed of these diaries, however, offer accounts beyond the broad strokes of the significant historical moment; they introduce the specific political and environmental challenges of the transcontinental move. One of the leaders of the immigration, Johann Wiebe, for example, offers a veritable blizzard of minute details of his work in organizing the migration, with ample references to both religion and the environment. Wiebe details several scouting trips to Bolivia even before the move of the first settlers in 1967, a dozen trips to Mexico City to negotiate with government officials, and innumerable visits for similar purposes to Chihuahua City. His diary, almost desperate in tone, begins with a clear imperative: "And so in 1966, Mr. Abram Loewen, Mr. Cornelius Reimer, *Vorsteher* [colony mayor], Peter Neufeld, and Peter Fehr were sent to Bolivia to find a place where we could secure our religious freedom." In Bolivia, writes Wiebe, "everything could be obtained," such as the right to their own schools and enough bushland to guarantee a separation from the wider society. Upon the land scouts' return to Mexico, the *Bruderberatung*, the assembly of all male church members, heard a report on the Bolivian offer, voted to purchase the requisite land, and sent another delegation to finalize the deal. After the second delegation's return to Mexico, a few more Bruderberatung meetings helped to divide the 40,000-hectare parcel, costing but fifteen pesos a hectare. Signaling their desire for continuity, they named the piece Riva Palacio, after their own settlement in Mexico. The only question that remained was "how to get there . . . by land and/or water?" To work out this detail, a third delegation, comprising scouts and Old Colony religious leaders, was dispatched, and it approved the route by sea through Chile. In the end, though, most Old Colonists flew by "jet," a value-laden word that Wiebe renders in English, from Mexico City to La Paz and then on smaller airplanes to Santa Cruz.[63] Ironically, the airplane, the latest tool of modernity, was key to transporting an antimodern, anti-car people to their promised land in primitive Bolivia.

Johann Wiebe chose to travel with a group that took the more traditional land-and-water route. He tells of the trip in epic language of religion and geographic fissure. It begins with his own personal prayer at the point of departure: "Also send on my journey / Your angels for my sake / And speak to me your blessing / Lend it to everything I do / Lord, send me your strength / From your heavenly abode."[64] The actual leave-taking from the Wiebes' 165-hectare farm in Mexico on February 22, 1968, was a departure "for the last time," and their exodus from the colony itself, on February 28, entailed a final "farewell from our parents, perhaps for the last time in this life." The chartered bus trip to Panama, the sea voyage to Chile, and the overland trip through the mountains to La Paz and eventually Santa Cruz each take on an epic quality. The voyage on the Pacific, as Wiebe describes it, was an arduous one, with the migrants encountering enormous, seasickness-inducing waves. Their arrival in high-altitude La Paz at night without any food turned into the miraculous: word spread in the city of the foreigners' need, and a throng of "adults and children" descended on the Mennonites' seven-bus caravan with "bananas and bread." The crossing of the 4,100-meter-high peak beyond La Paz, a passage fraught with danger, came next. But then the descent to Cochabamba, at 2,570 meters, and ever farther down to Santa Cruz ended, upon the bus company owner's insistence, with a victory lap around the city's central square.

The migration portion of the account ends as it begins, with a reference to land. As soon as the Wiebe family had been examined by Bolivian doctors, and the women and children settled in rented facilities in Santa Cruz, Wiebe and his eldest sons headed to the site of the new colony to report to *Ältester* Bernhard F. Peters, the bishop. Then they set about clearing bush for their own household. The story of the group's settlement in a strange environment is as detailed as the story of the land purchase and the physical relocation. Wiebe at once saw the difficult task ahead of him: "As the first villages [settled by an earlier group of Old Colonist migrants] and those closest to the [main] road were already full, we knew no other way than to begin a new village." Adding to the difficulty, the new village needed to be grounded in the bush well beyond the current end of the cleared survey lines. A few members from the Wiebe group had the means to purchase farms established in the previous months by the earlier group, but most did not. And so, returning to their temporary quarters in Santa Cruz, Wiebe discussed settlement options with the other household heads in his group, and together they decided "to go back to the colony with our boys, and if [survey] lines had not yet been cut open, to cut the lines and create a street by hacking out the bush by ourselves."

When the men and boys returned to Riva Palacio, they discovered, to their dismay, that the street through Steinbach, the village just before their own future village of Schoenwiese, was only half completed. Another five hundred meters of street needed to be cleared before work could even begin on their own farmsteads. So the Wiebe-led group hacked down the rest of the trees to complete the streets for both villages. Only then, writes Wiebe, did "we begin with the driveway of our own building site and begin cutting out plots of 15×15 meters so that we could have wells made." The work was done cooperatively, first the lot for one family and then the lot for another, until each household had its place in the forest. At that point they hired the Bolivian company Equipetrol to drill five wells, each seventy meters deep, in the village. Then, having constructed rudimentary "hen houses," which would double as initial dwelling places, the men and boys returned to Santa Cruz for April 12, Good Friday. There they assembled with the women, girls, and small children to sing a few songs "as there was no preacher in the city," and on Saturday they moved the families to the colony. Thus "on Easter Sunday we were nicely settled in the colony." Ironically, this day in which they could celebrate the first steps of settlement coincided with the anniversary of the 1952 Easter Revolution, which had enabled the opening of the lowlands to settlement in the first instance. Importantly, Wiebe's narrative makes no such reference. The detailed account, from the departure from Mexico to the settlement of the new farm village in Bolivia, is bookended with religious, not political, description.

Because this was an intentionally religious community, Wiebe's focus was not ultimately on his own farm. He also had duties to the village. Although Wiebe reports making good progress on May 2, hiring Bolivian workers to hack out two hectares of woods "so that we could begin planting," just three weeks later he received word that he had been elected Riva Palacio's *Vorsteher*, the colony mayor. Wiebe was distraught, as this news came on the very day that his wife became very ill, the beginning of a seven-month struggle for her life. Nevertheless, she insisted that he accept the result of the open election, and over the next two years his responsibility became overwhelming as one "difficult situation" after another faced the colony. "Each dealt with land, sometimes with no [written] agreement," writes Wiebe, alluding to the problem in eastern Bolivia in which the absence of a formal, systematic cadastral survey led to a system of many overlapping titles.

Without elaborating, Wiebe writes that his first survey-related problem had to do with "making villages" within Riva Palacio, presumably dealing with boundaries between villages. But then a second, larger problem arose when the neighboring Santa Rita Mennonite colony was forced to purchase 10,000 hectares since what had been believed to be a 40,000-hectare parcel, sufficient land for both colonies, turned out, in the absence of the cadastral survey, to be only 30,000 hectares. In the midst of a standoff between the two colonies, the government threatened to impose import tariffs unless the Mennonites settled their internal conflict. Then a third issue arose when the government's Enstituto de Colonicaio pressured the colony to pay for their land earlier than they had agreed. Emergency village assemblies were called, but the settlers simply did not have the required eighteen pesos per hectare. Fortunately, the state's creditors relented. Then, in a fourth upheaval, "a man from Santa Cruz came and said that he had purchased 5,000 hectares of our land in 1952," a claim that led to an intense year-long legal struggle before a settlement was reached. And finally, there was an internal problem when in 1971 the Mennonite farmer Johann Guenther, of Schoenthal, "unexpectedly incurred a large debt," landing him in prison and requiring the colony both to auction off his farm and to impose an internal tariff on all cattle sales to raise the money to cover his debts. As Wiebe puts it, "Oh, the situation was bad." And yet Wiebe never writes of wavering in his leadership role. The construction of an agrarian community in the bush was nothing less than a religious imperative. There would be no return to Mexico.

Another early text of the settlement years, a Texas A&M master's thesis in anthropology, tells the story of environmental adaptation from a fundamentally different perspective. The author, James Lanning, a non-Mennonite, made several visits in 1969 and 1970 to Riva Palacio, where he interviewed older members of the colony in English, a language they still remembered from their pre-1922 days in Canada. In this text, Lanning is the outsider, analyzing the newcomers' ongoing adaptation to Bolivia's tropical lowlands. He emphasizes that the "sea of green awaiting them is a shocking contrast to the barren wind-blown prairie characteristic of the mother colony in Mexico." He lauds the Old Colonists' response to the physical environment, noting how they exchanged their Mexican "oats, barley and rye" for Bolivian "corn, kaffir sorghum," and how they successfully planted potatoes, carrots, radishes, cabbage, peas, and beans as well as new crops—peppers, squash, melon, and papaya—within the tropical environment. He describes the settlers' homes, which for many young families were simple wooden structures, often having but an earthen floor, with furnishings constructed from local mahogany. Everywhere Lanning saw adaptation to a new environment. Even the children participated in a task designed for them in Bolivia's distinctive environment, "the collection of snail skeletons . . . dispersed on the surface of the earth," which were then ground and fed to their laying hens as a supplement.[65]

As an anthropologist, Lanning also emphasizes the cultural meaning and religious values that undergirded the rural settlement. He notes specifically that this land was meant to secure the Old Colony Mennonites' tradition-based faith, indeed serving as a place where "old ways could be restored." He even predicts on the basis of the colony's 1971 census, with a population of 2,172 persons, that it might well replace Mexico as the center of Old Colony religiosity.[66] He speaks of the linkage between farming and health but also of "the rigors of pioneer life," and he insists that "simple . . . living habits have blessed men, women and children with a degree of good health." He describes the Mennonites' self-confidence. They had disrupted large landholding estates and now served as national food producers. Lanning reproduces a 1968 op-ed in the national *El Diario* in which the federal head of immigration praised the pacifist Mennonite farmers as a "national asset," for "the more Mennonites we have tilling our lands, the greater number of nationals can go to war." Lanning recounts, from overnight visits with the religious leader Reverend Jacob Wiebe, the satisfaction settlers derived from speaking about pioneer hardships. He seems to respect the Old Colonists' insistence on steel wheels on tractors, although he is dubious about their practice of wearing "full heavy dark clothes that absorb the heat of the sun." Toward the end of his thesis, Lanning quotes the farmer Gerhard Enns, aged sixty-eight and father to fifteen grown children, who is grateful that "a new frontier opened up" in "a new underdeveloped country."[67] It was a frontier guaranteeing the Old Colonists a future in Bolivia.

The Wiebe diary and the Lanning thesis provide a detailed story of one of the locales that was part of a global narrative. Written from markedly disparate points of view, both nevertheless highlight an environmental adaptation shaped by the particular culture of a group of settlers. Both emphasize the farmers' singular focus on their own challenges. Although from two vantage points—the subjective internal and the social scientific external—both narratives assert that agricultural success will secure the farmers' place in the new land and fulfill a religious imperative.

Conclusion

The three farm communities in the Global South stand at odds with the four in the Global North. Each of the southern communities, located variously in South America, Africa, and Asia, has a story of colonial reach from the North to the South. But each is also characterized by resistance to Western capitalism in some fundamental way. The two farm villages in Java and Matabeleland were founded with full support from colonial masters, and yet both groups of farmers would assume some agency within those structures. Protected in part by the white missionaries, they were also propelled by the local farmers' own pre-Christian approaches to the environment, which highlighted values of humble yieldedness and self-sufficiency. The community in the Bolivian Oriente, founded in another century and following another logic, was a site of resistance in its own right. First, the settlement was linked to the agenda of a postcolonial government seeking within the Green Revolution the instrument to gain economic independence from Western dominance. Second, Mennonite farmers found within the eastern lowlands the resources for an agriculture that resisted the logic of full-blown capitalism, its attending technologies, and its economies of scale. Each of these communities was also founded on overtly religious principles: the communities in Java and Matabeleland were mission farms meant to safeguard religion through agriculture; the Bolivian community was an antimodern project, the culmination of a series of migrations. Each of these places contested the contours of modern agriculture.

Still, the story of these three communities in the South tells us less about the overall nature of agriculture in the Southern Hemisphere than it does about the remarkably diverse, local expressions of religion and land management. Religion may have worked itself out in a particular fashion in the integrated geography of Friesland, and in yet another way in the grassland diaspora in Canada, the United States, and Russia, but the linkage of religion and land was different again in the Global South, where it was inherently experimental, shaped by the idealism of both Mennonite missionaries and antimodern stalwarts. As the following chapters reveal, even this characterization does not fully account for the dynamic of religion in the lives of these farmers. Profound technological change, lived religion, inescapable government intervention, evolving gender relations, debates on climate change, and far-reaching transnational linkages all pointed to ways that local cultures played themselves out on the land. The common religious identity of these seven Mennonite communities certainly highlights the effect of religion on everyday farm life. But this narrative also demonstrates the limit of religion in shaping rural community. Indeed, it accentuates the immense diversity of agriculture at the local level, within this earth's varied climates, ecozones, and environments.

CHAPTER 3

Something New under the Mennonite Sun

A Century of Agricultural Change

In January 1923 the *Kalona News*, the foremost newspaper in Washington County, Iowa, carried a front-page story on the funeral of an Amish widow, Katie Miller. What the reporter found noteworthy in Miller, a farmer in Arthur, Illinois, 280 miles from Kalona, was her 600-acre farm, valued at $151,000, making her "the largest woman land owner and the wealthiest Amish lady" of her community. It was status enough, according to her bishop, to make this funeral the "largest" in the county's Amish history.[1] Hers was an old status marker; a person with farmland, no matter their gender, had the instant respect of folks from across the region. A similar marker was evident in the September 1922 report of a lightning strike on Edwin Hershberger's barn in Washington County itself. The hit, resulting in the devastating fire and loss of the barn. was accentuated by the report that his had been a "good barn" containing "a lot of hay, grain and other stuff," although fortunately "no live stock was destroyed."[2] The "good barn," like that of Miller's abundant land, was the measure of a farmer who had made good.

But in the 1920s another status marker was on the rise: technologies powered by fossil fuels. Two years before Katie Miller's death, a January 1921 story in the *Kalona News* reported on the auction sale of the local Mennonite farmer Emanuel Yoder, which earned "$1500 more than he expected." The Yoders now had the means not only to move off the farm and retire in the town of Kalona but also to plan a springtime escape by "auto route to the west [coast], where they hope the change of climate may prove a benefit to Mrs. Yoder's health."[3] The car, along with the distances it could travel, was a new mark of rural success. And so too was the gasoline engine in everyday farm life. Auction sale notices in the *Kalona News* reveal that Mennonites themselves were embracing the fossil-fuel craze: in November 1921, for example, the farmer Moses Gingerich's closing-out sale listed not only livestock and commodities but also "gas engine, quarter-horse electric motor, electric washer, tank heaters."[4] An August 1922 advertisement in the *Kalona News* typical for its time boosted the Fordson, "the Universal Tractor," not only "ready for any one of the 101 jobs" but now owned by 170,000 other farmers around the world, all enjoying a "drudgery-saving" life. Not everyone could own a 600-acre farm, but it didn't matter, as the new standards of success were the more easily acquired automobile and tractor. These fossil-fuel-driven machines also happened to recalibrate human interaction with nature, shortening physical distances and easing fieldwork. The machine would transform agriculture as Mennonites and even the Amish of Washington County knew it.

The story of agriculture in the twentieth century is an account of fundamental change. Mennonite farmers, in many places somewhat skeptical of modern ways, were particularly aware of new technologies, both mechanical and chemical, that propelled rural transformation. For most Mennonite farmers, the new agriculture was nothing less than "something new under the sun," to borrow the title of J. R. McNeill's important book; here was an "expansion of agriculture," echoing elements of an "agroindustrial revolution," as farmers learned to "manage" nature in new ways.[5] In the process farmers often also radically reshaped that nature. At the level of the farm this transformation could be multifaceted. It meant the decoupling of animals from the land and an alienating commodification of them.[6] It also entailed an embrace of crop monocultures, further drainage and deforestation, and the introduction of synthetic fertilizers and chemicals to fight weeds, insects, and pathogens. Each of the seven Mennonite farm communities in this study encountered these changes in some fashion. Crucially, each also experienced them in a distinctive way, some in a pronounced and sustained manner, others as a process that reflected but a modicum of change, and even then only late in the century. The particularity of national economies mattered, and so too did legacies of colonialism, civil and world war, and population densities. More localized factors were also important, including regional climates and soil types, degrees of local poverty, and the community's religious practice, including the abiding sense of a communitarian, moral economy.

This chapter tells the story of transformation at the local level, from the local perspective, based on local sources. The task is complex, as no two communities among the seven possessed identical historical records or even a similar volume of such texts. Their creation depended on each community's history of social stability, culture of recordkeeping, and prevailing levels of education. Nevertheless, from a variety of newspapers, memoirs, diaries, local histories, letters, account books, and institutional records, a story of the seven communities over the course of an entire century emerges. This chapter is based on a particular ordering of the seven places. It begins with those places that experienced agricultural "expansion" most systematically and incrementally (Iowa and Manitoba), proceeds to those with significant degrees of mechanization but interrupted by war and revolution (Friesland and Siberia), visits a community that intentionally resisted progress (Bolivia), and ends with those given to only sporadic and belated opportunities of development in colonial and postcolonial contexts (Java and Matabeleland). For some communities this ride into modernity was remarkably smooth, and for others it was inherently, even treacherously bumpy. In all cases, though, it was a story of change over the course of a century.

Iowa: From Rural Community to "Ag" Society

Iowa fits the paradigm of agricultural change in the most classical of ways: a slow but seemingly sure unfolding of modernity from the first decade of the twentieth century to the last, stalled only temporarily by economic downturns such as those there in the 1930s and the 1980s. Local Amish and Mennonite farm diaries and account books, many held at the Mennonite Museum and Archives in Kalona, Iowa, tell this story.

Those from the first half of the century reveal only tentative steps toward modernity. The account book of the thirty-nine-year-old Amish farmer Dan (Denny) Miller for 1939, for example, lists items sold at an auction sale just before he and his wife, Katie, and their seven children moved from Ohio to a 120-acre farm in Washington County, Iowa.[7] The auction sale list shows a farm edging toward modern ways: the items included a "dung hook," a neck yoke, 5 harnesses, 2 wagons, 975 bushels of oats, 400 of corn, 9 tons of hay, and at the end of the list 1 "gas engine." Clearly, this mixed-commodity farm still relied on animal power, but a small step toward modern technology was evident in the stationary gasoline engine, a sign that even the Amish had come to accept the fossil-fuel-burning engine as a necessary tool of agriculture. Meanwhile, the midcentury diary of the middle-aged Mennonite farmer Erlis Kinsinger, of Iowa County, abutting Washington County, reveals the contours of a larger, 200-acre farm beginning to concentrate on the commercial production of hogs and corn.[8] As his June 1954 diary attests, this farm too had embraced an aspect of modern agriculture, not so much in mechanization as in farm specialization, in this instance a focus on hogs. On June 15, for example, Kinsinger "took 5 hogs to Parnell," enough product to indicate the farm's main source of income. And yet, when, four days earlier, on June 11, Kinsinger had "moved sheds," giving the hogs fresh pasture to do their rooting and natural fertilizing in open air, he signaled the continuation of the old link between animals and the land. Then on that same day he "ground feed," and on the sixteenth he "got feed" from a neighbor, actions suggesting the basis of an old moral economy undergirded by household self-sufficiency as well as communitarian, neighborly ties.

The second half of the century fundamentally altered the pace of the change. For this time period, a single diary, that of the Mennonite farmer Duane Peterson, who was married to Janet Yoder in 1960 and served as a partner in the large Yoder family hog and turkey farm, reflects the very contours of modern agriculture.[9] Peterson's remarkable diary possesses an entry for each day from 1956 to 2000, the year he died tragically in a farm accident. The arbitrarily chosen date of April 13 for each of these years reflects a single annual reference point by which to measure change. The April 13 entry for 1966 tells us that Peterson "unloaded [a] carload of shavings" and "disked till 9 pm." The first task is clearly related to a large shipment of litter for the farm's large, confined turkey barns, and the second to a new era of the light-equipped tractor operating in the fields an hour after sunset. The entry for 1971 notes that Peterson "put [the] sprayer on the 806 [International tractor]," a new springtime task on a farm newly dependent on herbicides to fight weeds. On April 13, 1983, he writes that he received "8000 [turkey] poults this morn" and then "moved birds after from Pole [barn] 2," the first task reflecting the turkey farm's growing economies of scale, the second the "pole barn" system of keeping turkeys in massive, semiconfined barns with automated feeding and watering systems. His entry for 1991 reflects the rising wealth of the Yoder farm, allowing for a vacation to England's famed Lake District: Peterson marvels at the "left hand side" driving and the picturesque "stone fences," but he highlights a visit to Windermere's village pub "to talk . . . with the farmers of the area," from which he discovered that their "problems are the same as ours." Here was a modern farmer: his animals were confined to barns, his crop management was shaped by chemicals, and his reward was an international vacation.

The fuller context of the transformation signaled by these diaries is apparent in the *Kalona News*, which provides an uninterrupted record of Washington County news from the very beginning to the very end of the twentieth century. Unlike any other source, this weekly newspaper traces an almost seamless transformation over the entire century. The farmers with recognizable Swiss–South German Mennonite and Amish surnames—especially Yoder and Miller but also Brennerman, Gingerich, Hochstetler, Kreider, Mast, Mollinger, Nicely, Ropp, Semler, Snider, Swatzendruber, and others—are frequent subjects in this drama of agricultural modernization, which is based on ever-increasing degrees of technology and expanding local worlds.

The story of the Amish and Mennonite farmers at the beginning of the century is a simple one: a close, animal-based tie to the land. Most noticeable is the image of an inveterate reliance on horsepower. On April 27, 1900, for example, the newspaper reports horse thefts at both the Jonas Yoder and D. K. Yoder farms but also the near-fatal accident of young Moses Miller, who shattered his left arm after attempting to shoot squirrels from a moving horse-drawn plow. The same issue features ads by Mennonite entrepreneurs offering horse supplies and also cutting-edge horse-drawn cultivation implements: J. E. Miller lists a wide selection of "Light and Heavy Harnesses: Collars, Sweat Pads, Curry Combs, Brushes and Whips, till you can't rest," while Snider Brothers advertises three brands of the "Riding Cultivator" and three types of riding plow, the "Deere, X-Ray and Case." These horse-drawn implements are, of course, also linked to animal power itself, a source of energy that crescendoed in the second decade of the twentieth century, before the Fordson tractor became ubiquitous. Indicative of both animal power and a modern drive to seek constant improvement are ads in the May 22, 1913, issue for the best local stud services: Dave B. Hochstetler advertised the services of "Duke," a "grey steel Norman" stallion, offered to "stand at $8.00," guaranteeing a "colt to stand and suck," the final sign of successful breeding; Jos. C. Gingerich's "Obstine" was billed as an "imported Percheron Stallion . . . fit to win the sweepstake of any show-ring,"[10] an odd promise for a farmer merely wishing to build up a good team of horses but apparently an effective message nonetheless. Here was an old agriculture infused with the language of modernity.

As the middle decades of the century unfolded, the Mennonite farmers of Iowa encountered ever-new agricultural technologies accompanied by new mentalities. If newspaper reports had Mennonites embracing the car and the tractor in the 1920s, reports in the 1930s and 1940s had them accepting crop specialization accompanied by an array of new machines, alongside new ideas on agronomy and chemicalized agriculture. Crops like soybeans and corn were given a boost with new machines like the combine harvester, which Roy Yoder, of Wellman, used in October 1937 as he "combined beans for Sherman Miller" over a few days in the early part of the week. Both crops were also enhanced by new biological knowledge. An ad from June 1938 announced that Emery M. Yoder, at the Joetown Store, was selling "Hi-bred corn" seed, as well as three strains of soybeans together with the inoculate for them.[11]

Then the Second World War altered farming even more radically. An October 1942 front-page story titled "Desperate & Expensive War" declared that in fighting Hitler it was "Washington County's part to raise livestock, grain and produce lots of all of them." Mennonites might well think they were doing their "part" as good citizens by being kind and supportive neighbors, quietly helping one another as they did in November 1943, when "2300 bushels of corn was husked and cribbed last Friday for [ailing] Sam Mast by 76 friends and neighbors with 28 teams and wagons." But this old approach to agricultural production was being put to the test. The bar for good production, as the *News* reported it, was increasingly chemical in nature. Some of these ideas predated the US entry into the war, with one 1941 piece speaking of the arrival of "chemical fertilizers" for "the smart farmer [who] is looking ahead." But the new logic of chemical-based agriculture took off especially after the war. By 1947 local ads heralded "that magic compound" 2,4-D, effective in killing most weeds. In 1949, Washington County farmers were told that 2,4-D was now used on five hundred thousand acres of the nation's soils and that Iowa should get on board, for "Iowa's small grain fields can be wiped clean of weeds."[12]

The postwar decades saw ever-new ways of engaging a more industrial agriculture, and Mennonites walked in lockstep along this pathway. Local Mennonite-owned businesses, such as the large Maplecrest Turkey Farms, which included a "USDA-approved" killing plant, fully meeting "the new compulsory inspection law," pointed to new forms of farm specialization. And Mennonites were adopting a range of new fossil-fuel-powered technologies. In April 1950, the farmer Elwyn Brenneman received a public shout-out in the *News* for providing the "best" answer to the question why "farmers need a scientifically designed, all-steel, prefabricated [dryer]." Having done so, he earned a sizeable credit on the purchase of a new, propane-fueled Hi-Dry Hay Dryer. Even more consequential was a similar technology, the farm-based, propane-heated corn dryer. Such a tool, declared Washington County's extension agent, Tom Robb, in September 1960, allowed for earlier corn harvests, as moisture content became less of an issue. It would also render obsolete the old, massive corncribs that relied on sunshine to dry the kernels naturally over time. Then there were the ever-larger sources of farm power. Nothing spelled mechanical muscle like the "New Generation" 4-cylinder diesel 2010, 3010, and 4010 John Deere tractors, with cushioned seats and easy transmissions, announced in 1960. A September story from that year about Ernie Ropp, the prominent Mennonite owner of Farmers Supply Sales, leaving by "plane for Dallas where he previewed the new line up of John Deere tractors" lent the additional glitter of air travel and metropolitan city to the new tractors.[13]

Farmer Ken Gingerich and son, Mark, planting treated soybeans in Washington County, Iowa, with a front-wheel-assist Ford tractor, May 2015. Photo by John Eicher, Altoona, PA.

Alongside reports on new machines in the *Kalona News* were more references to chemicals. In January 1960 one commentator, anticipating cutworms
in corn, counseled farmers "to use 2 lbs. actual Aldrin or Heptachlor per acre
broadcast and disked in. . . . Ahead of planting . . . [as] smaller amounts won't
do the job against cutworm[s]." Insecticides also invaded the increasingly
larger barns. In a June 1961 piece, "Washington County farmers who milk one
or more cows" were urged to "'stop, look and then spray' before starting their
annual battle against flies" and to spray "absolutely no DDT, Lindane, Chlor-
dane, Toxaphene or Dieldrin."[14] Ironically, the warning actually signaled the
acceptance of such chemicals, at least on some local farms. There were also
signs everywhere of increased use of synthetic fertilizers. In a telling springtime warning during the 1973 commodity-price rally, the local county's Agricultural Stabilization and Conservation Committee, the ASCC, counseled
farmers to stockpile synthetic fertilizer in anticipation of a "more than normal" demand, given "favourable farm prices this year." The old organic-based
farming, attuned to the animal world, was now increasingly left to nostalgia.
In 1964 a news column, "Flint Ridge," by Don W. Yoder masquerading as
Bill Jr., began offering folksy wisdom, including weather forecasts: a typical
outlook was one from March 1973 that predicted rainy, mild weather because
"frogs have been croaking, the night crawlers are moving . . . and the moon has
been turned upside down."[15] The old ways of knowing had now become
merely cause for amusement.

The 1980s and 1990s marked more difficult times. In the early 1980s especially, myriad articles spoke of high interest rates, rising input costs, low commodity prices, and a cost-price squeeze devastating to increasingly helpless
local farmers. In September 1984 the Mennonite columnist Chester A. Miller,
in his column titled "Chet Says," voiced the farmers' new frustration: they
were no longer "the goose that laid the golden egg" but rather victims of "re-
lentless pressures on the traditional family farm" arising from "the cruelties
of the export market." Mennonites now joined their Catholic and Methodist
neighbors in public forums that focused on the ethical dimension of an agriculture leading to record foreclosures.[16] Now also came an implicit critique
of chemicalized farming.[17] In February 1980 the Washington County–based
Iowa Mennonite School sponsored a public seminar on "profitable farming
without pesticides" as part of its Profitable Farming System Series. Increasingly, Mennonite and Amish farmers began to sell certified organic foods: Joseph T. Miller advertised "organically grown sweet potatoes" in the "Buy
and Sell" section in May 1980, and Virgil Miller listed hormone- and antibiotic-

free beef in July of the same year. And yet fossil-fuel-based farming held the day. In March 1993 the local Mennonite corn and hog farmer Tom Yoder was quoted as noting in a Midwest survey of farm concerns that his foremost worry was not "poor farm profits" but a new "energy tax" effectively curtailing soil tillage and corn drying. His strategy for dealing with this challenge signaled the height of modernity, a take-it-on-the-chin approach, that is, "trying to be more efficient, tighten things up."[18]

The story of the Mennonite farmers' place in Washington County's agricultural modernization from a global perspective is relatively neat and almost uninterrupted. True, the economic upheavals of especially the 1930s and the 1980s disrupted the pattern of incrementally larger and more sophisticated worlds. But overall, the century-long narrative is one of increasingly complex technology, allowing for greater economies of scale.

Manitoba: From Monocrop to Monoculture

Of the Mennonite communities in this study, the one in Manitoba was most similar to that in Iowa. Indeed, the agricultural history of the Rural Municipality of Rhineland, comprising the eastern half of the historical Mennonite West Reserve, has usually been told in terms of mechanization and commercialization. The Manitoba story, however, was distinctive in several ways.[19] Unlike Washington County, Rhineland Municipality was a monocrop wheat-growing region on open prairie. Certainly all Mennonite farms here aimed at food self-sufficiency, with their dairy barns attached to houses and their separate pigsties and henhouses, but their focus during the first third of the century was on wheat. Census records tell of a certain stability in cultivated land per farmer, at 196 acres in 1921 and 210 in 1956, but also of a slow increase in the total acreage as pastures gave way to cultivated land, that is, from 144,000 in 1921 to 175,000 in 1951 and 200,000 in 1971. But the most significant changes came with regard to commodity selection, with wheat falling from 77 percent of total acreage in 1891 to a mere 19 percent in 1956 but then rising sharply again to almost 50 percent in 1976. In the meantime, sunflowers arrived, signaling a new, more intensive, row-crop-based agriculture, while oats fell from 23 percent of the acreage in 1921 to 17 percent in 1961 and just 4 percent in 1976, reflecting the end to a reliance on horsepower. Finally, rapeseed, or canola, rose from 248 acres in 1961 to more than 13,000 in 1971, reflecting the rise of oil seeds, now epically depicted as large fields of neon yellow flowers interspersed with waving fields of green, and then golden wheat.[20]

Diaries provide a fuller, local-level account of these practices and changes. They speak of farmers spreading manure on snowy winter fields, acquiring or cleaning their seed in April, waiting until the first weeds emerged by early May and then cultivating to kill those weeds, and finally seeding by drill. September was the harvest month, when the grain was cut, dried in either stook or swath, and then harvested, at first with a stationary threshing machine and later with the combine harvester. As soon as possible after the harvest, the land was plowed or deep tilled with a heavy cultivator, turning the black clay soil so that the next spring the sun's warmth could turn the frozen ground once again into a fertile plain. Every six or seven years, farmers black summer-fallowed their land, tilling it numerous times to kill weeds, leaving it absolutely bare.[21]

If diaries tell of nature's cycles and change over time, memoirs often provide a more nuanced picture of the community as a whole. In his personal history, the farmer Menno Klassen, from the village of Neuanlage, recalls the 1930s, when he was a teenager.[22] Although the farm village had been founded in 1876, with "a number of farmsteads on both sides," by the 1930s it had succumbed to the forces of modernity, with most farms moved to their own quarter sections and the white farmhouses now separated from the red barns; indeed, only four farmsteads remained in the village. Klassen's short memoir "visits" his neighbors' farms, "starting from the west end," and in the process offers a portrait of four distinctive households interacting with the land and modernity in ways reflective of their relative wealth.[23]

At one end of the village was the technologically advanced David Schellenberg farm, rich in hardworking boys; this farm was the "biggest" operation in the area, powered by "3 four-horse teams to work the fields" and notably graced with a "disk and land packer." These new technologies allowed, on the one hand, weeds to be uniformly disked into oblivion and, on the other hand, the land to be packed to preserve moisture. The other farm families had their own mix of modern and traditional. Jake Schellenberg, whose farm was next, embraced the modern in his own way, receiving a shout-out from Klassen for having served as his 4-H Seed Club leader and helping him win a trip to the Toronto Royal Winter Fair one year to exhibit his agricultural achievement. The next family was the very shy, female-headed Fitz family, with five single grown children; Lutherans in a sea of Mennonites, they were lauded nevertheless for living "very simply from a mere ten acres of land," a farm made up of a "big garden, some small fruits, a flock of chickens, and two Jersey cows" grazing on the farm's "lush pasture." The Fitzes had an especially primitive hay-gathering system, still making use of a scythe and a wooden hay rake. They also bartered their one farm commodity, eggs, and not owning a car, they walked their eggs to the local store in a "basket purchased from native Indians."

At the far east end of Neuanlage were the Mennonite Gerbrandts, another large, poor family, living on rented land. Their farm was located on a patch of "poor land, poorly drained, heavy gumbo soil, that kept the farmer poor as well." Indeed, their main accomplishment was that "they were very good, God fearing people." Where Klassen heralded the well-to-do David Schellenberg farm for its effort to mold the stubborn Red River clay soils with new technologies, the poor Gerbrandt farm was congratulated for molding itself to the exigencies of those same difficult soils. In both cases, though, farming in Neuanlage in the 1930s was still based on animal power and manual labor, household self-sufficiency and a barter economy.

Quite a different story develops in the pages of the English-language *Altona Echo*, later renamed the *Red River Echo*, the community's first weekly newspaper, launched in 1941.[24] Like the *Kalona News*, the *Echo* offers a play-by-play of the community's embrace of modernity with front-page headlines, weekly columns, and flashy ads. Although the *Echo* had a more distinctly Mennonite readership base—typical Dutch–North German Mennonite family names in this newspaper included Elias, Enns, Heppner, Klassen, Martens, Neufeld, Penner, and Wiebe—it hinted at an ethnoreligious community in conversation with a largely non-Mennonite-driven agricultural hegemony.

A range of stories charted a rural transformation. Among the main stories from the 1940s was the arrival of rural electrification in 1947. In January of that year, a story on the local screening of the Canadian National Film Board's film *Farm Electrification* heralded the coming of hydropower. Served on a province-wide electrical grid system, it promised reduced labor needs with "electric milkers, separators, and feed mixers" and increasing farm income with "chick brooders, milk coolers and honey extractors."[25] Other stories signaled the arrival of the chemical. One account from February 1952 linked it to the most successful of farmers, George G. Elias, a prizewinning barley producer of the Haskett district who addressed Manitoba farmers at the annual Canadian Barley and Oilseeds Conference at the luxurious Marlborough Hotel in the provincial capital of Winnipeg. The Mennonite farmer Elias had achieved the acclaim of a modern-day prophet, extolling good crop rotation and humus-producing clover but also chemical herbicides like 2,4-D, to be applied with scientific precision, "at the rate of two or three ounces of ester per acre when the crop is five to seven inches tall."[26] Still other stories and advertisements outlined the benefits of synthetic fertilizer. In one ad from July 1957, local farmers were urged to use the "uniform granules" of "Elephant Brand High Nitrogen Fertilizers" not only in the springtime but also in the fall after the harvest, speeding the straw's decomposition and depending on "melting snows" in spring to "carry" the fertilizer "to the root zone."[27] Yet another set of reports reimagined farmers' relationship with the environment, with accounts of new machines besting western Canada's short growing season. The combine harvester, not well suited to this climate, could finally be made workable with a brand-new invention marketed by Wiebe Bros. Equipment of Plum Coulee. The miracle machine was the hydraulic-controlled, Versatile-brand "self propelled swather," a fifteen-foot-wide machine that allowed farmers to cut grain still on the green side and allow it to dry in "swaths" with a certain ease; its "variable drive for canvasses," "steer-o-matic transmission," and "electric starting" made it truly "versatile," the mantra of modernity.[28]

The *Echo* of the 1960s and 1970s made these machines and chemicals seem commonplace. Stories of simple agriculture were now associated with the exotic. One article from August 1971, titled "Early Indians Left many Traces," reported on the local amateur archaeologist George Enns's extensive collection of Indigenous artifacts, which included evidence of prehistoric agriculture in the form of a corn hoe and a corn-grinding stone; the story quoted Enns as suggesting that even more tools were likely buried in "an old Indian mound only about two miles west of Altona." The mound was an anomaly in Rhineland's cereal and oilseed fields, where news of yields, whether bad or good, was now announced with reference to the chemical. For example, the lack of chemical application explained the mediocre crop in the Altona-Gretna area in 1966, when excessive summer rains kept farmers from getting "on their fields to spray when it was necessary," resulting in weed infestations and "high dockage." In contrast, the eye-popping bumper crop of 1967, producing fifty-four bushels an acre of "number 1 Manitou" hard red spring wheat on the Martens farm at Kleinstadt, east of Altona, was attributed by Mrs. Martens (in a telephone interview) to summerfallowing after a six-year rotation, an application of "some hog manure" in fall, capped off with "drilling . . . 11-48-0 fertilizer . . . in with the wheat in spring."[29] Herbicides and fertilizers had become publicly acknowledged, incontrovertible keys to the successful farm.

The 1980s and 1990s pointed to significant new horizons regarding chemicalized agriculture, in particular glyphosate, sold under the name Roundup.

One upbeat report from the Western Canadian Flax Growers Association annual general meeting in 1992 noted that local farmers, including Art Enns, of Arnaud, attending afternoon "presentations by chemical companies . . . happened upon a gold mine." Enns learned that Roundup could be used as a preharvest desiccant on flax, snuffing out fast-growing quack grass and thistles at a vulnerable point in their growth cycle. A report from the local Manitoba Department of Agriculture "Ag Rep," Ahmed H. Kahn, spoke of rising "herbicide-resistant weed strains . . . present in many weed populations" resulting "when the same herbicide" was used "over a number of years." By happenstance a fortuitous report on the opposite page filed by the Manitoba agricultural columnist Les Kletke announced a remedy for chemical-resistant weeds. It came from Monsanto Canada and its multimillion-dollar research project on a genetically modified "Round-Up tolerant canola." Roundup would address "chemical resistance," wrote Kletke, as "no weeds have yet to show resistance to glyphosate, Round-Up's active ingredient."[30] The chemical of all chemicals was about to become accepted as a central tool in Rhineland agriculture.

Most voices in the *Red River Echo* seem to support this four-decades-long march of the chemical, and oftentimes those most opposed were either peripheral to the community or arose from quarters outside the farm itself. Letters to the editor of the *Echo* from folks beyond Rhineland itself, for instance, were the most critical. In August 1995 Brenda Bourgeois, who identified herself as a mother of small children from Stephenfield, a hundred kilometers northwest of Altona, wrote to the *Echo* to say that "aerial spray planes" had dropped Furidan and Diathane on her yard. She called on the local member of Parliament, Jake Hoeppner, a Mennonite farmer, to stop the planes buzzing her yard like "busy little bees"; to make her point, she ended on a sarcastic Carsonesque note: "Oh yeah, the bees? They are gone too." But voices of concern also came from the unlikely quarter of local town corporations within Rhineland.[31] In 1995, for example, the large grocery store in the town of Altona, Penner Foods, boldly announced a program to "save trees" by purchasing a ten-thousand-dollar cardboard baler to recycle paper; the company was heralded as Altona's blue-box champion. As the manager, Richard Neufeld, stated, it was all about being a "responsible corporate citizen," making this town a "little greener," but it was also about "business efficiency." A new lexicon of environmental ethics was in the making among Mennonites in Manitoba.[32]

Friesland: Reclaiming Land from the Sea and Yesteryear

Of the remaining five Mennonite farm communities, the one that most embodies incremental modernization is Friesland, which was also shaped by technological innovation and attending economies of scale. But this change in Friesland was distinct in some fundamental ways, mostly in that it was sudden and concentrated in the decades following the Second World War. In broad strokes, the story of Friesland hinges on two main events during the second half of the century: postwar reconstruction and the introduction of the European Common Market. The story of the first half of the century still highlighted the small-farm householder, less given to mechanization than its North American counterpart but also seemingly more in tune with nature.

In contrast to the Mennonites of Iowa and Manitoba, those in Friesland continued to be less concentrated geographically. Thus, there was no weekly newspaper from a so-called Mennonite community, and given the fluidity of church boundaries, specific primary sources were also less identifiably Mennonite. Several other sources, however, do tell the story of Mennonite farmers in Friesland for the first part of the century. One is a rich collection of letters written from 1898 to 1940 by recent Dutch immigrants in the western United States back to the Netherlands, specifically addressed to Johannes Hoogland, a Mennonite farmer from St. Annaparochie, in northern Friesland.[33]

A sample of these letters, written from Minnesota in 1915, 1919, and 1923, offers much more than a picture of farming in the United States; it provides a comparison with more traditional forms of agriculture in the old homeland of Friesland, upon which the writers in the United States seemed fixated. Indeed, the composite picture painted in these letters is that of a Frisian agriculture with relatively rudimentary technology. In a letter of January 5, 1915, for example, the recently arrived Sjoerd de Jong writes to Hoogland from Crookston, Minnesota, with both the new and the old world in mind. De Jong is unequivocal: "Everything is different here, Johannes, than it is in Holland; if you are spreading manure here, you don't have to [fork] it off [the cart], you just need to pull at a spring and the cart [manure spreader] empties itself." In another letter, from May 15, de Jong describes his work on a Norwegian-American farm and again indirectly describes a relatively primitive Friesland: "Today I planted potatoes, but not like you do in Holland, no, I had two horses to do that. . . . That is actually my life, to work with horses."[34] De Jong doesn't describe the mechanical planter those horses would have been pulling in

America or the planting by hand in Europe; again, a dichotomy between the new and old homelands is assumed.

Several letters from 1919 make a similar comparison between the labor-intensive Netherlands and mechanical America. In one from May, a relative of de Jong's, presumably his married sister, Sijke Hoogland, describes her 300-acre farm at Eldred, Minnesota. She writes of the pleasure of working with modern machinery, equipment that is "very different than in Holland," including a "new grain binder . . . and a mowing machine," and adds that "we have a new one too." In a December 12 letter she repeats the words "here it is very different than in Holland" and explains that "we have all the livestock in the barn, horses and cows." Perhaps the barn simply protected animals from Minnesota's "severe winter," but the implication is that agriculture in America was somehow more sophisticated. The new farming methods seem to have come with other advantages too: a strong economy even during wartime and easy credit. Sijke Hoogland adds, for example, that she often thinks of Friesland "because here we feel nothing of the war and never had lack of anything." It was more difficult to farm in Friesland because of unstable supply chains but also because its society was more stratified. Farming in Friesland, writes Hoogland, was an ideal vocation, but only for "the one who has money," presumably an allusion to the price of scarce land.[35]

Despite this somewhat negative depiction of Friesland, a number of writers pine for its old, simple ways. The Minnesota letters from 1923 allude to Friesland's fertile land and abundant cheap labor, both of which make for a more relaxed life, especially for the women. In one September letter from Adrian, Minnesota, the male immigrant Sjouke Hoogland writes that oat threshing in America is especially efficient: with "a threshing machine, here, it goes faster than [in] the old country." But then he adds that the "oats are not as heavy [here] as in Friesland and much shorter." In addition, the "cows here don't give much [milk]. . . . The meadows here are not that good . . . [even though] it is heavy land here, with a thick, fertile layer." Further, he has a new appreciation for Frisian canals and ditches, writing that in America "there are no waterways . . . to stop the livestock, everything is [en]closed with barbed wire." He does not seem to be especially impressed with this infamous American agricultural invention and laments the never-ending cycle of work, which is especially burdensome without the availability of cheap labor. He notes, for example, that in the United States the farmwife is burdened with both the housework and farm duties that include milking the cows and tending the layer chickens.[36] Farm life in Friesland may be less mechanized, but from afar it appears superior in other ways.

Poetry provides a glimpse into the rural mind-set behind this economic equation. Indeed, it reinforces the idea of a lush and relaxed Frisian countryside. Two such rural sketches are provided by Frisian Mennonites. One poem, from 1924, by the Mennonite pastor Reinder Jacobus de Stoppelaar, from the village of Warga, describes a scene from daily rural farm life. In it, he celebrates the "most beautiful species," the cuckooflowers, which "stand along the wettest borders of the ditches," and also "the marsh marigold," the "flower of April, the pearl of pasture."[37] Here was a sketch by someone passing by the fauna of the ditches and pastures of Friesland and taking note of nature. A poem from a lesser-known writer, an active farm householder, Renske Visser-Oosterhof, a mother of three, offers a similar portrait, but from one laboring within the Friesland environment. As a dairy farmer in eastern Friesland's Wouden region, a place of reclaimed peat land that features the famous Frisian Holstein cows, Visser-Oosterhof offers a perspective from the farm itself. In one of her many poems, "De Koetsebeibeam" (The red currant tree), from the anthology *Meniste Jistertinzen*, she describes the early morning moment before heading out to the *Jister*, the outdoor, grassy milking corral: "Each morning when I open the back door / While rain is pouring down this summer / Then the first thing that I see / Is, as always, the red currant tree / How flamingly red is your beauty / You, tree treasure of our Woods region." She also describes ambling in the meadows after the morning milking, "When walking along old footpaths / Or strolling in my silent dreams," where she again sees the currant tree, now "from a distance, amidst green tree-covered dikes"; it has "survived all winds and water."[38] The Frisian farmer in these depictions is deeply in tune with nature, a common characteristic of the small, mixed-farm householder.

This poetry, as indicated, predates the postwar transformation of agriculture in Friesland. After the Second World War Friesland, like Iowa and Manitoba, embraced technology with abandon. Its tractor count doubled in the 1960s, its fertilizer consumption tripled from the 1950s to the 1980s, and its landscape was transformed after the famous 1968 Mansholt Plan introduced Common Market policies. New regulations promoted specialized agriculture, especially in the dairy, hog, and poultry sectors; they financed new roadways and mandated the consolidation of small parcels of farmland.[39] Mennonite farmers, especially noted for their commercially marketed potatoes in Friesland's north and west and dairy in the south and east, welcomed and cooperated with this wave of rural modernization.[40]

This economic impulse is especially apparent in a biography of the Mennonite farmer Siebe Peenstra, of Akkrum, in south-central Friesland, produced from his extensive diary covering the years 1947 to 2012.[41] After Peenstra married Pietje de Jong—on "a beautiful day" in the Akkrum Mennonite church in 1947—he and his bride moved onto a small dairy farm. Reflective of the primitive rural milieu of postwar Friesland, the young couple's first farm was rented, had no electricity, and produced little expendable income. They sold bacon and butter to pay for the household furnishings and borrowed money from Peenstra's father to buy their first livestock. The farm lay a distance from the public roadways, so the couple transported all their belongings by boat, a *schouw*, and then hauled them along a small path connecting two pastures. At the farm itself, an agrarian routine set in; Peenstra describes daily milking by hand, seasonal haying and spreading of manure, and even rhythms of socializing with the community's "many coffee drinkers." The work was hard, but the community was tight-knit and the farm was life-giving.

This seemingly slow progress gave way to early signs of modern agriculture. In October 1948 Peenstra bought a "new tractor, a gasoline-powered Ford Dearborn," which he used to improve the farm's drainage. Then that very December he and Pietje enrolled in a milking-machine course and in March 1949 milked "for the first time with the milking machine." Two years later the farm took an even more costly step and "connected to PEB's electricity grid," that is, to Platform Electrische Bedrijfswagens, a step that led to acquiring a vacuum cleaner and a washing machine but also sleepless nights with Pietje lamenting to Peenstra, "What have we done? [these machines] we cannot pay for." But the farm survived as at every turn Peenstra attempted to make it more efficient. In March 1952 he purchased a shipment of potato fibers from a local potato flour mill, a new feed source for the farm's cattle. Later that year Pietje attended a regular meeting of the Plattelandsvrouwen (Countrywomen Society), while Peenstra himself attended a special assembly of the Landbouw Economishe Instituut (Agriculture Economic Institute), both organizations known for promoting rural improvement. Sometime later Peenstra received a visit from two "briefers" from that very meeting, who did a "cost price calculation" of the farm's milk production, enabling him to aim for even greater efficiency.

The major preoccupation of the 1950s, though, was procuring farmland and making it profitable. It was a difficult process and often entailed disappointment. For example, in 1956 Peenstra placed a bid for the lottery to win a share of the newly drained, massive Noordoostpolder, off the Ijsselmeer, but was unsuccessful. In early 1957 he visited four different farmers rumored to be selling or leasing their farms; he was turned down by two owners, and he rejected a third as "too sandy" and a fourth, offered on a twelve-year rental term, as "too light and peaty." But then in March of that year the farmer Johannes Zwaagstra, a fellow Mennonite from Wommels, accepted Peenstra's offer for his land. Peenstra described it as "a good-looking farm, well situated and [with] good soil," although he took a soil test and sent it to a soil lab to determine its nutrient value. On April 25, a cold day with "strong, southeast wind," Peenstra and his neighbor, Fokke, had another "look at the land." With the lab reports in hand, they determined how much synthetic fertilizer, *kunstmest*, literally "artificial manure," it would need and then phoned in their order. They also went to the "K.I. [i.e., *kunstmatige inseminatie*, or artificial insemination] station in Arum to look at some bulls" and began plotting a herd-improvement program. The foundation of a modern farm had been firmly established.

During the 1960s Peenstra continued to strive to be a professional agriculturalist. Three events in early 1962 alone pointed in this direction. In January he took a dairy-management course from the Coop Zuivelfabriekeren (Cooperative of Dairy Factories) in Oranjewoud, twenty-five kilometers to the southeast, followed the very next week by a tractor-maintenance course in Franeker, forty kilometers in the opposite direction. Then in March, the day that Peenstra heralds the local skating star and future Olympian Sjoukje Dijkstra, he also crows that his sow "about to give birth is taken into the new sow cage." The new process of animal confinement yielded twelve healthy weanlings. In April, Peenstra and his wife engaged in the annual *Jûnpraters* gathering; they visited another farming couple and staying until past midnight, socializing with sweets and an alcoholic drink. But the main object was to tour the neighbors' farm and together, as farm couples, compare the host farm's "cost-price numbers" and evaluate the farm's "mechanization" strategy. New machines, innovative methods, and a never-ending process of critical self-evaluation shaped Peenstra's vocation.

Still, the 1970s marked a time of an increased sense of well-being as the farm couple benefited from their hard work. Peenstra now sat on farm-association boards, although their agendas of "mergers, collecting debt and freeing of debt" often frustrated him. The farm was prospering, and Peenstra now took time for leisure, and given his love of nature, he also invested in getting close to it. In April 1972 he bought a Ford Corsair, "a white car with seatbelts, a radio and tow-bar," and later a "lovely" new camper trailer, equipped with a TV, and he and Pietje began regular camping trips within the

Netherlands and farther away in Germany and Switzerland. Then they hired a regular weekend milking hand, whose services allowed the farm couple "to sleep in and go to church" on Sundays. However, life did not suddenly become easy. For one thing, Friesland's weather was always challenging. The year 1972, for example, proved an exceptionally wet one: in early July the land was "terribly wet because of all the rain"; in August they tried to save the hay by compacting it with the tractor and then, employing a new method, they covered it with plastic to turn it into useable fodder, silage; November brought a terrifying hurricane-like storm that shattered the roof windows of their farmhouse. Despite nature's capriciousness, however, a new thought came to Peenstra that year: perhaps it is the human who is capricious, and not nature. In April, while helping his neighbor Willem van der Weg de-horn sedated cows with a massive shear, it occurred to Peenstra that this was "not a nice job" and that perhaps "cows should have horns." A new environmental consciousness had seeped into his mind-set.

In 1977, the year he turned 54, Peenstra sold the dairy and concentrated henceforth on commercial hay production and renting out the farm's pasture. This enterprise occupied him for the next ten years, until 1987, when the Peenstras sold the farm itself and moved into a house in Sneek, a town located on a canal, allowing him to indulge his passion of boating. Now that he had much free time, he accepted the position of vice-chair in his local Mennonite church. He also became the elder observer of ever-new farming methods. One day he noted that "men from the P.E.B. [the Provinciaal Elektriciteitsbedrijf] have dug power cables into the ground and connected the farm to the net," meaning that old electricity poles could be removed. A previous sign of modernity was erased from the landscape as a new connection to the wider world was made.

In his extensive memoir, Peenstra made few references to events such as wartime farm destruction, the creation of the Common Market, or land-consolidation programs that reshaped Dutch agriculture. But his story of a progressive agriculture stands in contrast to the farm culture depicted in Visser-Oosterhof's poetry or the letters written home by the Hoogland family in Minnesota. After the Second World War Frisian agriculture took off, and Mennonite farmers applauded the process.

Siberia: A Land of Elusive Optimism

The story of Siberia is also one deeply entrenched in science and modernization. But this story is even more fundamentally ruptured by war and centralization than Friesland's. The Bolshevik Revolution of 1917, collectivization in the early 1930s, Stalin's 1937 terror, the Second World War, and postwar ethnic repression all disrupted the Mennonite village of Waldheim (officially renamed Apollonovka) and its satellite communities.[42] The story here is told by way of memoirs of onetime farmers but also reports and articles from several farm organizations. The first set of documents stems from a Mennonite farm society begun during Lenin's mid-1920s New Economic Program (NEP), and the second from the executive committee managing the massive state farm, the *sovkhoz*, of which Apollonovka was part after the mid-1950s, at first Medvezhinskii and then the even larger Novorozhdestvenskii. It was a rougher ride into modernity than anything experienced by the other three communities in the Global North.

Memoirs tell the story of the postrevolution, terror-ridden transformation in sweeping terms. The memoir of the farmer Johann Epp recounts a very short lived, six-year time of agrarian peace in the Siberian birch forest from 1911 to 1917. There had been signs of possible disruption earlier, when the First World War "disrupted the quiet" of Waldheim in 1914, but the much "greater upheaval" of 1917 introduced marauding "Communist troops, moving through the country," coming "also to Siberia, and to our village, to do their damage." No one dared stand in their way; the Mennonites were forced to watch as the troops' actions crippled their local agriculture. Epp writes cryptically, "What the Reds wanted to do, they did." One cold autumn day soldiers descended on the Epp farmhouse and demanded that his mother cook them potatoes and that his father take them by horse and sleigh a few hundred kilometers to the north. The Epps survived the Bolsheviks, and with Lenin's progressive 1921 NEP, rural life improved. Once again "the farmers diligently served their fields and gave the Creator of sky and earth praise [and were] blessed and thrived."[43] Life almost returned to normal, and Epp's father, now "full of hope," spoke of a future of real agricultural progress. But there was a crop failure in 1921, just as fields were once again being cultivated after the revolution, leading to hunger in some places. Yet, on a diet of potatoes and millet gruel Waldheim survived. And even though the Communist state took over the village school in about 1927, hanging a photograph of Lenin on the wall, overall, writes Epp, the 1920s marked a "blessed time: all went well, in the fields as well as with the cattle."[44]

The 1930s, however, marked a return to terror. Epp writes that it began with collectivization in 1930, when "all land was designated as common property to be worked jointly by all farmers."[45] When Waldheim was ordered to collectivize in 1930, it resisted for a time, but ominously, on March 31, 1930, authorities arrested two of Waldheim's leaders—Isaac Toews and Johann Klippenstein—and their families and evicted them from the village. To procure their release, Waldheim acquiesced and fell under a new order: the state "dictated the [farm] production plan, how much of the yield had to be paid to the state, and only what was left over the farmers could divide among themselves. . . . Everyone had to surrender their horses, wagons, plows and other devices." At this time, several of the neighboring Mennonite settlements were added to the new collective, commonly known as a *kolkhoz*, including at least part of Wissenfeld (renamed Pokovyrovka), the family estate of Berezovka and, it seems, the family estate of Bekkerschutor as well.[46]

The year 1937, the time of Stalin's purges, was even worse, as nineteen Waldheim farmers were arrested on trumped-up charges. They were taken away, and most were never heard of again. "Where their graves were, no one knows," writes Epp. Now the churches were closed, the villagers were traumatized, and farming changed inalterably as "many old and experienced farmers, who had worked the soil all their whole life . . . were gone." A radically new farm culture ensued. On the collective, the Mennonites were forced to "push God aside," and farmers came to rely on nothing but their own "strength and knowledge." Eventually they adjusted to the new system; when Epp married, he and his bride learned to be content with what the collective allowed: a "little house . . . one cow and a pig."[47]

This twenty-five-year agricultural narrative, from the beginning of one war in 1914 to another in 1939, was graced with one particularly bright beacon of hope, the Allrussischer Mennonitischer Landwirtschafter Verein (Pan-Russian Mennonite Agricultural Association), or AMLV, and its meaty newsletter, *Der Praktischer Landwirt* (The pragmatic farmer). The AMLV signaled the Mennonites' hope that the NEP era would allow for the return of the Mennonite commonwealth of the prerevolution years, complete with agricultural innovation and cooperation. In the very first issue, of May 15, 1925, the editor, P. F. Froese, promised a Mennonite periodical for the "benefit of our farms," while another contributor spoke of being inspired by Johann Cornies, that "powerful driving force for the perfection of Mennonite agriculture" in the nineteenth century. The May issue reported optimistically on the state of crops in Mennonite communities across the Soviet Union and offered indirect criticism of the government's agricultural policies. In one story, under the heading "Landwirtshaftlicher Verein zu Omsk," in the August 1 issue Waldheim, in Siberia, was singled out for diseased and underfertilized wheat, but the implication was that the failure arose from an inefficient Soviet system. Yet most reports were positive. The July issue reported on the successful introduction of the German Red cow to Siberia, indeed as successful as "it had been in the south," in Ukraine. Throughout the newsletter as well were articles promoting modernization and a more scientific approach to agriculture, bearing titles such as "Of What Use Are Statistics?" or "To What End Does the Farmer Need Meteorology?" The latter piece argued that it was not enough to know "the chemical workings of the soil" and urged all newsletter readers to obtain a Russian translation of W. S. Harwood's progressive 1923 book, *The New Earth: A Recital of the Triumphs of Modern Agriculture in America*, and read it "thoughtfully."[48]

The newsletter invited Mennonite farmers to fully embrace modern agriculture, and yet it continued to idealize the Mennonite village. The issue from October 1925, for example, carried an anonymous traveler's enthusiastic report of the Mennonite villages, including Waldheim, described as located in a somewhat remote region north of the Trans-Siberian Railroad station of Isil'kul. These scenic, clean villages, set among birch forests, were a "painter's paradise," wrote the traveler: "On the street are strikingly green poplar trees and behind them are houses, freshly white-washed each springtime. Behind the houses are the nicely planted vegetable gardens. . . . On the yards and in the gardens one sees hardworking people."[49] It would, however, be the last report for a long while of idyllic farm-village life.

No one could imagine the terror that 1930 introduced or the grinding existence that resulted on the collectives, now ever larger in size. From 1930 to 1951 Waldheim, or Apollonovka, ironically named after a prerevolutionary landowner, together with the three smaller Mennonite satellite villages, constituted its own *kolkhoz*, or collective. In 1951 the Apollonovka-centered *kolkhoz* was amalgamated with yet another three villages—Voron'e, Gavrilovka, and Teter'e—and ordered to join the larger Kolkhoz Kirov. But then in 1957 Kirov was further amalgamated with the Medvezhinskii, a *sovkhoz* or state farm. The 1957 Sovkhoz Medvezhinskii "Amalgamation Report" notes that a five-person commission that included the Mennonite Isaac D. Loewen, was created to begin the process and that "accept[ing] the property of Kolkoz Kirov" was an eighteen-person subcommission that included the Mennonites P. I. Tevs (Toews), A. Tevs (Toews), and A. A. Jansen among its members.[50] The monetary value of Kirov's farm machinery and stock is indicative of the relative strength of this amalgam of onetime Mennonite farms. It was reported that Kirov owned 99,000 rubles' worth of motorized vehicles and 168,000 rubles' worth of farm machinery, the very symbols of modern agriculture. And yet Kirov was also reported to have more value in its 238 horses and 39 oxen, "working cattle" worth 299,000 rubles, than in its machinery. Evidently, the collective had been strongly reliant on animal power despite the Soviet fetish for tractors.[51] Other figures suggest a high degree of specialization in dairy: significantly, while the collective listed 1,146,000 rubles' worth of "productive cattle"—cows, heifers, and bulls—it only had 197,000 rubles' worth of swine (349 animals), 93,506 rubles' worth of sheep (797 animals), 14,000 rubles' worth of chickens (1,000 birds), and 2,400 rubles' worth of bees, the latter categories seemingly meant for household subsistence. That the collective also declared 286,000 rubles' worth of seed grain suggests the importance of its cereal section, which was perhaps at least as important as its dairy division.

Although the total acreage for Medvezhinskii is not available for this study, a 1956 "Land Balance Report" produced by the highest regional authority, the Executive Committee of the Regional Council of People's Deputies for the District of Isil'kul, notes that of the 239,416 hectares within the region's various *sovkhozes*, some 144,000 hectares, or 60 percent, were devoted to cereal production.[52] Other figures showing a total of 5 percent hayland, 11 percent grazing land, 13 percent trees (3 percent of which was "protective belts"), and 6 percent marsh, water, ravines, or gullies indicate the varied native landscape of southern Siberia. But the mere 4,900 hectares, or 3 percent, devoted to "gardens and personal land use of the individual collective farmers" was a stark reminder that life on Medvezhinskii was a far cry from what it had been in pastoral Waldheim.

The era of the 1960s to the 1980s marked a time of further agricultural expansion at the *sovkhoz*. A 1963 list of "the lead workers and specialists of the *sovkhoz*" indicates a commitment to scientific farming: the fifty-nine-person leadership team included the *sovkhoz* director, of course, but also a chief agronomist, a chief zoologist, a chief engineer, and a chief veterinarian, each with select subordinates.[53] These leaders reported annually to the Isil'kul Executive, which in turn offered its critiques of the *sovkhoz*. In 1961, for example, Isil'kul criticized "the extremely irresponsible attitude of the chairmen of [both] collective farms [and] state farm directors" and their experts for causing a breakout of brucellosis among livestock, and it named Apollonovka as one of twenty "points" with the problem. Then in a 1967 report Isil'kul cited its various collectives and state farms more generally for "poorly developed crop rotation . . . low productivity and high mortality of livestock . . . and slowly renovated farm machinery."[54] In 1968 it sharply censured Medvezhinskii's director, Comrade Gapon, for his "frivolous attitude" regarding "unsanitary conditions" at his *sovkhoz*'s three cattle-breeding farms. Specifically it named random manure piles, a lack of fences, and unwashed dairy utensils and dairy cows. It also cited the farm for failing to disinfect livestock and to follow the instructions from "the chief veterinarian and the chief sanitary doctor of the region."[55] No Mennonites were named, but it would seem that the larger farming units now defining their workspace had lost an old sense of agricultural vocation.

In 1973 Medvezhinskii was disbanded, and the part that encompassed Apollonovka, plus eight other villages was amalgamated in an even larger *sovkhoz*, Novorozhdestvenskii. The officially filed aims of Novorozhdestvenskii at this moment linked modernity with lofty nationalism. Those goals included "the implementation of national plans for the production and sale of grain crops and livestock products," as well as the use of the "latest achievements of science and technology in agricultural production . . . [directed by] financial and economic discipline, [including] efficient consumption of fuel and energy resources," heightened productivity, product quality, and a "steady increase of the prosperity of farm workers."[56] Part of this sweeping agenda was the incorporation of chemicalized agriculture. A 1979 Novorozhdestvenskii report on pesticides and herbicides, for example, listed a wide variety of chemicals used, including Thiram to treat wheat seed, Formalin for barley, oats, and grasses, Hexachlorocyclohexana to dust corn, and Metri for beets.[57] A second report noted the use of Formalin to combat smut, mildew, and root rot in grains.[58] The reports from Novorozhdestvenskii were upbeat, and there was no inkling in either the 1973 amalgamation report or the 1979 chemical report that in 1989 the state farm system would collapse.

Although fewer statistics for the last decade of collective agriculture were available for this study, the 1960s and 1970s already appeared as the apex of industrial agriculture. The seventy-two-year period between the Bolshevik Revolution and the end of the Soviet system represented an incremental modernization of sorts, but one shaped by sudden points of rupture and a forced push toward economies of scale that separated nature from the ordinary Mennonite farmer. The lofty Soviet goals of the 1970s seem to have been as elusive as those of either the Epp family or the AMLV in the 1920s.

Bolivia: Changelessness in a Changing Environment

In several ways the experience of Old Colony Mennonite farmers at Riva Palacio in Bolivia resembled those of the farmers of Apollonovka and the Kirov collective in Siberia; both groups bumped into forces outside their control, and both engaged in modern agriculture in fits and starts. For the Old Colony Mennonites of Bolivia, however, those historical moments did not result from a centrally planned economy. Rather they arose first, from a creeping modernity in Mexico, threatening old worlds of agrarian simplicity, and then from a surprise postcolonial economy offering a lifeline to those very worlds. Of course, Bolivia too had its revolution, the 1952 Easter Revolution, which ushered in the progressive Paz Estenssoro presidency. But this political rupture was bent on modernizing the country's economy with the tools of state-supported capitalism: it built the highway linking the mountainous western Altiplano to the fertile Oriente, the eastern lowland, and it oversaw this fertile region's development by inviting immigrant farmers to settle it.[59] The Old Colony Mennonites from Mexico who responded enthusiastically to Bolivia's invitation would be ambivalent partners on this modernization path. On the one hand, they were committed to overtly simple, communitarian ways. But they were also farmers determined to adapt to a radically new environment that seemed to demand agricultural expansion, new technologies, and a bolder, more assertive mentality. It was an evolving approach to the land apparent in several distinct records: issues from two German-language newspapers, one set from 1922, the other from 1977, and two personal diaries, one for 1963, the other for 1994.

Letters written to the *Steinbach Post*, the privately owned Canadian biweekly newspaper published in Steinbach, Manitoba, cast light on the Old Colony Mennonites in Mexico well before their settlement in Bolivia. Letters from the early 1920s especially reveal the Old Colonist adaptation to the semi-arid valleys of Mexico's northern Sierra Madre after their arrival there from Canada. Clearly, the Mennonite settlers thought of themselves as technologically superior agriculturalists. One young visitor from Manitoba, P. K. Doerksen, announces in a May 1923 letter that "I had never thought that Mexico was so backwards, some of the people say that it is a hundred years behind. . . . Here one rides on donkeys, walks bare feet, eats . . . and sleeps on the ground." And yet most settlers approached the new environment and its native farmers with a sense of respect. Reverend Abram Goertzen, for example, reports in a June 1923 letter that he had planted "some corn" on land broken only four months earlier, in December, and learned the hard way that "the freshly broken soil did not have moisture"; indeed, his plants remained stunted, while "what the Mexicans . . . planted in the old soil grew very well." Similarly, the farmer P. B. Zacharias's letter of December 1924, which narrates a year-long study of Mexico's climate, reports that the key to farming in Mexico was to watch the "Mexicans who had 'old' land under cultivation and planted their wheat in the middle of April really deep where it has the moisture to grow."[60] The Mennonites concluded that they needed to invert a practice they had learned on Manitoba's heavy clay land and return to the ways of their grandparents in Russia's light, semiarid soils. In Mexico preserving moisture was imperative, and planting more deeply and only after rainfall the new strategy. Moist soil was to be treasured.

A comparison of these settler letters from the 1920s with letters written by their grandchildren in Bolivia in the 1970s, published in a second Mennonite newspaper, reveals an increasingly bold approach to the environment. The *Mennonitische Post*, first published in 1977 under the auspices of the North American NGO Mennonite Central Committee (MCC), was similar to the *Steinbach Post*: it linked the Low German Mennonite diaspora across the Americas, especially in its letters section.[61] Usually written by farmers, these missives formed a rich record of Mennonite relations to the environment in Bolivia. Indeed, in the *Mennonitische Post*'s first issue, published on April 21, 1977, the farmer Gerhard D. Klassen, from an unnamed Bolivian village, describes a new agriculture with new seasons and new commodities. "The weather is very dry for the late seeded" kaffir, corn, and soybeans, he writes, "and we very much await rain" even though "the early seeded [soybeans] are ready for the harvest, and the kaffir is already being cut." By August 1977, letters from Riva Palacio itself began to appear. In the first of these letters, a writer signing off as "The Searching Pilgrim" begins by denouncing Mennonite assimilation in Canada but then segues to the environment. He is exuberant: "Here in Riva Palacio colony we have had a rich harvest overall, God be praised, at places three tonne per hectare." He then explains for northern readers the inverted seasons: "July and August are here the coldest months," but of course it was a relative coldness: "When the south wind blows from Paraguay and Argentina, then one happily sits indoors in the warmth."[62] The message is clear: Bolivia held the moral high ground, but it also met the environmental bar of a promised land.

A second writer from Riva Palacio, the schoolteacher Wilhelm S. Braun, also emphasizes the unique qualities of Bolivia and the Mennonites' ability to succeed in the new land. In a letter dated October 8, 1977, he offers an astronomy lesson to readers in the north: the Big Dipper appeared low on the horizon in the south, and the "North Star can never be seen." But then in another letter just two weeks later, Braun describes in rather positive terms Bolivia's agriculture: a welcome rain had fallen, the planting season was upon them, and even an unexpected hailstorm had left the Mennonites unscathed. Significantly, Braun is even more upbeat about news he repeats from a newly established colony east of Bolivia's Rio Grande at Nueva Esperanza, where settlers "have cleared five hectares of bush and want to do more." And as part of the program, they had already set up a sawmill for "cedar, mahogany, ocho-o, green 'lager,' and other kinds of trees."[63] It was an enthusiastic embrace of an invasive and brash project of turning forest into farmland. The 1922 letters from Mexico suggest a much more tentative approach to the environment than do the 1977 letters from Bolivia.

A second textual comparison of environmental interaction by Mennonite farmers in Mexico and Bolivia underscores this increasingly expansive mindset. It is one apparent in two personal diaries, one from the 1960s in Mexico, the other from the 1990s in Bolivia. The diary from Mexico, for 1963, is that of the farmer Jacob Enns, from the well-established Nord Kolonie, the parent community of Riva Palacio in Bolivia. The diary was penned just four years before most of Enns's neighbors moved to Bolivia, and shortly before the Enns family moved north to Canada, whence they eventually returned south, first to Belize and then to Bolivia. The second diary, from 1994, is that of Jacob Enns's son Abram Enns a few years after his 1988 migration from Belize to Riva Palacio in Bolivia.[64]

Both diaries are rudimentary farm household records. Indeed, Jacob Enns's note on January 6, 1963, that it had been "a difficult day for me" and the junior Enns's entry on April 10, 1994, noting that he "felt bad" are uniquely personal entries for their respective years. For the most part the diaries offer no comment on emotion; they are records of farmwork and environmental interaction. Both refer to the ancient *Strassendorf* system of rural landowning transferred from one country to the next and describe the ways in which it ordered farm activities. Jacob Enns in Mexico, for example, references local trade with farmers in neighboring villages, while Abram Enns in Bolivia refers to his three fields with the German word *Kagal*, a reference to small fields within the open field system:[65] there was first the *Hauskagal*, the home field; then the *Butakagal*, the outside field; as well as the Blumengart *Kagal*, located in the neighboring village of that name. Like his father, Abram Enns describes business deals in neighboring villages, including Steinreich (Stone Regime), Grossweide (Large Meadow), and other villages bearing old names that reference local environmental features.

Both the 1963 and 1994 diaries also report on weather and its effect on how farmers intervene in nature in their effort to achieve household self-sufficiency. In this respect the diaries are almost identical. Jacob Enns's 1963 diary reflects this story for temperate northern Mexico. His entries for January describe days that are often "cold and sunny," times he spends on aspects of farm household maintenance: on January 3 he builds a calf corral; on the ninth he obtains firewood; on the twelfth he sells oats and purchases flour; on the thirtieth he hauls water. He makes very few trips to the city, and those he does make, such as the trips on February 8 and March 12, occur just after bean threshing, evidently to market the beans. The months of May, June, and July bring the first occasional rains, and then the imperative of seeding sets in: on May 9 the Enns household starts its plowing generally; then, with June 11 to 14 bringing plenty of rain, they prepare for seeding, including the spreading of manure. On July 5, a rainy day, they plant the first corn, followed by the first beans on the eleventh, the garden on the thirteenth, and oats on the eighteenth. The summer months are easier, but by October the harvest is ready: the corn and oats are cut and bound into sheaves, and the beans are hauled home. On November 14 they "butcher the bull" to replenish the larder for winter. By December 7 the task of bringing in the sheaves is complete and cold weather sets in, with a minus 8°C frost on the eighth. But it is also the postharvest season, and from December 14 to 19 Jacob Enns and his wife travel to visit siblings in the new Casas Grandes settlement, near the US border; indeed, they "visit very much." On December 20, a "very nice day," they thresh 1,150 bushels of oats, and on the twenty-first the first of the beans. The year ends as it began: on December 31 Jacob visits two nearby villages to "pick up water." His world revolves around seasons of a particular environment and an agrarian, communitarian self-sufficiency.

In this respect his entries and what they represent resemble the 1994 diary of his son Abram Enns in Bolivia. Except for the inversion of seasons, the concerns of winter, spring, summer, and fall are similar. The January and February entries, for example, record the farm's preoccupation at high summer with a scarcity of rainfall: there are only three inches of rain in January, but fortunately the January 30 lament of "waiting for rain" is countered on February 22 with a report of a five-inch deluge. The rest of the year, the winter months from March to October and the spring months of November and December, have all the marks of a farm following the contours of the seasons. There is the cutting of lumber on March 14 and 15, followed by the first harvest of soybeans on April 7 and the first planting of winter kaffir corn on May 26. These are also the social months: at Pentecost, May 22, Enns's "daughters Elisa and Agatha are baptized," and then on September 9 they have a double *Verlobung*, engagement celebration, and a week later the double wedding. But the record of farmwork overwhelms these social notes. October is the harvest month for the winter crops, and November the planting month for the summer crops, soybeans and corn. Once again, Enns is deeply concerned about the lack of rain. The rains do come, however, and by mid-December the soybeans are growing quickly. No work occurs on December 25–27, the three sacred days of Christmas, and then because of a five-inch rainfall on December 29 and 30, no fieldwork occurs on those days either. On the thirty-first Enns records that he "got weanlings," a fitting year-end activity that promises to satisfy the household's meat requirements for the new year.

In an important difference between the 1963 and 1994 diaries, the latter also hints at selective adaptations to modernity. In some ways, of course, the farm's technology had not changed over the thirty years. Riva Palacio in Bolivia followed the same rules its parent colony had in Mexico: no electricity, only steel-wheeled tractors, and only horses and buggies for transportation. Nevertheless, Abram Enns's diary records the arrival of certain new technologies. A medical procedure, for example, is evident when on January 19 Enns vaccinates (*gespirtzirt*) cattle for four different neighbors, a task he undertakes again on February 12, when rabies (*Dollkrankheit*) is the targeted disease. There is also new mechanical technology, evident in frequent references to Enns's tractor: on January 1 he writes that the tractor has new "*Logeon*" (Low German for ball bearings), and on March 4 he takes the tractor to Neuendorf for a new "starter," a word he spells out in English. On the last day of the year, the day he takes delivery of his weanlings, he also "puts in a new pump ring," presumably in the tractor. Of course, given the farm's dependence on fossil fuels, Enns takes a shipment of "diesel" one day and a delivery of "oil" on another.

But the foremost sign of modernity on the Enns farm is the chemical. Indeed, the task of "spraying" insecticides and herbicides is referenced with the actual word *spraying*, again written out in English in an otherwise German language script. Abram Enns sprays the early soybeans on the *Hauskagal* on January 3; then on the twelfth he sprays for his neighbor Peter Peters; on the twentieth he sprays his own Blumengart *Kagal*, and on February 2 the schoolteacher's land. Then, having completed the first round, Enns commences with a second round of spraying: on February 17 the Blumengart field, on March 7 the *Hauskagal*, and on March 31 the *Butakagel*. But spraying doesn't end with this season: this is multicrop-per-year Bolivia. From June 22 to July 11 Enns records several days of spraying the corn, and on December 19 he announces that it is very windy and they are fortunate that the soybean spraying has been completed.

The story of agriculture at Riva Palacio reflects the adaptation of Mennonite farmers to Bolivia's postcolonial economic independence through Green Revolution policies. Ironically, this national agenda attracted traditionalist Mennonite farmers whose foremost concern was to build a self-sufficient and isolated rural society. They rejected economies of scale and technologies that encouraged such expansion, but they relied on their machines and medicines and increasingly on chemical application to secure the foundation of their antimodern community. Especially this ironic preoccupation with the chemical, less obvious than the horses and buggies and the steel-wheeled tractors, signaled an expanding agriculture at Riva Palacio.

Java: Mennonite Rice-Paddy Involution

Because of colonial structures that kept education rudimentary and technical and then a war of independence and a variety of natural disasters, including a flood, local farmers' records for Java were not available for this study. Those same forces also ensured that agriculture in Java developed more slowly than in the five places studied thus far in this chapter. The written story of Margorejo's agricultural history is heavily dependent on records of Western mission boards or development agencies whose workers resided in the village, and thus the narrative emphasizes changes introduced from the outside. Of particular importance for the history of Margorejo after 1925, once again, is the narrative of the mostly rural GITJ church by Sigit Heru Sukoco, a native of Java, and Lawrence Yoder, a onetime Mennonite seminary professor at Pati, near Margorejo, and later, in the 1980s, the "country rep" for Mennonite Central Committee.

Outside of this specific history, this study had access to a single locally produced source, a village census of 2010.[66] It represents the basic demographic outline of the 647-household community, compiled by sixteen local census takers, with households divided among them by alphabet. Because only thirteen of the record creators went to the work of asking the question about profession on the form, only 433 of these households can be analyzed with respect to economic activity, but it can be assumed that they are representative of the village as a whole. Significantly, fully 85 percent, 369 of these 433 households, listed at least one adult as a *tani*, a farmer, meaning that the village was still predominantly agricultural in nature. And the fact that about 58 percent of these farm households listed both husband and wife as farmers meant that most families were entirely devoted to farming. Also, because in only 5 percent, or 16, of the farm households was the *tani* a *buruh tani*, meaning a farm laborer, it also seems that the original vision of inhabitants owning land still largely held true.[67] Significantly, however, the average age of a farm laborer (29 in total) was 35.6 years, with an age range from 20 to 60, suggesting that landlessness was not a matter of life cycle but a more deeply rooted intergenerational issue of class. The statistics also indicate that no other category of profession challenged the dominance of the farmer, and most—the *ibu rumah tangga* (housewife), *swasta* (self-employed), *wirasuasta* (entrepreneur), *buruh* (laborer), *guru* (teacher), *sopir* or *pengemudi* (two categories of driver), and *pedagang* (trader)—appear to be professions that might well have served to support the agricultural base of the community.[68] The village was still predominantly Mennonite, with 83 percent of all the village households including at least one baptized member in the GITJ and 59 percent listing both husband and wife as baptized members. This does mean that in 17 percent of the households no member had been baptized as a Mennonite or baptized at all, and in 41 percent at least one spouse was not a full member of the GITJ, suggesting some fraying of old church lines.[69] All in all, Margorejo at the beginning of the twenty-first century remained as it had been at the beginning of the twentieth, a predominantly Mennonite farm community.

The historical research of Sukoco and Yoder supports the theme of rural continuity. The legal foundation of Margorejo was the seventy-five-year lease of the land from the state, beginning in 1882 and ending in 1957. Throughout this time the agricultural focus of the 179 hectares (443 acres) was divided between the low-lying rice paddies and the higher land with its crops like cassava and peanuts. True, significant changes were under way, including a slow indigenization of leadership, with a transfer of power from Dutch and Low German–speaking Mennonite missionaries to local leaders. Indeed, it was a change foreshadowed by an early twentieth-century internal evaluation of Doopsgezind missions citing the white missionaries in Java and Sumatra for overreach in serving not only as evangelists but also as "'governors' and landlords" in the mission villages.[70]

It took time, but in 1928 Nickolai Thiessen, as well as the aged Pieter Anton Jansz, agreed that Margorejo should choose as pastor their own Rubin

Martorejo. No changes, however, were made to the farm landownership. Indeed, Sukoco and Yoder's work deemphasizes Margorejo's early-century farm history, no doubt because little changed. Germany's June 1940 invasion of the Netherlands ended any communication with the Dutch Missionary Society and effectively granted leadership to the Javanese. Despite this management change, agriculture remained unaltered. By all reports, the Margorejo farms were successful. In preparation for possible independence, GITJ church leaders toured the various farm villages in 1940 and reported that Margorejo had an especially rich civil life, one that included a "farmers' organization," and that its church was financially solid. Indeed, the church alone owned more than half of the Margorejo land, 91 hectares, of which 27 hectares were devoted to rice paddies.[71]

The Japanese occupation of Indonesia from 1942 to 1945 did affect Margorejo's progress. One result was the rise of a local Islamic militia, the Ansor, which attacked Margorejo and threatened to subsume it, citing its farmers for historic collusion with Dutch masters. Local farmers sought to protect the village's foremost leader, Samuel Saritruno, the church board chair, and the hamlet head, Wirjo Soemareh, from the Ansor. But ultimately the farmers, "distraught with fear," were forced by the Ansor to burn down their own church building. Fortunately, the elderly Saritruno, as well as Mayor Soemareh and others, persuaded the Ansor militants not to seize the farmland, claiming that it "was not the possession of the Margorejo people" but held by Europeans.[72] Evidently, the mystique of colonial power was enough to ward off the militants. Ironically, a side benefit of the Japanese occupation was that the invaders insisted that tensions between Margorejo and its Muslim neighbors be resolved, indeed forcing the Muslims and Christians to apologize to one another and to resume farming peaceably.

With the Japanese surrender in 1945 came other significant changes. GITJ youth, for example, set aside historical Anabaptist pacifist teachings and joined the military forces of independence, with Margorejo becoming a rebel outpost known as "Post Four." In retaliation, a Dutch army unit occupied the village for a time, but the unit withdrew when the church elders persuaded them that Margorejo's youth were fighting without church sanction. Yet in 1947 GITJ synod leaders met with those very youth in Margorejo and arrived at an understanding with the next generation of leaders, "especially with regard to the agricultural settlement lands owned by the church." Presumably the accord foreshadowed the shift of landownership from the Dutch to the local congregation that occurred with the acquiescence of Dutch forces in 1949. Other changes were also afoot. A farm inventory that year, for example, took note of the village's overall "wealth" and its agricultural operations, expanded beyond its rice paddies and dryland crops to include six dedicated hectares of fishponds. The most significant postwar moment for the farm village, though, was when the old Jansz lease came due in 1957. Indeed, government officials insisted that a second lease of thirty-eight hectares should also expire in 1957 even though technically it was not due until 1969. This move, argue Sukoco and Yoder, caused Margorejo to come "alive . . . not . . . in a spiritual sense, but . . . with efforts of the people to . . . become owners of the land they had been living on and working."[73] The authors describe confusion as farmers scrambled to secure the required funds to pay the new owner—the government—for the land. Without elaboration they write that the events of 1957 instilled in farmers an unprecedented sense of landownership.

The 1970s and 1980s were decades of considerable change at Margorejo as Indonesia joined the Global South's embrace of the Green Revolution. No documents were available for this study to indicate the process by which Margorejo expanded its rice production with high-yielding "miracle rice" seeds, synthetic fertilizers, or herbicides. Sukoco and Yoder do note, however, that when the North America–based MCC moved into Indonesia in 1949 it established a clinic at Margorejo. Then twenty years later, in 1967, it amalgamated with three other European and local NGOs to form the Komisi Kerjasama Ekonomi Mennonit (Mennonite Cooperative Economic Commission, or MCEC), and this newly reconstituted agency promoted the diversification of agriculture in the village. One initiative offered ocean fishers from seven Javanese communities, including Margorejo, loans to obtain their own boats. Ultimately the program failed in most of the villages, and in Margorejo too, for a variety of reasons: a "lack of skilled talent," a lack of accountability, and two serious mishaps, the sinking of one of the fishing boats when it was hit by a ship and the tearing of a net when it was snagged by another ship.[74] Another effort by the MCEC, the creation of tofu factories, did succeed at Margorejo, although no agricultural consequences of the factory are offered.

These small documented steps at Margorejo signal broader changes that the MCEC was attempting to achieve in places around the village. At nearby Kudus, it oversaw the creation of a fish farm; at Pati, a mulberry plantation and silkworm operation; and at Jerukrejo, a dam and rice-irrigation project. The latter project eventually achieved rice production by "fine-tuning its upstream terraces so as to regulate the mud," allowing farmers to grow rice during the dry season. At other places the agency pursued pumping projects with gasoline or diesel engines that ultimately did not succeed. The fact was that local farmers often disagreed with the MCEC's top-down approach. In 1972, for example, when the old commission evolved into the MCC-backed Yayasan Kerjasama Ekonomi Muria (Muria Cooperative Economic Foundation), or most often simply YAKEM, a set of ambitious irrigation projects south of Margorejo were opposed by local farmers. One report described these farmers as "poor . . . uneducated and so traditional" that they "felt that it was sinful to leave old ways." This resistance to change was exacerbated by the difficulty of solving environmental issues that mattered to farmers, including irregular rainfall, "weaverbirds which ravaged the rice in the field," or "rats and various other pests which attack the rice." Sukoco and Yoder are critical of the fact that instead of addressing these issues YAKEM emphasized technology that promised to increase production. In 1975, for example, a new director from the United States imported a range of new equipment to increase rice production, including "7 'hand tractors' on credit, 3 highway truck trailers, 13 rice drying bins, 6 threshers, 5 blowers, 80 grass cleaners and 100 hand sprayers," at a total cost of US$53,000.[75] A sharp internal debate within MCC Indonesia and YAKEM ensued as the wisdom of pursuing a high-technology strategy among local farmers was questioned; in 1978 YAKEM was disbanded.

The rice paddies and peanut fields at Margorejo encountered this YAKEM-based push for modernization only indirectly. The high-technology strategies that MCC and its Indonesian associates sought for the small landholders in Java may have been questioned by local farmers. Perhaps the records until 1978 indicate a certain resistance to Western ways, but oral testimonials reported on in subsequent chapters do tell stories of an acceptance of miracle seeds, chemicals, fertilizers, and rudimentary engine-powered machines in the 1980s and 1990s. Agriculture until 1978 may have been characterized by a certain *involution*, to reference Geertz's classic conclusion about local poverty in Java, but a particular *innovation* was beginning to form.

Matabeleland: A Racialized Modernity

Of the seven places, Matopo in Matabeleland has the most troubled history with agricultural innovation. An oppressive segregationist system and then, more recently, a faltering national economy tempered the rise of a modern agriculture. Indeed, in the decades after Hannah Francis Davidson penned her extensive 1915 mission history, farm life at Matopo changed only slowly. Farmers on small plots used but primitive technology. Both men and women worked in the fields with oxen, producing a variety of vegetables, exported to the city in donkey-drawn wagons.

Like the written texts for Margorejo, those for Matopo are mostly missionary reports, which continued, as in the past, to describe the community's environment in the esoteric language of an outsider's gaze. Letters sent from the second generation of missionaries back home to the United States are illustrative. In his first letter from Matabeleland to Upland, California, in July 1932, the young missionary David Hall describes awaiting final deployment to Matopo. In the meantime, he was learning about his new environment. At one juncture in the letter he exclaims, "Mother, you should see this country . . . such beauty you never saw. Hills and rocks, big rocks, colored rocks, rocks on top of one another . . . Rocks that look just like an old wooden shoe from a quarter mile distance, others look like an elephant's head, others like tall cathedrals, surely beautiful." Hall shows a special eye for problematic wildlife. He refers to a Matopo missionary, Mrs. Taylor, "whose husband was chewed up by the lion," and to "six big kudus" caught eating in the missionaries' sweet potato patch.[76] In a subsequent letter from July 1935 Hall makes two allusions to the Matopo environment: first, "Cold! Say, but this is a cold place and the past few days have been very windy. . . . Dust, as blowing everywhere"; and second, seen en route to a missions conference, "a big troop of baboons . . . a wonderful sight to see, a few large families of them in the wild."[77] In both cases, a missionary is looking in on a strange environment. That environment is dangerous and difficult but also titillating and sensational; it is not a garden, field, or kraal.

A 1950 history of Matopo penned by two seasoned missionaries, Anna A. Engle, the widow of the founding missionary, Jesse Engle, and John A. Climenhaga, does offer insight into agriculture at Matopo. Like Davidson, they refer to Africa as a "vast stretch of land, long lying in darkness," as well as to its salubrious climate, annually birthing a promising agriculture. "Each returning spring," they write, "demonstrates one of Nature's great marvels. From out of dry hard soil, which has had no rain for months, there begin to spring up tiny plants. . . . Seemingly dead trees begin budding and busting forth in beautiful color." References to agricultural expansion before 1950, however, usually relate to the growth of BICC mission farms in wider Rhodesia. Matopo itself did not grow, but it was joined by four newer mission farms, including the 6,000-acre Gwanda farm, established in 1906, and the 7,000-acre Filabusi farm, purchased in 1924 for one dollar an acre. Specific references to agricultural performance, however, occur within upbeat missions reports. Engle and Climenhaga quote from the 1945 summation of the BICC mission superintendent, Charles Eshelman, that "farming operations were as usual and good crops were obtained." And then the authors add what mattered most to the missionaries: "We still feel that a good rural life for our people is essential" as "there is a close connection between salvation and the soil."[78]

Within Engle and Climenhaga's book are more specific references to a changing agriculture, albeit at the slowest of paces and within a colonial context. The first change affected tillage. True, the authors describe Matopo agriculture at midcentury as "primitive," still overtly patriarchal, with woman "still the servant of man . . . [for] it is she who digs the garden, grinds and stamps the grain, fetches the water from the river, gathers the wood from the veld." And yet some technological innovation has mitigated this gender imbalance. "Plows are common today," write Engle and Climenhaga, and now "the man plows his lands with oxen, often the woman or a child leads . . . the oxen." A second change was in the delousing of cattle, although it is announced in a passing comment. To illustrate the mission farm's transformation since its inception fifty years earlier, when there had been "nothing but the veld with its kopjes, vleis, streams, ravines and characteristic trees," they reference the mission's livestock-dipping station, located at the end of a lane of "stately eucalyptus trees." Third, they refer to the farm commodities themselves: the "beautiful grove of citrus fruit" and the "adequate vegetable gardens and large mealy fields," the latter deeply rooted in Matopo culture but credited to the missionary endeavor nevertheless. Indeed, write Engle and Climenhaga, it was the missionaries who ensured that "the almost barren veld has been made to yield its fruit in its season," a task completed "with the African helpers" who "tilled the soil, sowed the gardens, planted the trees . . . and developed the landscape generally."[79] Matopo had progressed, but its progress had been controlled by the missionary.

The Matopo Mission staff minutes from the 1930s and 1940s suggest an incremental, if uneven, extension of modern agriculture and science teaching. An entry regarding school discipline from February 1934 indirectly reports on progress on the farm. It orders the boys and girls to be separated on the mission grounds by referencing geographic boundaries that also serve to signal a modern sense of order. While the girls are instructed to stay close to the "old mealy garden," the boys are to stay within an area demarcated by "the eucalyptus grove to the cattle kraal" on one side and "the wire fence of the orange grove" on the other. Not without coincidence, the marks of sexual order at the school were reinforced by lines drawn in the landscape.[80] But another mark of modernity referenced resistance to modern discipline by some of the students. An entry from October 1937 reports on the settlement of a "strike" by the boys, refusing work, classes, and church attendance. In an exchange that saw the "boys beg pardon" of the missionaries and "settle down to obedience," the boys were promised "meat six times a week," plus "monkey, nut dressing or such other food as may be available."[81] Modern farm methods could be made acceptable so long as they were accompanied by elements of tradition.

Other mission staff minutes speak more directly to change, however tentative. An entry from 1936 reveals that two of the thirty books purchased for the school's library that year—*The Garden Science* and *Farm Building Construction*—relate to modern food production. The school's emphasis on agriculture study is also apparent in a minute from August 1940 about the board's decision to send the 1939 *Textbook of Agriculture*, by J. C. Brash, "to the educational board for approval as a text for Std's 5 & 6."[82] Other changes, ironically, empowered girls to do the very tasks their grandmothers had done, that is, to farm, although within a context emphasizing science.[83] A March 1937 entry demonstrates the school's commitment to creating "agricultural plots for the girls" but also to a program to "teach the . . . girls agriculture."[84] While the scientific base of the girls' agricultural education is merely implied, it is confirmed by a March 1942 order to extend a current "history-geography-science" curriculum to include the rural kraal schools, beyond the mission station. Now science would also be available to the farm children of Matopo more generally.[85]

The period from the 1950s to the late 1970s was a time of increased political tension at Matopo as independence movements swept the African continent and armed resistance to white regimes like Rhodesia's intensified. Significantly, reports of white missionaries emphasize progress on indigenization of leadership, but they skim over any of the bitter tensions rising from racialized landownership laws. Indeed, they deemphasize the mission farm itself. For example, a seventy-fifth-anniversary booklet on the founding of Matopo Mission, published in the midst of these tensions, in 1973, outlines the rising African leadership within the BICC church but refers only indirectly to the mission farm. It notes that Reverend Ndeabenduku Dlodlo was appointed as the "Overseer of Matopo District" in 1930, but with reference to the farm the booklet notes only that as "overseer" he lived "very near the [white] General Superintendent" and worked closely with him. Indeed, these missionary reports focus on the dogged imperative of mission work. For example, the missionary response to the forcible removal of many Matopo people from their farms in 1953 and their relocation to distant Gwaai District in western Rhodesia is described simply as one of "moving along and establishing a new mission outpost."[86] Other sources do mention the work of white agricultural extension workers pushing the BICC mission farms to accept modern ways. A short, unpublished memoir by the missionary George Bundy, who supervised the Mtshbezi mission farm, north of Matopo, after 1957, notes in passing that in 1958 he "established a farm plan with the government extension agriculturalist," the result of which was that "a lot of fencing took place that year." Bundy's reference to the end of white farm supervision in 1977 is also brief: he writes that in January of that year "we began hearing of the dangers of being on the mission stations" and that in June the BICC mission board ordered all white missionaries to "leave the rural areas."[87] The independence of Zimbabwe in 1980 spelled the definitive end of the white-run Matopo Mission even as the newly constituted African-led BICC maintained Matopo Mission as a church-based farming community.

The white missionaries' somewhat uncritical appraisal of life at Matopo stands in sharp contrast to the writings of native son Scotch Malinga Ndlovu thirty years after the height of the War of Liberation. Although he was a graduate of the BICC-run Matopo Secondary School and a member of the church, his 2006 book, *Brethren in Christ Church among the Ndebele, 1890–1970*, nevertheless offers a stinging rebuke of the US-based BICC missions. Ndlovu's

Farmer Moffat Mayo (*right*) with helper (*left*) near Matopo Mission in Matabeleland, using a stationary Honda gasoline-powered water pump to irrigate their gardens, 2015. Photo by Belinda Ncube, Bulawayo.

argument is straightforward: BICC's mission program had considerable merit, but its focus on personal piety, nonviolence, and rural isolation prevented the Ndebele from engaging on issues of land justice. He reaches back to the time when King Mzilikazi oversaw a "spiritual link with the supernatural world," and he celebrates the restoration of these traditional environmental teachings of the Ndebele after the 1980 independence. He argues that the ancient practice of *umbuyiso*, the shepherding of the dead into the afterlife, for example, occurring as it did at "a time of green leaves . . . just before the first rains are due," reconnected the Ndebele to the land. He recalls his own experience with this environmentally sensitive ritual in 1984, when his father died. As the eldest son, he was asked to return home to Zimbabwe to rekindle the family name. Back at his family's farm he experienced a blessing when a bull named Ubabomkhulu (Grandfather) gave Ndlovu, a Christian, a special bellowing and a rigorous head-shaking, thus signaling that the *idlozi*, his father's spirit, had passed into the animal. Ndlovu lauds the 1960s-era decision to have the administrative duties of the white mission superintendent assumed by a black pastor, a black principal, and a full-time black farm manager. But he is critical of the fact that the white missionaries "controlled . . . all the farms" for as long as they could. Ndlovu is angry that even after Liberation old ways were not fully respected, and he links this failure to environmental catastrophe. He recalls a dream in which the "the spirits were angry" because "once, our people held regular rain ceremonies and made sacrifices to appease *amadlozi* . . . but then our culture became diluted with blue jeans and color TVs and . . . the skies became empty and the soil thirsted."[88] Agriculture in particular had lost its traditional roots, and onetime farming families had become separated from both the land and nature.

The published texts that recount the history of agriculture at Matopo report on a segregationist, colonial structure in which progress was dictated by white leaders. And any technological advancement was very slow and certainly inhibited by the bloody 1970s War of Liberation. Even with independence in 1980, few Matopo farmers embraced modern agriculture; indeed, some of the most articulate voices championed nothing less than old spiritual ties to the land.

Conclusion

The twentieth-century history of the seven places in this study reflects the particulars of their environments, which are disparate in multiple ways. A comparative analysis of local records indicates a variety of perspectives on this century of change. But it also reveals a significant gradation of modernity that affected farmers' relationship to their environments, which differed vastly between the places of the Global South and those of the Global North and among the specific communities within these broad designations. For the farmers of Iowa and Manitoba, agricultural expansion was disrupted mostly by occasional economic downturns, especially in the 1930s and the 1980s, and even these hardships hardly affected a seemingly inexorable pace of change. For farmers in Friesland and Siberia, it was the horrors of war, both in the 1910s and the 1940s, and then powerful new forces of state-sponsored change that dictated a sharp rise in mechanization, land consolidation, and agricultural science. In contrast, the impetus for change was mitigated in the communities located in the Global South, even as the gradual introduction of Green Revolution technologies altered old livelihoods. In Bolivia, migrant farmers produced their own upheaval through costly migration and adaptation to frontier conditions, all the while resisting the logic of agricultural capitalism with their selected approach to technology. In Java and Matabeleland, colonial structures, revolution, and struggles for independence ensured that farmers would reap fewer of the benefits of the twentieth century, even as some farmers venerated old life-giving forms of agriculture.

The twentieth-century history of global agriculture cannot be told as a monolithic account. Certainly common elements existed, but the specifics of local soils and climate and of culture and social fabric apparent in the seven places examined here resulted in seven specific Mennonite stories. For the Doopsgezind farmers in Friesland the twentieth century represented an opportunity to continue the logic of agricultural reform as they had known it in previous centuries, seamlessly accepting change alongside fellow Frisians. For the Mennonites in North America, those changes compromised social boundaries of founding decades and turned farming from a guarantee of the separate community into a particular livelihood within an integrated global economy. For the farm communities in the Global South, founded upon very specific religious beliefs and set within developing economies, the twentieth century brought fewer changes: agriculture here complemented the values of humility and community cohesiveness. In all seven places, change differed from one community to the next, each trajectory affected by a dialectic between large forces of agricultural "expansion" and local conditions, including climatic, political, economic, and religious ones. The result of this dialectic shaped the farmers' ultimate relationship to their respective environments and made for seven rather localized accounts of global agricultural change.

CHAPTER 4

Making Peace on Earth

An Agricultural Faith of the Everyday

In 1929 Andries Lucas Broer, the Mennonite pastor at Harlingen, Friesland, published the first of his dozen books of poetry and essays on nature and cultural history. Titled *Open Vensters* (Open windows), it contained numerous poems describing the Dutch landscape. In "Neveld-age" (Misty days), Broer contemplated the beauty of autumn in Friesland: "I truly love those silently grey days / Now the earth takes a rest from labor / . . . / Now, after the full ardor of the summer sun's fires / My heart silently rejoices in the things of silence." It is as if he had an eye on the farms of his parishioners, celebrating with them that the summertime labors were completed. But he was also a person of deep faith and uttered his divine gratitude for the natural: "Thou, who folds the misty garments around meadows /And by whose mighty Word soon the sun returns / . . . / Thou hast woven the golden yarn of Thine secret / And I thank Thee that I'm allowed to trace her."[1]

It may seem unremarkable that a rural pastor would express this sentiment, yet ten years later in a 1939 essay, "De Natuur en ons Godsdien-stig Leven" (Nature and our religious life), published in the Doopsgez-ind magazine *Brieven*, Pastor Broer criticized his church's view of nature. His own love of the environment evidently was not shared broadly by other Doopsgezinden, the Mennonites, he knew. He contrasted them to the Calvinists, who, he asserted, spoke more easily about God as creator and indeed of creation itself as a sacred text. The Mennonites, wrote Broer, saw the divine only in their inner conscience and a personal, loving Jesus. Broer admonished Mennonites to learn to see the ways in which "nature manifests itself as strength, power, wisdom, order and beauty that highly exceeds what humankind is capable of" and therein "find the very footsteps of God." And he moved quickly from this general admonition to denounce modern haste and waste. As he saw it, the Mennonites had "desecrated nature . . . restricted, mutilated and destroyed her." He wondered how wealthy Doopsgezinden could pay for expensive paintings by Rembrandt, Vermeer, and other Dutch artists but then ruin "God's pieces of art . . . the moors, the marshes, the dunes, the lakes, the forests."[2] To Broer's mind, his own people had failed to preserve natural Friesland and in so doing had stagnated in their spiritual life.

Culture, and religion in particular, has guided human perspectives on land and nature from time immemorial. Sacred scriptures everywhere infuse nature with the spiritual.[3] And significant writings—whether specific essays such as Lynn White's now classic 1966 critique of Judeo-Christian values, "Christian Myth and Christian History," or Wendell Berry's corpus of writing with titles such as *The Art of the Commonplace*—argue for this centrality of religion. White saw within Christianity the ideological underpinnings of a culture of indifference to nature and easy exploitation of it, while Berry saw within it the possible grounding for a tacit reverence and gratitude for the land, even the grounding of an ecospirituality.[4] Indeed, a wide scholarship agrees that religion has shaped an environmental dialectic, for good or bad.

But Broer was right: historically, Mennonite thinkers and preachers have not interwoven their religious and environmental thinking. Certainly Mennonite scholars have demanded that their coreligionists do so, especially since the modern environmental movement took off in the 1970s. Calvin W. Redekop's 2000 collection *Creation and the Environment: An Anabaptist Perspective on a Sustainable World*, for example, features essays that denounce the historical lack of a Mennonite theology of creation and those that outline just such a theology. In his foreword, Redekop observes an uneven Anabaptist-Mennonite discourse, "a unique, though not totally consistent, philosophical and ethical position regarding the creation."[5] On the one hand, as the theologian Walter Klaassen laments, Mennonites had "done *no* thinking about the resources of our tradition of nonviolence in the human war against mother nature" until the late twentieth century;[6] and a number of essays, including Redekop's own about his family plowing the fragile semiarid plains of Montana, underscore this charge of "no thinking." On the other hand, several writers in the book do outline "an Anabaptist perspective" on nature: Theodore Hiebert sees in Anabaptist thought "the human as a servant of creation"; Dorothy Jean Weaver sees within its stance on nonviolence the promise of a "re-stored" relationship between humans and "the rest of the created order."[7] But these are writers who suggest what should be done, not what Mennonites have accomplished. True, some contributors celebrate specific historical actors who embody this link between Anabaptism and land—the Amish father who "gave more to the world and its inhabitants than he took," the urban Mennonite's organic "Shared Farming" business, the politicly engaged Mennonites fighting for farmland preservation—but these are more often the exceptions in the history of Mennonites and "a sustainable world" rather than the representatives of it.[8]

Since the publication of Redekop's collection, other works have advanced an even more hopeful idea of an Anabaptist theology of sustainability. The collection *Rooted and Grounded,* edited by Ryan D. Harker and Janeen Bertsche Johnson, for example, traces "the intersection of land and discipleship" as an extension of "the biblical vision of peace."[9] But this too is a volume that outlines a problem, "a detachment from the land," with a call to action, the need for "the restoration of the land that God created." And most of the book is a nuanced intellectual challenge to readers to avoid "the way of dominance" and embrace "watershed discipleship."[10] A smaller number of writers celebrate a historical Mennonite reverence for the land. There is Douglas Kaufman's account of an environmental application of "a theology of humility" in the 1760s; Ezra Miller's outline of a "symbiotic relationship" with the land within the Indiana "settlerscape" in the 1840s; Rebecca Horner Shenton's history of "faithful stewardship of the land" in the conscientious-objector camps in the 1940s.[11] Here are historical challenges to the readers of today.

Scholars in disciplines such as anthropology and earth sciences, documenting what has been rather than what should be, are more skeptical of a direct connection between Anabaptist religious practice and sustainable agriculture. Martine Vonk's 2011 comparative study of Anabaptist and Catholic farmers' relationship to the environment concludes rather decisively that "a direct causal relationship between religion, measured by denomination, church attendance and biblical beliefs, on the one hand, and environmental behaviour, measured by attitudes, concern and/or diverse action on the other hand, cannot be found." At best, she argues, this relationship has been "indirect and complicated."[12] David L. McConnell and Marilyn D. Loveless come to a similar conclusion in their 2018 *Nature and the Environment in Amish Life.* They argue that while the Amish in the United States do "see themselves as stewards of the earth," this is at best an indirect consequence of their theology of simplicity. Indeed, the Amish reject any idea of spirituality "embedded in nature," but they do teach "self-denial" and possess a historical predisposition to "parochial" rural life, which simply entails a particular "resource dependence" and closeness to nature.[13] Where they have practiced a sustainable agriculture, it has not been as a result of religious practice per se.

The viewpoints outlined above reflect the thinking of a foremost apologist for linking religion and land, Wendell Berry. In an early essay, "People, Land and Community," Berry argues that "the connections that join land, people and community . . . in a healthy culture . . . are complex," and old, moralistic dichotomies obscure such linkages. What matters is the culture that infuses the local, the visions that propel a farm's everyday agenda of work, and even the quotidian messiness that hard work in the soil entails. The culture of a "good" farming community, writes Berry, arises not from a set of moralisms but from "an order of memories preserved . . . in instructions, songs and stories." Religion may be central to rural culture, but not as formal theology. What has guided rural communities has been an everyday faith, not adopted from intellectually sophisticated wellsprings but rooted in "work and love . . . within a kind of community dance."[14] It is a culture known and embodied by farmers at the local level, in daily interactions with ecology, more often tacitly acquired than verbally proclaimed.

This lived reality describes the relationship of religion and land in the seven Mennonite communities under study here. Here are seven different sets of these "instructions, songs and stories" of the everyday, reflecting ordinary farmers as people of faith within the physical environment. The stories are messy, contradictory, often articulated by "work and love" rather than coherent philosophy. Two distinctive sources tell this story of the everyday. First, there are the local church's historical texts—sermons, congregational histories, memoirs—from each of the seven places, which most often address ecological issues only indirectly but set the stage for them nevertheless. Second, oral-history testimonials provide the perspectives on faith and land by one farmer or gardener in each of the seven communities. Importantly, the interviewees were arbitrarily chosen and were not vetted for being each community's most articulate, systematic, or hopeful. Perhaps these texts refrain from making a case for a specifically "Mennonite" approach to the land. But each offers at least one version of how Mennonites from around the world, as religious persons within the everyday, have imagined the biotic within the local community. Side by side, the locally written text and the quotidian memory speak to "complex connections" to the land, and they defy easy dichotomy.

Friesland: Peaceful Fredeshiem and Fabled Jorwerd

In Friesland, the Doopsgezind farmers, spread out as they were, a minority among Reformed and Catholic neighbors, developed an especially complicated sense of religion and the environment. These farmers moved easily among denominations; they intermarried, blended into neighborhood associations, and changed church affiliations at will. Then too, religion for most Frisian Mennonites, even by 1900, was remarkably open-minded and rational, with a focus less on correct religious doctrine or an emotional spirituality than on culture and ethics, expressed especially in acts of charity.[15] It also mattered that old Anabaptist ideas on pacifism or simplicity had long fallen by the wayside. But complicating this historical trajectory was the twentieth-century rediscovery of an Anabaptist ethic, old teachings on nonviolence and ideas on simplicity that had been lost for four hundred years. By the year 2000, Frisian church leaders readily attributed a certain sacredness to land stewardship.

Two different written sources trace this changing view of the land. The first is the lengthy 1955 article on the Netherlands in the *Mennonite Encyclopedia* by the Frisian pastor Nanne van der Zijpp and the Holland-based peace advocate C. F. Brüsewitz. They emphasize the prevailing "moderate supernaturalism" in place at the beginning of the twentieth century, with its emphasis on personal "virtue and enlightenment" and "the teachings of the good Jesus." It was at once a liberal and a personal religious system that addressed neither rural life nor environmental relations. It served well the needs of sophisticated middle-class society in both city and countryside, indeed so well that a precipitous two-centuries-long Dutch Mennonite decline was halted and by the beginning of the century there were some fifty-eight thousand Doopsgezinden in the Netherlands. But, as noted above, with but few traces of the old Anabaptist teachings of simplicity and nonresistance, there seemed to be no specifically Mennonite system of ethics, let alone approaches to nature and land. The last of any groups of Mennonites that strove for agrarian simplicity in Friesland, write Van der Zijpp and Brüsewitz, were the traditionalist Old Frisians at Balk, who had emigrated to Indiana in 1853. Dutch Mennonites would continue to make their mark as a liberal branch of Christianity, an identity reinforced when in 1911 they became the first denomination in the Netherlands to appoint a woman, Anne Zernike, as pastor, placed, remarkably, in the rural Frisian community of Bovenknijpe. Van der Zijpp and Brüsewitz do emphasize that as the century progressed the Doopsgezinden rediscovered some of their old teachings, including the historical Anabaptist idea of nonresistance allowed for by the 1925 Dutch Law of Alternative Service.[16] It was a small step toward a religion that in time would make room for the extension of peace to the land.

A second, more recent church history, the 2006 Dutch Mennonite account within the Global Mennonite History series volume on European Mennonites, *Testing Faith and Tradition*, makes a stronger case for the link between peace and land. But the authors of the chapter on the Dutch experience, the pastors Annelies Verbeek and Alle Hoekema, see such a tie developing only slowly over the course of the twentieth century. They too lament the Mennonites' enchantment with enlightenment and personal virtue at century's beginning, which overrode old teachings on nonviolence and humility and left them spiritually "stagnated." When the Mennonites recalled their Anabaptist past, as they did in 1879 with the erection of the Menno Simons monument on a farm just outside Witmarsum, in Friesland, the impulse arose not from his "radical" ethics of the everyday but "out of nationalistic and romantic feelings."[17] But in time the Doopsgezinden did indeed rediscover historical peace teachings. The hosting of the third Mennonite World Conference in 1936 in three Dutch locations, including historic Witmarsum, provided the Doopsgezinden with an enhanced Anabaptist identity. At the same time, a "Quaker spirituality" and "Bible-oriented" ethics, newly popular within the wider church, also gave new credence to old teachings.

A new association of religion and nature seemed to follow.[18] In this spirit Mennonites created several countryside retreat centers, including a 1929 Frisian initiative, Fredeshiem (Ground of Peace), near Steenwijk. These rural retreats contemplated the lost teaching of nonviolence, certainly emboldening a wartime critique of Nazi occupation of the Netherlands but also generating criticism of the Dutch colonial war in Indonesia in the late 1940s. Indeed, this overseas war also gave birth to the Doopesgezinde Vredesgroep, which organized a minority of conscientious objectors resisting the draft to fight in this colonial war.[19] Ironically, as Doopsgezind church membership declined precipitously once again in the second half of the century, from 38,000 in 1956 to 10,000 in 2006, the Dutch Mennonites' historical awareness of their Anabaptist heritage increased. The 1974 founding of the Doopsgezinde Historische Kring was significant for this reason. And with this consciousness also came a sharper critique of unchecked rural capitalism and modern agricultural ways. Verbeek and Hoekema conclude their chapter with reference to Geert Mak's famous Frisian-based 1997 work, *How God Disappeared from the Village of Jorwerd*, a signpost, since "several rural Men-

Mennonite (Doopsgezind) church at Pingjum (Netherlands), the village in which
Menno Simons served as a Catholic priest from 1525 to 1531. Constructed in the late
1500s as a *hidden* church, it complied with Dutch laws forbidding Mennonites from
constructing publicly recognized church buildings. Photo by H. H. Hamm in 1950,
Mennonite Heritage Archives, Winnipeg.

nonite communities have experienced this fate." At Jorwerd, full acceptance
of farm mechanization, commercialization, and economies of scale undermined the ethical foundation of rural society, while secularization produced
a particular cultural ethos that made "Christianity itself, superfluous."[20] With
their work, Verbeek and Hoekema may have served the Doopsgezinden as historians, but they were also pastors at a moment in time, calling for a return
to simplicity, communitarian wholeness, and respect for nature.

This commitment to a religiously based rural ethic is apparent in the oral
history of actual farmers in Friesland. Jelke and Roelie Hanje, who farm near
Joure, articulate it despite insisting that little has distinguished the Doopsgezinden from their Reform or Catholic neighbors.[21] They point to Jelke's mother,
a Reformed woman who married a Doopsgezind man: "It didn't matter at
all to [my dad] what someone believed," says Jelke. In fact, for a time his parents attended both the Mennonite and Reform churches, each one every
other Sunday. This didn't mean that religion and rural life were not connected;
the association was simply unarticulated and shared among the Mennonites,
Catholics, and Reformed. Roelie recalls stories about social justice from Jelke's

grandfather, who, suffering from rheumatism, relied on employees to run his farm and was widely known as "the first one who paid the union wage to workers." In fact, he was seen as so generous that the neighbors rumored that "Hanje is going bankrupt, because he pays his workers such high wages." But it was a familial ethic and not a religious practice. A similar generosity was revealed when Jelke's father welcomed a number of "famine children" during the Second World War, including one boy who "actually couldn't stand, he was so thin and when he left, he was as big as a bear." Again, Jelke doesn't see this act as linked to a specific denomination: it was something Frisian farmers in general were known for; it was a time when "farmers were held in high esteem." Communitarian ties that created this particular agricultural ethic are now a thing of the past, says Jelke. He recalls the time before the land reforms of 1962, when farms still shared a communal pasture, when all the farmers—Reformed, Catholic, and Doopsgezind—each with only half a dozen cows, would meet at the *melkherne*, the milk corner, and call out their own animals for milking. After completing their work, the thirty or so farmers would head back to Joure in a group, by bicycle. "Well there's nothing left of that now," says Jelke. The land reforms brought electricity, and with it came the milking machine, sequestering community farmers in their own barns. Jelke misses the interchurch communitarian nature of agriculture of yesteryear.

And yet, Jelke says being Doopsgezind did matter in some narrow, mostly unspoken ways. His Reform and Catholic neighbors, for example, do refer to the Doopsgezinden as "peace apostles," and it is true historically, for "if the neighbors had a problem, [we] just sent my brother there and it always came up right"; his brother "was never angry." And in their farm household, any sibling squabble was settled before bedtime: "We were never, ever allowed to go to bed angry." Indirectly this peace ethic also affected their approach to land, for what mattered was the community and the extended family, not the individual. For example, the Hanjes valued the intergenerational family farm. Jelke says he acquired a farm because "my father and my great great grandfather were all farmers, on my father's side, my mother's side too"; indeed, "my father made us all [four brothers] farmers," providing each one with ten cows: "We wanted very much to become farmers." And this farm lineage, says Jelke, created a wider sense of belonging and indeed a Doopsgezind identity. The farm and church in the Hanje family had a close, if unarticulated, link.

Certainly the Hanjes' farm has been guided by an ethical dimension, but only indirectly is it linked to their own evolving Doopsgezind ideas on peace.

Jelke Hanje's father was known for his nonviolent approach to life, but the link between nonviolence and environmental care is a more recent concern. Pastors Verbeek and Hoekema in 2006 highlighted what had gone wrong at highly commercialized, fictional Jorwerd, not what had already developed at Joure and other Frisian districts in which one might find Doopsgezind farmers.

Iowa: Geography and an Ethic of the Everyday

Quite a different version of Mennonite life ensued in the North American diaspora in general and in Iowa in particular.[22] An early church history from 1939, *Mennonites in Iowa*, by native son Melvin Gingerich, uniquely discusses the Amish and Mennonites' relationship to agriculture.[23] At every turn, Gingerich depicts the Mennonite farmers as hardworking and thrifty, people of faith treasuring the health of Iowa's rich soils. They "are closely attached to the land," writes Gingerich, "and they obtain a large amount of their pleasure from watching things grow." Moreover, they are model farmers, planting legumes and rotating their crops as in the past but now in the 1930s also using the latest techniques, including "artificial fertilizer." Neither this nod to modernity nor their general avoidance of state-based soil-conservation units equaled a cavalier view of the soil; Iowa's Mennonite farmers simply thought it wrong to "join any organization in which non-Christian people may be in the majority." Indeed, an indication that the community linked religion and the land was its perennial search for inexpensive farmland in places farther west for the next generation, helping establish new rural congregations and, of course, keeping young couples from the reach of harmful city lights.[24]

If the Mennonites associated faith with land, their Amish neighbors did so even more intentionally. Gingerich lauds the Iowa Amish commitment to agrarian simplicity, seen, for example, in their steel-wheeled tractors, which kept farms small. Without elaboration he also praises them generally for being "sympathetic to any plan to save the soil." He highlights the Amish yieldedness to nature as a spiritual strength, and he cites their rejection of lightning rods based on the belief that God alone should control the lightning strike. Gingerich insists that this act of Christian humility contradicts the controversial pre-Christian folklore of the Amish, of which he disapproves, especially "pow-wowing," paying "much attention to the different phases of the moon in their planting of trees or crops," and "even the slaughtering of hogs and construction of buildings . . . according to the signs of the Zodiac." The Amish, writes Gingerich, also attempted to heal sick children or to stop animals from bleeding with charms. But he insists that by 1939 these old

European-based practices had largely been eradicated and relegated as "superstitions."[25] This Iowa story heralds rural life as a bulwark against the wider world and healthy soil as the foundation of rurality. As Gingerich saw it, these early-twentieth-century tillers of the soil were indeed faithful Anabaptists on the land.

A 1994 work by the local writer Franklin L. Yoder, *Opening a Window to the World: A History of Iowa Mennonite School*, describes a similar ecclesial relationship to land in the second half of the century. At the outset, Yoder links Iowa Mennonite School (IMS), a Washington County high school established in 1945, to rurality. This association was reinforced daily with a simple view of the Deer Creek watershed from the school windows, a vista of "rolling fields and farmsteads . . . an image of abundance and tranquility . . . the crops and fields creat[ing] a mosaic of order and beauty." Together these farmsteads and the surrounding land, argues Yoder, served the old Anabaptist religious ethos of "humility, nonconformity and separation from a sinful world" in the cities beyond. During the fervently patriotic Second World War, for example, these farms marked a haven from the wider world, a place where the Mennonites could be left undisturbed. Fifty years later, in 1994, the farms had changed in their "search for efficiency and economies of scale." Yet the farmer's humble outlook on life remained, affected by the vagaries of the weather, a commitment to the land, and loyalty to local congregations. Moreover, the historical mystique of Deer Creek remained: IMS was a private, religiously based Mennonite institution linking a particular physical environment to a specific religious identity and mind-set.[26]

This nexus of faith and land is also expressed in the oral history of Washington County farmers. Gerald Yoder and his adult son Brent, for example, readily speak of faith as intersecting with agriculture.[27] In many ways the Yoder farm is a typical Iowa operation. Located in Washington County, just outside Wellman, it has served the family for generations. Gerald was born in 1942, just months before his parents purchased the 80-acre (32-hectare) farm where his mother had grown up, to which they added a smaller piece his dad had purchased before the Depression. Gerald recalls the mixed farm of his boyhood: they kept a couple of cows, a small flock of chickens, a few hogs and sheep, and some geese and ducks, and they raised the corn and hay needed to feed them. The animals were all kept outdoors, including the hogs, out on the fenced-in range.

This is the story of their farm, but the Yoders are also very conscious of their Mennonite affiliation and relate it to farmland. Indeed, as a couple Gerald and his wife, Kathy, have never questioned their Mennonite association: "That was where I was baptized at, so I was [Mennonite], my wife is Mennonite"; he grew up in the East Union church, she in the Lower Deer Creek congregation, ten kilometers apart, both of Amish-Mennonite lineage. Moreover, the sense of being Mennonite was shored up with a particular and evolving approach to the environment. In the past, says Gerald, Mennonites valued the idea of working together, helping one another: "When one of them had hay to make, they would just help each other out," especially in times of need. In time this religiously informed ethos of service also crossed community boundaries. Gerald has served with Mennonite Disaster Service, founded after the Second World War, and recalls its vital assistance when a tornado hit their community in 1984. He also has participated in canning meat for Mennonite Central Committee to feed the hungry overseas, another, more recent service opportunity, since "when I was a kid there wasn't so much of that."

Reflecting this new ethos, Gerald also says that as a farmer he has been guided by values of faith even in his cultivation practices. Remarkably, he gives as an example his rejection of the plow, which had shaped Washington County Mennonites' approach to the land since the 1840s. "In 1990 we started 'no till' [farming]," leaving the corn stubble and soybean residue on the ground; in fact, he rues the fact that "we used to plow everything. . . . I'd say we're better stewards of the land [today than in the past]." Not plowing means less soil erosion from wind or water runoff but has made the Yoders reliant on chemicals. Gerald's son Brent has been critical of organic farming that relies on stirring the soil: "When you look at stewardship and farming it's difficult. . . . You could take organic farming and say 'I'm the best steward because I have an organic farm' [but] they're losing a lot of soil. . . . So to me we don't know the effect of the move [to chemicals] yet. . . . I wish they would have never invented that thing [the plow]." Ironically, he questions the ethics of organic agriculture, commonly practiced by Mennonites and Amish enunciating soil stewardship. The Yoders have their own view of the meaning of faith-based soil conservation.

When father and son reflect more deeply on how religion has affected their approach to farming, certain generational lines emerge. Religion matters for Brent in an especially personal but less tacit way. Using evangelical parlance, he says that farming has been a spiritual vocation, and one applicable in the everyday: "I would say that there is a faith interaction with my business. . . . When I accept Christ as my savior, that's everything. I can't give him parts and pieces. . . . If God called us to do something else . . . you would have to say 'yes'!" In this respect farming has a certain sacredness, but mostly in that farming was a vocation that he himself had been called to. For the more elderly Gerald, farming has been sacred in a different way, as a vessel of ethics related to old teachings on humility. Indeed, they have determined his view on land: "When land was $3,000 an acre, I wanted it like everybody else. But . . . we've never [been] dog-eat-dog to get any land. . . . There was land come up and I prayed about it, but I never got it." When he ponders how their farm business and religion have been intertwined, he says, "We've been blessed," and he repeats that "we've never been dog-eat-dog either." Gerald sees faith as a historical check on rural capitalism.

But it's been more complex than this singular equation. Religion has determined the Yoders' farming culture in several interlinked ways: it has defined certain geographic boundaries, given meaning to their vocation, added imperative to soil conservation, and challenged greed. The way the Yoders have interpreted this ethic of the everyday may well put them at odds with some of their Mennonite and Amish neighbors, who have come to different conclusions on how to achieve peace on land. The ethical culture related to land even within a single Mennonite locality has hardly been homogenous.

Manitoba: Land Reserve, Faith Fracture, and Mystic Garden

In Manitoba, the wider world was also held at bay by geography. Because the eastern half of the West Reserve, or Rhineland Municipality, was home primarily to members of the Bergthaler Mennonite Church, Rhineland's history focuses mainly on one group of settlers from a specific geographic entity, Bergthal Colony in Russia. Hence, this congregation has a particular physical identity. Indeed, the local historian Henry Gerbrandt's 1970 history of this church, *Adventure in Faith*, emphasizes its material boundaries: the Bergthalers resided mostly on the difficult black clay soils of the eastern half of the West Reserve, and the more traditionalist Old Colony Mennonites on the lighter, more easily cultivated western half.[28] Then in 1892, when a schism over schools tore the Bergthaler people apart, established geographic boundaries of the Reserve contained the split. The progressive, pro-public-school Bergthaler church rump, led by Bishop Johann Funk, remained in place, but now they lived interspersed with the breakaway, more traditionalist and communitarian Sommerfelders, named after the rural village of their new bishop, Abram Doerksen. This schism marked a separation within Rhineland.

Indeed, it was a sharply divided place. *Church, Family and Village*, a collection of writings edited by three historians of the West Reserve—Adolf Ens,

Jacob E. Peters, and Otto Hamm—outlines the ways in which the progressive Bergthalers differed from the communitarian Sommerfelders. While the former deemphasized farming as *the* Christian vocation, embraced public education, and allowed their youth to work in nearby Gretna and Altona, the latter remained ensconced in the rural farm villages. Then, in the 1920s, recommitting themselves to agrarian simplicity, many Sommerfelders emigrated from Canada to avoid education in assimilative English-language schools. In 1922 some 600 Sommerfelders followed Abram Doerksen to join the Old Colonist Mennonites in Chihuahua, Mexico, and in 1926 another 400 joined similarly minded East Reserve Chortitzer Mennonites in a migration to the Paraguayan Chaco; then in 1948 yet another 500 moved into the rain forest of eastern Paraguay alongside like-minded traditionalists from Saskatchewan.[29] In each of these places, Sommerfelder settlers found agricultural land to be a check against the modern reordering of rural life they had observed in Rhineland.

The distinctiveness of these Bergthaler and Sommerfelder histories is reflected in sermons. Significantly, a New Year's address by the Bergthaler bishop Johann Funk in 1892 articulated a spirituality in which environmental discourse served as religious metaphor: "We are only pilgrims . . . who have no permanent place on this earth . . . and journey towards eternity." His was a personal spirituality: "Even though, the roads outside are covered with the frost and snow of this winter, God's mercy-paths . . . shall never be covered with snow nor be frostbitten."[30] Ironically, sermons by the more agrarian Abram Doerksen contained fewer allusions to nature but ultimately served to keep farmers on the land. Certainly Doerksen preached a general piety of "repentance and forgiveness," but overall his sermons admonished and challenged rather than expounded or theologized. In the twelve-week period preceding the annual springtime communion service, Doerksen preached one sermon based on 1 Peter's call "that you have sincere love for each other," another on 1 John's challenge to "walk in the light," and a third on 1 Corinthians 11's demand that everyone "examine himself" before they "eat of that bread, and drink of that cup." Analyzing Doerksen's sermons, the theologian David Schroeder highlights their warning "against believing with the mind only . . . and not having a change of heart" but argues that at their core they "emphasized character . . . rather than ethics."[31] The equation was straightforward enough. A change of heart for Sommerfelder Mennonites spelled humility, which shunned the capitalistic impulse, and ultimately character led fifteen hundred Sommerfelders to relocate to demanding environments in the Global South.

This emphasis on a communitarian gospel set the Sommerfelder apart not only from the Bergthaler in 1892 but also from a third congregation, the Rudnerweide, or Evangelical Mennonite Mission Church (EMMC), which separated from the Sommerfelder in 1937. In that year twelve hundred members, or about 20 percent of the total church, broke away to establish a new ecclesial body with a decidedly evangelical bent. The West Reserve native Jack Heppner's comprehensive history of the EMMC, *Search for Renewal*, describes a highly personal and emotional religiosity. He recounts an experience of the church's founder, Isaac P. Friesen, who spoke of a compelling moment when "the room was full of light" and "all fear vanished and I was able to say, 'Jesus has forgiven my sins.' . . . Oh how happy I was."[32] The focus of faith had shifted from the communitarian to the individual. The writer Mary Neufeld, in describing the religious thinking of her father, Wilhelm Falk, the EMMC's first bishop, tells of a church moving away from its agrarian roots and becoming more attuned to the demands of urban living. In 1937, for example, Falk outlined a five-point pathway for the new church, including "certainty" of belief, personal morality, youth instruction, "spiritual awakening," and foreign missions.[33] Here was a confident individual piety ready to negotiate a world beyond Rhineland.

Of the three church groups in Rhineland, the Sommerfelder Mennonites were the most committed to agrarian ways as religious expression, in the form of everyday action as much as that of the spoken word. A similar equation is seen in the oral history of a daughter recalling her gardening mother. It recounts the everyday life of Norma Giebrecht within the transplanted farm village of Neubergthal and in and around a house barn that her immigrant grandparents from Russia built in 1901. Her story bridges the theme of faith and soil, but indirectly and almost without words.[34] She recalls her mother at work in the family's perennial garden, bordered by lilac hedges and located just behind the summer kitchen, itself draped in grapevines. Norma's mother was in tune with the seasons and patiently tended her plants and the soil accordingly. As Norma puts it, her mother "let everything grow, let everything come up, and everything always did come up." Norma recalls reading stories of her great-grandparents bringing seeds from Russia, following requests sent back home by the first group of immigrants: so the "next batch . . . would bring seeds, carefully packaged. . . . So I imagine her mother before her and her grandmother, they will have brought whatever they could," perhaps "coneflowers, and hollyhocks," from their own places of origin in the diaspora. Norma finds this environmental connection to an imagined old homeland personally compelling.

Norma is also moved by her own mother's direct tie to the land. She describes her mother's immense knowledge of plant life despite only a rudimentary education. She knew "all about seeds, when to plant, when to hoe and when to harvest." And she knew about growing food organically. Norma's mother had the inherited knowledge to keep plants "healthy without the magic of fancy garden books or commercial fertilizers," and she knew that "prize tomatoes and corn thrived on manure, from the barn and the chicken coup." Her garden was also closely linked to sustenance. The vegetable garden, she says, was the foundation of a farm's self-sufficiency; there were beans, peas, and corn, of course, but especially potatoes: "We had terribly many potatoes. . . . They were a staple."

But then Norma takes the image of seeds brought from Russia and the fullness of the garden and utters a single line that instills those memories with meaning. She knows that her mother "felt the nearness of God in her garden and there drew peace of mind. It was a bright spot in the days of drabness and demanding work. She took humble pride in sharing a task with the 'Master Gardener.'" The spiritual emotions of "joy and love," says Norma, were indelibly related to the orchard and garden. It was a particular peace on earth—the experience in the big gardens behind the house barns of the village of Neubergthal. And in these gardens church lines mattered little. In fact, Norma makes no mention of whether her family was Bergthaler, Sommerfelder, or EMMC. Their varied theologies, no doubt, reflected certain approaches to the land, but when viewed through the filter of the everyday, conditioned by local constructions of gender, their differences diminished in significance. All three congregations within Rhineland, after all, included farmers as members, all preached nonresistance, and all venerated interdependence in some form and a particular intergenerational modesty on the land. What mattered for Norma was a certain wholeness of being within the garden, one filled with "instructions, songs and stories" linking the past to the present, and both to the land and sustenance.

Java: Colonialism, Decolonialism, and the GITJ

The colonial context of the Global South affords yet another set of Mennonite religious expressions on land and agriculture. The history of the Gereja Injili di Tanah Jawa (GITJ) at Margorejo during the twentieth century is, of course, rooted intrinsically in the coupling of farm village and religion. This theme marks the beginning of the twentieth century, continues during the immense upheaval of the Second World War, survives the church growth during

Indonesia's anti-Communist campaigns of the 1960s, and outlives the refashioning of ecclesial testing in environmental hardship in the 1970s. In telling this story, once again Sigit Heru Sukoco and Lawrence M. Yoder's rich comprehensive history of the GITJ is crucial. Indeed, it is much more than a linear church history. It alludes to a dynamic between a Western cosmology and Javanese traditions that affected the very way local farmers related to their environment. At the beginning of the century certain "unbaptized" village members, hesitant about Christianity, as well as some baptized members, practiced Javanese life-cycle rites, including puberty-related rituals and prayers for the dead, the *selamatan*.[35] These were the recorded events, but the coexistence of the two cultural streams also resulted in a spirit of acceptance or submission, even taking both the good and the bad as coming from God.[36] At a more empirical level, the strength of Javanese culture eventually compelled the Europeans to relinquish control and the church elders to recognize the nationalistic, even militant impulses of the youth. By happenstance, it was at about this time that the church dropped the words *Tata Injil* (Gospel Way), from their nomenclature, a signal that the church wished to be a less exclusive body. Certainly the old Anabaptist idea of pacifism was increasingly seen by the village's

The GITJ (Mennonite) church at Margorejo, Java, constructed after the original church was destroyed on the orders of Indonesian nationalists in 1942.
Photo by Danang Kristiawan, Jepara.

youth as a tacit nod to colonialism. In any case, Margorejo now aligned itself with Indonesia's cause of independence.[37]

This rising sense of self-determination among Javanese Mennonites also was evident in the way church members related to land and the environment. During the Second World War, GITJ affiliates at the various mission farms simultaneously maintained a mystical Javanese approach to the land and accepted Western-derived food aid. This connection could best be seen at Margokerto, on the west side of Mount Muria, opposite Margorejo, where a former herder, Wirsamta, took over the farm village, declared himself to be "supernatural," and then redistributed the congregation's land, livestock, rice, and coconut groves in ways that he considered just. At Margorejo itself, the village youth's political action group also demanded a reassignment of church agricultural land profits and a redivision of farm labor. And then the Margorejo farmers themselves, propelled by a Javanese sense of justice, turned against their own church leaders in the 1950s when they agreed to become the food distributors for the US-based Mennonite Central Committee's food-aid program; local farmers saw such activity as an acquiescence to colonialism and to lost honor.[38]

In the 1960s this link between Javanese mysticism and land justice strengthened. At this time Margorejo, along with various other mission farm villages, attracted new converts from among local, mystically inclined peasants who were driven by a new, anti-Communist national law to join any recognized religious body and thus seek shelter from the state's violent program to squelch Communist insurgents. Indeed, a number of these converts had earlier been associated with the PKI (Partai Komunis Indonesia) Communists and the BTI (Barisan Tani Indonesia), the Indonesian Farmers Front, with a strong commitment to land justice. And yet, concurrently there was a rebirth of Christian teachings on charity with an environmental implication. In the 1960s and 1970s, for example, the local Diaconal Service Body (DSB) joined forces with MCC and other NGOs to address drought in 1962 and 1963. Then terrible flooding in 1967 and 1968 was met with food aid of rice and meat for more than 180,000 persons. As Sukoco and Yoder put it, locals were impressed that especially in 1972, as "water, wind and fire played on the disaster stage and . . . as . . . the soil itself [seemed] . . . not [to] want to see its friends playing without joining in," the DSB mobilized, saving many Javanese villages.[39] Christian and Javanese cosmologies were both apparent at Margorejo in the second half of the century.

Once again, oral history with a single farmer fleshes out this story. The account of the elderly farmer Sukarman, who traces his birth to about 1940, reflects a decidedly Christian view on land but with a certain linking to a traditional Javanese view on nature.[40] He begins by connecting Christianity and the land, telling of the time during the Second World War when his Muslim parents came to Margorejo and converted to Christianity because they "would be given land and didn't need to buy it." But at the get-go his story, while Christian, bridges strict dogma, for his testimonial exemplifies interreligiousness, with no judgment passed on the veracity of one faith over the other. It is a simple fact that Christianity gave Sukarman a chance to farm at Margorejo. As a young adult, having grown up at Margorejo, he was in a position to purchase 1.5 hectares of dryland after building up a ten-head cowherd, and then later he inherited a piece of wetland from his parents. His story of farming begins with a religious act.

He also relates religion to the story of his later success. Although he insists that "if a farmer doesn't start with a cow he won't buy land," he believes that success is associated with Christian forgiveness. He describes the theft of his very first cow, which forced him to practice a life of religiously informed submission: "[Someone] borrowed my cow, sold it, loaned out the money and then didn't return the money to me. So I left it alone. . . . I said, 'It is no problem if he doesn't pay me; God Almighty will repay me.' . . . Wealth and possessions in this world, all come from God." At the root of his farming vocation has been an acquiescence to the monotheistic divine. This same submission has shaped his approach to land and crops. But it has been a complex act as the boundary between Christianity and traditional Javanese religiosity is blurred. Sukarman says that "as a farmer I surrender to God. When I plant I ask the Lord to give us a blessing . . . as my Father. As a father, when a son asks for something, doesn't he give the best to him?" Thus, Sukarman has prayed at both planting and harvest, but he has been mindful to accept whatever occurs within nature as coming from God. The "main point," he adds, has been to "surrender."

With this statement, which may well reflect both a Christian idea, and in particular an Anabaptist sense of yieldedness and humility, as well as the core of traditional Javanese teaching on acquiescence, he segues unabashedly to the latter. Sukarman describes his historical reliance on the old Javanese calendar, which has intersected with a mystical view of the land. He maintains that he has never been a shaman: "I just observe what . . . happened in the past, as the older people taught us." They followed an eight-year cycle, the *sewindu,* based on the ancient Javanese calendar, in which particular years coincided with specific days on the calendar. In time, says Sukarman, most villagers stopped following the old mystical ways, as faith in a monotheistic God grew. Just after this disclaimer, he nevertheless explains the function of the Javanese calendar: "Last year . . . I said, 'Be careful, this year is *Rabu Wage*, it will rain heavily, with heavy thunderbolts that will be frightening.'" He recalls that many people ridiculed him for his predictions. But then it all happened as he had foretold; in some regions, in fact, people were killed by floods and by lightning.

The Javanese calendar has also helped Sukarman accept the cycles of life, the good and the bad, an attitude he sees as complementing central Christian tenets. Indeed, he moves seamlessly from the Javanese belief system back to the Christian as he insists that "everything is the Lord's; God controls everything; humans actually need to accept everything and do their part." But pride has stood in the way. He recites the Christian doctrine of human sinfulness, of Adam and Eve challenging God in the Garden of Eden, "tempted to be more powerful than God." Humility has made for a repentant heart and the faithful farmer. Sukarman ends by referencing the ethic of yieldedness once again: "The human life span is not more than a hundred years, so don't be too ambitious. As long as we are able to eat, that is enough, isn't it?" Disease, in either plants or humans, is natural, he explains, and thus he as a farmer has needed to accept it as his destiny. He has treated the plants with fungicides, but he has told himself that even "if we leave them alone, the disease will be overcome." Perhaps the yield will be reduced but will suffice nevertheless.

Sukarman's story is interwoven with a number of ideological strains: the reverence for a monotheistic creator that Christians and Muslims share is foremost, but an environmental mysticism of Javanese origin, with possible strains of Anabaptism, is evident too. This mix of ideas has shaped the everyday world of the farmers at Margorejo. It is a lifeworld that has evolved since the time of Tunggul Wulung, whose own hybrid religiosity was the inspiration for Pieter Anton Jansz's Christian village.

Matabeleland: Religion, Race, and the Return of Black Africa

The twentieth-century story of Matabeleland's 3,000-acre Matopo Mission farm, under the direction of BICC missionaries presents yet another localized linkage of religion and land. It is a three-part account, the first iteration of which puts the white missionary and a Western cosmology at the center. This story, told in Anne Engle and John Climenhaga's 1950 missions history, accounts for white missionaries' success in a physically arresting geography. They tell a story of incremental indigenization of leadership, paying tribute to the first black deacon at Matopo, Mazilbopela Ncube, who was instated in 1922; they also describe the ordination of three Ndebele preachers, including Reverend Manhlenhle Kumalo, a graduate of the school at Matopo and grandson of King Mzilikazi. As white missionaries they also celebrate the firm establishment of a Christian cosmology that shaped local ideas on the environment: they highlight a sermon at Matopo in the 1930s by the Reverend Kumalo, who openly critiqued his Ndebele people's old teachings on the environment by proclaiming that only the Bible could serve as "the medicine" of the people and only heaven constituted a "sacred cave." The authors retell stories of how the Ndebele came to believe in a monotheistic God in charge of nature. One account, dating from 1946, tells that "the wife of King Lobengula's chief witch doctor" possessed "a simple faith in God," evidenced by the time she drove devouring locusts from her garden by kneeling in the soil and praying.[41] The message was clear: the Hebrew God of the missionaries, not the Ndebele spirits of former times, could best manipulate nature. In Engle and Climenhaga's telling, at Matopo one view was in certain decline, the other in clear ascendancy.

A much more recent version of this narrative of the Christianization of Matopo is told by African BICC members themselves. It is somewhat critical of the white missionary, even as these writers celebrate the BICC itself and its ultimate evolution into a church body that searched for land justice. As part of the 2002 Global Mennonite History Series volume on Africa, *Anabaptist Songs in African Hearts*, Bekithemba Dube, Doris Dube, and Barbara Nkala lay out a narrative that is at once postcolonial and Christian. They describe the establishment of the white-run mission in less than laudatory phrases, as an attempt to "incarnate . . . itself in the Ndebele society," and impose a Christian worldview to address "disease, death, drought and fertility."[42] They also implicitly critique the white missionaries' moral code, which undermined an environmentally based, native religiosity. They recount how in 1914 BICC missionaries blacklisted the *utshwala* beer, a traditional reward for workers tilling the soil; they describe how in 1916 the missionaries forbade "traditional singing and dancing," even though these celebrations ordered work on the farm; and they recall the time when missionaries ended the traditional seed blessing by an *inyanga* (African doctor), an ordinance that many converts ignored in times of drought.

Yet these African BICC authors applaud elements of Christianity that benefitted the Ndebele. They highlight the missionaries' opposition to polygamy and the related exploitation of women as beasts of burden on scattered farms.

They point to the missionaries' opposition to belief in the *tikolotshes*, the "half human and half spirit" said to be responsible for droughts and deaths, and the spirits that terrorized local farmers. They also celebrate the BICC mission schools, including the agricultural and animal husbandry courses taught to the boys and the gardening subjects offered to the girls. Finally, they praise the continuity of Ndebele-led BICC mission schools after the 1980 independence, including the schools' "intensive cultivation of vegetables under irrigation."[43] A white-led past may be challenged where it undermined a life-giving culture in a fragile environment but affirmed where it advanced an equitable agriculture.

Oral history provided by BICC farmers echoes this dualistic environmental outlook, at once Christian and Ndebele. Josephine Siziba, a 70-year-old widow with a deeply rooted, rural identity—born in "Ntabazinduan under chief Khayisa"—farms in Matopo with hired help. It is the same farm she came to when she married in 1968.[44] She has grown maize and some rapoko in the past, and also sweet potatoes, which she has fed to her goats. But most recently she has grown the leafy green viscose, a form of kale, which she exports to the market in Bulawayo. She has never needed much income, because she has sufficient food from her own fields and gardens. But the farm has also strengthened her psychologically. She recalls needing to farm after her husband died in 1997, and in time the farm came to define her, being "the one thing that has kept me up." It is now in her blood: "I don't know how I will separate myself from the soil, because even if my children ask me to stop farming, I find myself going for irrigation as early as 5 a.m."

Yet Josephine's connection to the land has also become deeply religious. Her own view of this link evolved over time, for when she arrived at Matopo in the 1960s, old Ndebele religiously oriented environmental practices were still popular, and she believed in them. At the time, she says, Matopo was imbued with animistic ideas about the soil and the environment. The "place was surrounded by traditional healers, traditional drums would be playing. . . . Beer brewing was very common, and I used to brew it often. . . . Most areas in Matopo were infested by traditional healers; in this very house we used to have traditional drums." These practices and artifacts were connected to a particular view of securing fertility and precipitation. She recalls the ritual that accompanied the annual planting, how she used to "take the *Ama Vagazi* [tweak roots and bark] and wait for the first rain, and then . . . mix the seeds with rain water; like I said, this place used to be full of traditional healers."

Her very description hints at a transformed view of the religious and the land. But she also says it unequivocally: she herself has left the old ways. She has stopped brewing beer, and the drums have been left to rot as no one used them any longer. Her approach to the environment has paralleled her acceptance of an evangelical faith. She says that since "I was born again, I no longer use *muthi* [the traditional medicine] on the farm"; she has come to see the farm merely as a way of earning an honest living, as the foundation of an upright life: "If you teach your children to farm they will grow up knowing that one has to labor to put food on the table." Spiritually, she thanks the creator God for caring for the soil and for "the food He provides through the soil." She also sees God as an ordering, divine force in nature: "The rock you saw at Ntunja used to be [up, at] the centre of the mountain, but people . . . removed it because it was threatening [to fall] and disturb our place for weddings." It was an act by people fearing a capricious nature. The fact is, says Josephine, that "it is God" who kept the famous Matopo rocks balancing on tips of granite outcroppings from falling, and "no one else."

The farmers at Matopo, as members of the BICC, have heard sermons that accentuated the role of a divine creator. They nonetheless hold memories of an animistic past that lies just below the surface. Josephine knows where she has come to stand, but it is clear that she also believes that her neighbors might not all agree with her.

Siberia: Choosing between Lenin and Jesus

The two white settler communities founded in the twentieth century—in Siberia and Bolivia—were remarkably different from the places in Africa and Asia discussed above. They were also sharply different from each other. The story of the older of these two, Apollonovka, formally Waldheim, is an account of a deep Mennonite Brethren pietism that was undercut and pummeled by a Communist regime.

This story is told by the Apollonovka native Peter Epp, author of the centennial history of the German Baptist Union, which includes the story of the Mennonite Brethren church at Apollonovka. The main theme of the book is captured in its title, *100 let pod krovom Vsevyshnego* (One hundred years under the roof of the Almighty). In a 2012 summary of the account, Epp tells of euphoric growth after the initial Mennonite Brethren church in Siberia was founded in 1901, including a rich ecclesial life—German-language schools, short courses for preachers, choral music, large baptism services, exuberant interethnic evangelization. But this religious culture ended abruptly with "the inhumane campaign of dekulakization" in 1931, and Epp cites a heart-wrenching letter from the Mennonite Brethren member Luisa Bekker, of Apollonovka, in 1932: "My heart is breaking from anguish and despair. From morning on I feel that God has deserted us." By 1937, writes Epp, the congregation at Apollonovka had been destroyed; Stalin's terror had resulted in the mass arrest of the village's men and the execution of all the ministers. Yet the Apollonovka Mennonite Brethren never lost faith. Under the ruse of "birthdays, holidays and weddings," they gathered for worship even if they had no minister. In time they found a leader in their own prodigal son, 24-year-old Jacob Reger, who had once left the village and Kolkhoz Kirov, of which it was a part "to live a free unrestricted life far from mother's admonitions." In 1950 he returned to Apollonovka, having "heard the word of God" in a "strange land," and at once became a local sensation. In a single night his testimonial led to a "sudden awakening," with seventy-six villagers rediscovering the old faith at an intensely emotional, secret meeting that lasted until past midnight."[45] The foundation of the Mennonite Brethren congregation has been restored.

Local authorities responded with repression, instituting "a new campaign of 'tightening the screws,'" including featured lectures on atheism. Then in 1953 they arrested four men from the village who had "found God"—Jacob Peter Reger, Ivan Abram Wall, Ivan Vasilyevich Epp, and Jacob Franz Dirksen—and sentenced them to twenty-five years in prison. But the men's surprise early release and the founding of the subversive Omsk Association of Mennonite Brethren and German Baptist Congregations in 1957 created a new alliance, deemphasizing old denominational lines but strengthening local faith. Even Nikita Khrushchev's new antireligious laws, which led to the badgering of the local Apollonovka congregation and the arrest in 1972 of three men and two women, Elizabeth Panina and Maria Toews, for organizing a Sunday school class, did not break the Mennonites. Certainly, these times were very difficult; a letter by one of the arrested men, Jacob Franz Dirksen, written after his trial and addressed to "my dear Katya," his wife, spoke of the "burning" pain he felt on hearing his five-year-sentence read at the courthouse. Now on the train to the prison, Jacob "had the same pains and feelings as on the day of my arrest," a debilitating "soul sickness." It was an emotion arising from familial fissure, itself resulting singularly from religious persecution.[46]

Yet despite such difficulties, writes Epp, the believers prevailed. They met in private homes throughout the 1970s and 1980s, they sang from songbooks preserved by the elderly, and they pursued active children's programs. With the beginning of Glasnost in 1987 services became public, a church building was constructed, and a rich associational life erupted, including youth programs and evangelistic meetings. The main threat to religious life now came not from hostile authorities but from a mass emigration of Mennonites to secular Germany and, for many, a secondary migration to far-off Canada, mostly to southern Manitoba to settle within the predominantly Mennonite communities there. Those who remained in Apollonovka were committed to Russia and worshipped in Russian even though they still spoke Low German at home. Their three-generational story, as told by Peter Epp, is a history with a singular focus on the survival and even rediscovery of an old faith. It leaves no room for other aspects of religion, including its link to nature or agriculture.

Yet Epp's account sets the scene for the stories of oral history. In the version told by Margarita Janzen Drei, born in 1952 on Kolkhoz Kirov, religion and agriculture do intersect.[47] The third of seven children of Low German–speaking agricultural workers, Margarita recalls vividly her father as a "field brigadier" tractor operator and her mother as a milk technician. She describes daily life at Apollonovka, especially after the 1973 establishment of the massive state farm Sovkhoz Novorozhdestvenskii, with religion as her foundational narrative. As a "Baptist" and a "German," interchangeable postwar terms that often replaced *Mennonite*, especially in Mennonite Brethren communities, Margarita says that she and her fellow believers "were oppressed." After completing the eighth year of school, she and her "German" girlfriends were compelled to quit their education and work on the *sovkhoz* as milk technicians because of their religious beliefs. Working to reduce the stigma of being "believers," they "tried hard, always went to work and honestly, cleanly, to the end." They wanted to please their Russian bosses and to prove that "believers" had had a good, indeed superior upbringing. For Margarita, faith was certainly a matter of the heart. She says that "when we grew up we understood that this [Communist] world didn't bring happiness, and then we went to services and spoke about God, how Jesus suffered for us and about that happiness that Christ creates in your heart: that he frees us." But faith had a social dimension, which she says the Russians respected. It meant that she and her friends always fully completed their assignments, something Margarita insists "the Bible teaches . . . therefore Baptists are real farmers. The Russians say, where there are . . . believers, there is order." A "real farmer" was an honest, hard worker.

Margarita's life as a "believer," however, ultimately brought her into conflict with the state farm's administration. Although her close friends provided strength, united in their belief that you "leave it to God to help, work and agriculture, all is in His hands"; this is how "you begin to understand what is the meaning . . . of life; it is to belong to God." They knew that this faith also impeded their professional advancement. Margarita recalls her vivid interest in animal husbandry and in the correspondence courses she could take, earning her the designation "Chief Expert in Dairy," and then she attained the position of "dairy brigadier substitute." As a result, in 1982, at age thirty, she was asked to become Novorozhdestvenskii's lead brigadier of dairy operations. But the appointment came with a hitch: at the *sovkhoz* center, twelve kilometers from where she worked, the director, under "a big portrait of Lenin," asked her, "Which God will you serve, the one on the wall, or the one above?" Margarita was put in a quandary, for she pined to be a respected dairy professional. She could only pray "'God give me strength so that I can answer correctly'; I wanted to work cleanly, wanted it to be intelligent . . . but my conscience wouldn't do it [and so I] firmly said, 'That God' [above], and he said, 'Leave!' It was the time of persecution." But she was joyful: better to be a simple milk technician and feel a deep peace. In time, she did earn the respect of the director and even won a few national milk-technician awards, traveling twice to Moscow to compete in the Soviet Master of Milking Machine Operators competition. She has good memories of the competition—"We had white clothes, white dressing gowns, white pants. . . . All the cows stood correctly; it was very beautiful"; she even enjoyed the outings to Red Square and Lenin's mausoleum. The crucial point of the story for Margarita is that she kept her faith. It lay at the root of her vocation in the *sovkhoz* dairy.

But Margarita also talks about being "connected to nature everyday" through a linkage of agronomy and religion. She says that her father, who bequeathed her this appreciation for nature, possessed a God-given knowledge of wind patterns that preceded rainfall and other signs of impending weather change and offers that "God has given me that talent" too. And then to add emphasis, she adds, "I see Him in nature, how it changes" and offers accounts of how God has spoken to her through the environment. She recalls, for example, how at age 41, having been single all of her life, she felt drawn to marry a recently widowed man and thus become the mother to his seven children. She found clarity one day when she "went through the forest, early in the morning, the birds were singing, the sun was out, how beautiful and how well the birds sang, and suddenly a low voice said, 'it is your fate, it is your path.'" She recognized the divine message in the birch forest and so married widower Woldomir.

At the end of the interview, she leaves the topic of religion and links not only her love of nature and farming but even nature and the industrialized state farm on which she grew up. She says that "sometimes, when there's lots to do, I tell Woldomir, 'take me to the forest,'" and there she finds peace. But she adds that she "really likes the fall, when the combines are going, when there's grain, when they mow," and she and her husband "take pictures . . . and really love it, and then we go home." Her view of nature may be typically Western, separating agriculture from the divine and finding the creator's voice as distinctive from creation itself. But in Siberia, this dualism was made more explicit by the state's industrialization of farming and its persecution of religion. Religion on the state farm at Apollonovka certainly had a social consequence; the link of religion and nature itself became a private matter, out of the sight of the authorities.

Bolivia: School, Church, and Humble Agraria

The final account of Mennonite religious life as it pertains to agriculture differs from the others in significant ways. The story set in Bolivia is based on a succession of migrations, each reinforcing a commitment to a simple agrarian and deeply communitarian world. In literary-text form, this story is told most clearly in a 1997 document titled "A Retrospective and Answer from the Reinlaender Mennonite Church," addressed to the Bolivian government and authored by Mennonite ministers from the three Low German–speaking colonies transplanted from Mexico to Bolivia in the 1960s. One of the three leaders was Bernard F. Peters, the *Ältester,* or bishop, of Riva Palacio, the largest of the three colonies. The document's clear aim is to tell the Bolivian officials that Mennonites were a "farm" people as required by the 1962 Law 6030, the Mennonites' "Privilegium," but the authors take the extra step of linking agriculture to Mennonite history and their Christian faith. They thus begin by describing "where our ancestors come from and how the Mennonites have developed and how they plan to proceed." The writers mention Menno Simons, "our antecedent *Ältester,*" but then they note broadly that "our religion" is based on "God's Word and the teachings of Jesus Christ and his disciples."

From this point, the ministers proceed to connect the Old Colony Church to a broad Christian understanding of religion and nature. They announce their belief in "God the Father, Son and Holy Spirit, who created the heavens, the earth and all seen and unseen things," and support it with lengthy scriptural passages that reiterate the Old Colonists' hope for eternal life. Thus the earth itself is not sacred, but within the Old Colonist's cosmology it is a site for a temporary sojourn before eternity. But as sojourners they are farmers. It is a connection the ministers make repeatedly as they outline in detail their

Anabaptist ideals on nonresistance, community, and simplicity. They begin by noting their historical pacifism, as "our Lord commanded us not to repay evil with evil," and they outline the communitarian nature of their church, based on voluntary vows to obey its rules.[48]

But here they introduce the centrality of agriculture in their faith world and their dogged determination to avoid any form of urbanization: "Our children are taught to faithfully pursue farming," the ministers declare, "and thus to make their living." They explain that as a community all members are assisted in attaining the position of a self-supporting and self-sufficient "humble farmer." Indeed, this aim is at the heart of their church-run, German-language school system, which teaches nothing but "that which is required for a life on the land, namely God's Word, in the High German or Low German languages, reading, writing and arithmetic."[49] Agrarian simplicity is at the heart of their faith.

The close link between Old Colony faith and farming becomes even more apparent as the writers shift to historical narrative. They outline the Mennonites' various migrations, beginning with their move as Dutch Mennonites to Poland in the sixteenth century and then in the late eighteenth century to Russia, where they attempted to live "within the gospel of Christ," especially as it pertained to "a humble life on the land and also with livestock." But after eighty-five years, when the Russian government repealed their absolute military exemption, they migrated again. Western Canada in the 1870s was a place where "parents could teach their children to pursue a humble life in agriculture and advance their livelihood as caretakers of livestock." And for forty-eight years they succeeded in this aim. But then provincial governments in western Canada passed new school laws in 1916 that ended the Old Colonists' ability to teach their children "according to the gospel." Thus in 1922 they moved again, this time to Mexico, being promised freedoms of "church and school" and the permission for parents to "direct their children to a humble life on the land and with livestock." In the 1960s, when an encroaching modernity threatened this freedom, Bolivia's "generous government" offered them the terms to restore their rural life "as a humble agricultural people," and so they migrated once again.[50]

The ministers end their submission by reiterating both the Mennonites' distinctive features and those they shared with all Bolivian Christians. In the first instance, they explain their peculiar use of the horse and buggy and steel-wheeled tractors. They cite the scriptural passage 2 Timothy 3, which taught them the importance "of staying with that which you have been taught." Thus, they use only "tractors that do not drive on rubber tires" or have electric lights and, not being allowed to "own cars and trucks . . . within the colony we drive by horse and buggy." Then in the last paragraphs the ministers strengthen the bridge to wider Bolivia: they salute its government and make an ecumenical pitch based on "one baptism" and a faith in God, who is "over all and through us all and in all."[51] With this particular ending, the Mennonite ministers have not only created a local place for the separated, humble farmer in Bolivia but made an appeal to a common cosmology in which the divine orders nature.

Oral history possesses little of the overtly religious language of this 1997 document, intended for Bolivian officials. Indeed, the Old Colonists' articulation of faith comes not by way of words but by action, with the use of the physical artifacts—the horse and buggy, the steel-wheeled tractor, and the electricity-less house—to announce their submission to God. Thus, when Johann Fehr, a 57-year-old farmer from Riva Palacio, talks about his farm it is mostly in temporal language.[52] His narrative focuses on the Mennonites' arrival in the Bolivian bushland, their work to clear-cut it, the extermination of horrific bugs, and the building of houses with cement to withstand termites and mold. Johann outlines the process of clearing the forest, the rise of the soybean market, and the ensuing sandstorms, which made them regret their greed and "return the soil to pasture." Agriculture has been rescued, and now he focuses on the next generation. He hopes to give his children a start with land. His wife brought twenty-five hectares to their marriage, and his goal is to do nothing less for their children.

This statement of agrarian intent is as close as Johann comes to reflecting the core of the official 1997 text on Old Colonist goals. Life is about being humble on the land and being content with little, says Johann: "25 hectares is perhaps the average; it's not a lot, but you can work [with it] . . . you can plant something. And little by little you can save money and buy a little more land." Only when he is asked more specifically about religion does he speak about it, and even then in short, crisp phrases. Religion is crucial, he says, because "we are in this faith, to always live here as farmers." But his faith has also taught him that with respect to nature Mennonites must yield to the divine. Regarding weather patterns, "this we don't know, it all depends on God. . . . In Bolivia . . . one year the rains will arrive late, another year they will arrive early. It's different every year." Nevertheless, he daily prays for divine intervention on the farm: "We always pray and ask for rain . . . , everyone in the church, we do this. . . . Yes, really, every day, we pray for this, that's certain; everything depends on God. We can't do anything. . . . Here in Bolivia, we

Mennonites, we believe very strongly in God; well, the Bolivians do too." Iron-ically, as personal piety goes, the Old Colonists share a lot with their Catho-lic neighbors; it's in the lived-out part of their faith that they differ.

Significantly, Johann declines to talk about the way of the horse and buggy and steel wheels, and he attributes the Old Colonists' 1967 migration from Mexico to a lack of land and drought. He leaves out the religious reason for their emigration, the threat to simplicity from members who gave up on the horse and buggy and steel-wheeled tractors. These symbols of the "humble farmer," artifacts that have kept the farm small, serve as their own text. Only within their exclusive church services and when required by government, as in 1997, have the Old Colonists chosen to express their faith in words. But whatever the text—written words or artifacts of agraria—religion and land have come together in Bolivia as the very genius of Old Colony culture.

Conclusion

The interviews with these farmers and gardeners tell a global story of Men-nonites and the land. Being at peace with the land has been a core religious goal. This aim has entailed stewarding the earth, and the farmers acknowledge the times when they have not or when they have been alienated from the soil. But most often these aims are unspoken, embodied in everyday activity geared to sustenance and to generational succession. And despite their common Ana-baptist roots, their approaches to meeting these aims have been remarkably disparate. Their stories possess no single Anabaptist vision regarding farmland, no common extension of nonviolence to the land. They have followed their faith within the daily moments of farm life, affected by the vagaries of weather, the imperatives of survival, and the peculiarities of local church history. And listening to the nuances of their stories, it is also clear that neighbors within these Mennonite communities may well hold different views.

Environmental history offers a complicated account of the past. For farm-ers in Java and Matabeleland, a mystical, animistic worldview and a Judeo-Christian cosmology have been in a contest, with some members being more accepting of pre-Christian ideas than others. In Friesland and Siberia, farm-ers have come to terms with dramatic challenges to religious life in diverse ways even though any effects on agriculture were circumscribed by a remark-ably acculturated civic society in Friesland and by a repressive centrally planned economy in Siberia. In settler societies like Manitoba and Iowa, farm-ers have articulated both antipathies to and nostalgias for a simpler past close to nature; in the former, this link was expressed most clearly in emigration to rural enclaves in Latin America, in the latter by the continued strength of "old order" groups like the Amish. In Bolivia, the farmers at Riva Palacio have expressed an implicit relationship between simple technologies and a humble life on the land, most often without words.

To find the link between the divine and the land within the local is to find myriad expressions, even from members sharing a common religious label. It is to see how religiously informed visions have been shaped by everyday reality, but also how they have informed work on the land, even if in indirect and complex ways. At one level, these stories all feature the silent, hard work of historically aware gardeners and the quiet pondering regarding soil preservation in the larger fields by concerned farmers. And yet, to understand the life of Mennonite farmers is also to see many forms of "community dance" expressing "work and love."

CHAPTER 5

Women on the Land

Gender and Growing Food in Patriarchal Lands

Krismiati has farmed peanuts and watermelons in Margorejo since she married Yoga in 1992.[1] Her life story centers on her family—her husband and two children, a boy and a girl—and relates the accomplishments of each member. She tells of the time Yoga worked for a year in Sumatra, where he saved enough money to get married; how he rose in social stature, recently becoming vice chair presbyter in the local Javanese Mennonite (GITJ) congregation; and how her son became a student at Satya Wacana Christian University. But her narrative also includes her own story of accomplishment and status. Krismiati completed junior high school in Bopkri, for example, thus obtaining a higher level of education than her husband. She also inherited land, giving her and Yoga a chance to farm. As among the village's poorer farmers, the couple has had to *mocok*, that is, supplement their household income by working for more well-to-do rice farmers. Still, as a landowner Krismiati has status, and she speaks about planting crops on "my land." Although the field she inherited was not the choice wetland of the well-to-do rice producers, even as less expensive dryland, it was still farmland, and it was hers.

Krismiati's life has straddled the house and the fields. She has always been in charge of the house, even while tending the crops alongside Yoga. But she has also been a farmer and speaks about her agricultural decisions with authority. She describes what makes farming in Margorejo distinctive: "In the rainy season we plant rice, then in the dry season we plant watermelon." But she easily moves to another level of knowledge. She recalls the arrival of new genetics, "super seeds," which allowed rice to reach maturity in only four months, resulting in four rather than three crops a year. She speaks about the introduction of synthetic fertilizers, "urea and ZA," which she adds to manure to obtain the correct level of fertility at an affordable price. She also describes technological advancements: how they used to harness their plow to a cow but now rent a hand tractor at a rate of three hundred thousand rupiah per quarter hectare.

And she situates this knowledge within her own history, learned from her father: "In the beginning I followed my father. . . . But [he] passed away so I cultivate his field. . . . I learned to be a farmer from my father. I know how to plant rice." Significantly, despite this association of agriculture with a male, she puts herself at the center of the farm and once again references her inherited land. Krismiati has always had only one criterion for land ownership: those who cultivate it should own it. Folks who leave Margorejo for the city should divest themselves of their holdings. Absentee owners haven't cared "for our land" nor realized that "they will have nothing if our land is damaged." Krismiati's position as farmer is reinforced by landownership, received knowledge, and memories of innovation.

Mennonite women have always been farmers. They have worked the land, owned land, made agricultural decisions as farm householders, and, like all farmers, approached their work according to the way they have imagined nature. They have also contested unjust systems of landownership within extended families and voiced their views on correct crop management and how to care for the soil. True, their worlds have been shaped by patriarchal structures: as women they have been challenged by wealth and technologies controlled disproportionately by men, by the programs of male-dominated farm organizations, by male-led marketing cultures, and by laws privileging men. Also, more often than men, women have found themselves on the outside of cultures that exploit nature rather than nurture it. And yet the women in the stories that follow have developed ways of exercising agency on the farms. They have drawn upon deeply rooted lore, claimed their right under local inheritance customs, organized their own sisterhoods, and marshaled status within their communities by being producers. And certainly both the obstacles and the opportunities in their lives as female agriculturalists have been conditioned by local factors.

In scholarly explanations of the relationship between gender and nature, ecofeminism has been a particularly important approach. Clearly very few Mennonite women represented "a feminist rebellion within . . . radical environmentalisms," one definition of the term.[2] But the wider definition has promise for an understanding of agrarian, sectarian women's worlds nevertheless. First, ecofeminism raises the issue of power and focuses specifically on ways in which women experienced forms of environmental degradation generated by patriarchal privilege. Second, ecofeminism asks how the relationship between environment and gender is constructed and how it has evolved dialectically. And this dynamic means there is nothing *essential* in this equation. Perhaps women like Rachel Carson and her stinging 1962 critique, *Silent Spring,* inspired the modern environmental movement, but the relationship between women and nature, indeed between gender and land, is complex and in constant flux. Third, ecofeminism has begun focusing on *difference,* giving room for the intersectionality of a variety of subidentities. As Greta Gaard puts it in her 2011 piece "Ecofeminism Revisited," the relationship between women and nature always "intersects" with many "boundaries," including those "of race, class, gender, sexuality, species, age, ability, nation."[3]

These variables mattered also for Mennonite women's relationship to the environment. They may have worked the land within patriarchal structures and experienced their share of male-led environmental degradation, but their experiences were filtered by such factors as race, class, age, ability, and nation and certainly by religious cosmology. Especially in Java, Mennonite women's role on the land was affected by class and relative wealth. In Bolivia, age or generation and the way women related to animals on the farm affected their status. For women in Iowa, it mattered that the wider nation held to a particular civic imagination and possessed a highly developed economy. In Matabeleland, race was paramount in shaping women's subjugated locations on the land, while in Siberia, ethnic, religious, and class identities within the context of political repression directed a particular relationship to the land. Often these linked phenomena diminished the range of options for farm women, but often they also were the very resources women employed to assert their agency within agriculture and contest patriarchal structures.

Like chapter 7 in this book, this one focuses on five of the seven places, mostly to more clearly outline the argument. These five places—Bolivia, Iowa, Java, Matabeleland, and Siberia—represent five rather distinctive communities as identified by economics, civic culture, and degree of integration. Bolivia's households, within a profoundly communitarian society, allowed for the clearest study of gender and generation, Matabeleland provided the sharpest focus on race, Java on class, and Siberia on ethnicity. Iowa represented a world of increasingly sophisticated technology in a progressively integrated community, not unlike Manitoba and Friesland, the two communities not considered in this chapter. A focus on five communities allows for an overall focus on the idea that no one sweeping interpretative tool can account for the way women devised strategies to sustain a semblance of agency and meaning in patriarchal societies.

Java: Divided by Class, United by Faith

A central factor determining the well-being of the Mennonite women of Margorejo has been social class, that is, the chasm between the landowning women and the laboring women. But it is a particular version of the linkage of class and gender, distinct from classical studies conducted in South Asia. Such studies have argued that even as socialism transformed rural worlds in countries like China, women's worlds were still tied to both high fertility rates and rural poverty, leaving them "doubly marginalized."[4] The Javanese Mennonite women's story reflects a rather different set of constraints; indeed, this story is based on the reality of local class differences, but mitigated by a culture of relative gender equality in the Javanese countryside. It parallels, for example, works that argue that rural Javanese women within female-run local collectives and self-help organizations—where they pooled capital, built rural economies, and found educational opportunities—had real "bargaining power."[5] It also reflects studies on rural Java that observe that while patriarchal structures are dominant, women have owned land at the local level and thus enjoyed "more freedom" than women in other Asian or Islamic societies. This equation is true, these studies argue, even if gender in the Javanese countryside has been ordered by the "exercise of state power, of religious authority, and of [certain] cultural norms and values [that] reinforce . . . inequalities."[6]

Certainly Margorejo has been led by men over the years, from the time of Peter Anton Jansz's founding of the village in the 1880s to the heroism of Samuel Saritruno in the 1940s and the male-led Margorejo Farmers Union in the 1990s. Still, the worlds of the women, as exemplified by the farmer Krismiati, have benefited from a long-standing culture of relative equality among genders. A foremost expression of this phenomenon is that both men and women have owned land, raising women's status within the community. While this means, of course, that landless women have had less status, even the poor women have known a modicum of gender equality, derived from working alongside their husbands in the fields. Then too, the village's religious homogeneity means that both rich and poor women have attended the same GITJ congregation, both claiming inherent social belonging. Finally, the very possibility of landownership and the village's limit of seven hectares of land per household have mitigated the effect of social class.

Landownership has certainly shaped the lives of Sutarmi and Sumarmi, twins born in Margorejo in 1954.[7] Both women are well-to-do by local standards; indeed, they are among the largest landholders in the village. Sutarmi, who is widowed and lives with her children and grandchildren, says that she is the sole owner of the farm now that she "cultivates the land that was left by my husband." Sumarmi, on the other hand, is married, but she too speaks of land "owned by me." And when they reflect on their personal histories, both women say that they have been farmers since childhood, learning the craft from their parents and then having it reinforced by marriages to farmers. This lineage has granted both women the status of village landowners, an identity they have passed on to their children. Sutarmi, for example, has met the cultural goal of giving each of her children, both the boys and girls, a field similar in size to the others'. "Yes, the same," she says. "They are all my children; there is no difference."

With land has come status, and Sumarmi apologies for even talking about it: "I'm like a conceited person if I talk about my property." But then she admits that "I bought a field of 1.5 hectares and inherited 1 hectare, giving me a farm of 2.5 hectares," half a hectare in the wetland district, the rest in the less profitable higher and drier area. Again she apologizes, as "it's not my intention to show off," but landownership has mattered. Sumarmi admits that because of her status, villagers have addressed her in the formal Javanese reserved for officials. She insists that "actually I have had compassion for the poor people" and always felt sympathy for them, "so if I had a project [on the farm], I invited them to work for me." Perhaps the twins have acted charitably when they hired the poor, but their farms have also depended on wage labor, reinforcing class lines in the village.

As farmers, the two women also understand that knowledge, and in Margorejo agricultural knowledge in particular, has spelled power. Both say that they learned farming from their parents and their neighbors, and they assume that other farmers have learned from their successes. Indeed, Sutarmi has overheard male farmers comment on her cassava fields, "Oh, my friend plants cassava, I will plant cassava too." The sisters have derived confidence from such comments. They assert that "we know [agriculture] because we are farmers," and they claim that no outsiders, not even government officials, have taught them how to farm: "it all depends on the farmer when to plant what." They display an almost cavalier attitude toward agriculture. Sutarmi says that she has "no problems; I'm happy in farming, so no problem." Sumarmi may complain that the "water is shallow here," but about the flood the previous year she says, "That was normal. . . . We lost the plants . . . [but] that was normal. . . . Nobody was crying, . . . of course not!" As landowners, employers, and skilled farmers these women hardly acknowledge the existence of patriarchy.

Another story of womanhood comes from another set of Margorejo female farmers, the very poorest. By happenstance, this story is told by two men, one who describes his impoverished mother making good to describe his own good fortune, the other a well-to-do landowner who speaks about the cycles of poverty his female workers experience. Ironically, class difference between women is accentuated by the very attempt by these men to downplay it.

The first of these men, Suyatno, tells of his mother's poverty at his birth in 1979 and then of her remarkable economic ascendency.[8] He and his wife, Sriyuliati, have two children and earn most of their livelihood as farm laborers. But they do own some land, and Suyatno credits his mother for his economic start. He explains that after being abandoned by her husband, she became a part-time wage farm laborer. But she was very entrepreneurial, and after she purchased a bull she became a cattle breeder, demanding payment in kind for her bull's services and thus eventually creating her own small herd of cows. Suyatno recalls how during his school years he helped his mother cut grass along the roadway for animal feed and how he eventually obtained his own cows, which provided the capital for his first house. Eventually he also began growing cassava on rented National Forest Department land. The farm is at a point now that his wife works in the fields only if the children don't need her: "If she is free, she helps me in the field." A culture of relative gender equality has helped bridge the social chasm Suyatno knew as a child.

A less hopeful account of impoverished women comes from the landowner Sutari.[9] He says that in the past, labor-intensive forms of harvest gave poor women an opportunity to work in the fields and benefit from *ngasak*, gleaning the remnants of the rice harvest. The ancient tradition of *ngasak* was an integral part of the low-technology *dos* system of harvest, which relied on having ten to fifteen women workers and gleaners in the rice field at a time. But *dos*, and the work culture around it, has been threatened by machines: "Nobody can *ngasak* if we use the blower to harvest, because it is very clean, so that no remnant of rice remains." Sutari says it's unfair to hire women to plant his fields but not harvest them, causing them to forgo the chance of gleaning. He blames the rice merchants, who historically have decided which harvesting system to use, the old *dos* system or the new blower system. Sympathizing with these women, Sutari has asked the overseer of the new blower system to compensate the women with extra rice. In addition, Sutari has accorded these women dignity by "not watching them," concerned that they might feel uncomfortable about accepting something that "is free for them." As the landowner, he has been these women's benefactor, a position from which charity is offered, but charity that is nevertheless imbedded in power.

Clearly, the women of Margorejo share a history in which they have been divided by wealth. The owners of wetland, in particular, have had status; those owning some land but needing to work the lands of others as laborers alongside their husbands have had less status; the very poorest have been entirely reliant on the needs of the landowners, female or male. But locally, status has

Mennonite women, led by Ibu Munarti, planting rice seedlings at Margorejo, Java, 2014. Photo by Danang Kristiawan, Jepara.

not been fixed. For one, social standing has been complicated by religion, as common GITJ membership has downplayed class differences. Furthermore, the poor women say it matters that they have been allowed to work alongside their husbands in the fields and help build viable rural households. But land has been especially important. Women who have inherited even a little land talk of status. Class lines are real, but ironically they signal a form of female agency. The very possibility of landownership has allowed even the poorest women an expectation of success outside the labor market.

Matabeleland: A Racialized Earth and the Right to Land

Of all the seven places under study in this book, Matabeleland most clearly reflects the intersectionality of race and gender, core to ecofeminist inquiry. Indeed, scholars have highlighted the historical injustice in Zimbabwe, where women have labored in agriculture but typically have not owned the land. This equation has been made complex, and especially difficult, by political struggle, cultural underpinnings, and legal structures specific to Zimbabwe. Here long-standing tradition dictated that women were expected to grow "women's crops" on small-scale *tseu* or *isivande* farms, over which they had only limited usufruct rights. Significantly, a 1982 law gave women "legal adult status," and a law from 1985, the unrestricted "right to own land," but neither addressed historical male control of the land.[10] In Rhodesian times, patriarchy was intersected by race, that is, by white colonial control over all the land, creating a culture in which women were "left trapped in a vicious cycle of poverty."[11] In the more recent Zimbabwean era, Matopo women suffered again when intra-nation conflict pitted the dominant Shona people against the minority Ndebele. The infamous Gukurahundi massacres, which left twenty thousand Ndebele people dead in the early 1980s, laid bare this inequality. More recently, a "dead economy" arising from Robert Mugabe's mismanagement undermined women's well-being in southwestern Zimbabwe.[12] Together, these factors have destabilized rural society and disadvantaged women. But here, as in Java, women's agency has had several sources, including land from the mission farm, family networks, and religion, which these women invariably insist has been life-giving.

This story of inequity within Matopo is obtained from two sources, both of which point to the variable of race. The first is the white missionary Hannah Francis Davidson's 1914 history of Matopo Mission.[13] Significantly, Davidson identifies as a woman much less frequently than she does as a white missionary. True, as a woman she acknowledges the male leader, Jesse Engle, as the early mission's head and at places reflects a culture of gendered subordination: she writes, for example, that in the actual construction of the first mission house "we women assisted" as carpenters, roof thatchers, and masons. Yet on balance her narrative portrays an assertive woman often disregarding male authority.[14] In fact, at one point she places herself at the center of a legal fight against a bothersome white male farmer who blatantly encroached on Matopo Mission lands. She writes that to make the mission's case she and other missionaries undertook their own land survey: "[We] stake[d] out the land . . . climbed hills, went over precipices, and waded swamps under a hot August sun and made a diagram of the desired farm [boundaries]." They did not rest until the mission's ninety-nine-year lease was confirmed by a white judge. In colonial parlance, she writes that the injustice was not that white farmers owned Ndebele land but that this particular farmer had no plan to cultivate the land and wanted it only to extract fees from poor black inhabitants. As a woman missionary, Davidson would be undeterred by Victorian ideas of the "fairer sex" and defended what she asserted were the rights of the black convert within the colonial system.

At other junctures too Davidson was the protofeminist, often in particular solidarity with the native women. She clearly respects the Ndebele women for intervening in nature; for example, she praises them as "always busy and forehanded with their gardens, their grass cutting, and cutting and carrying firewood to stow it away before the rains come." She comes to their defense when women are subjugated as farmworkers: at one point Davidson complains that "more than one native has been heard to exclaim, 'these are my oxen,' pointing to his wives," with the chief difference, retorts Davidson, "that whereas the oxen get some time to rest and eat, the wife gets little, as she must grind and prepare the food in the interim of digging."

But ultimately Davidson is a missionary first and a feminist second. She writes about confronting strong Ndebele women about practicing their old, animistic religion. On one occasion Davidson challenged a female rainmaker, citing her for immorality, putting on airs, and duping gullible local farmers. "Being jealous for Him whose ambassador I was," writes Davidson, she stood up to the prophetess, accusing her for "deceiving the people and pretending" that she could make rain, when they both knew that "God only can produce rain." When in a sudden about-face the prophetess surprisingly pronounced a Christian view of creation to the assembled villagers, Davidson was skeptical and countered that the rainmaker was merely playing for the farmers' support. In the end, Davidson's disgust at both exploitative white men and a supposedly disingenuous, black woman are judgments from a position of white cultural privilege.[15]

A hundred years after Davidson penned her work, Matabeleland had been transformed into part of an independent postcolonial state. Twelve Ndebele women gardeners at Matopo interviewed in 2014 by Belinda Ncube, the spouse of a descendant of BICC converts, described a fundamentally new era. These women remember Zimbabwe's independence in 1980 as promising an end to any form of injustice. The white missionaries returned to the United States, and the land was to be redistributed. Women and men were supposed to experience a new form of equality. Some of the women did in fact speak of new land freedoms after 1980. Heta Mlilo, for example, born in 1950 in Manama but raised at Matopo, recalls being forced to live in a particular place by the white, segregationist regime.[16] But after 1980 she was told that "people could stay wherever they wanted to," allowing her to return to her parental farm. Heta also associates a new gender equality with independence in farming, asserting that in the postwar generation men and women began to undertake similar work and "farm together." She says that "long back" things were different: "men would put away their tools after plowing, then women would cultivate"; any man caught weeding was made to feel "disrespected." And then she repeats, "People now work together."

A more complicated version of this same story of post-1980 hope for justice on land is told by Eldah Gumpo.[17] She was baptized by a white BICC missionary, married in 1964, and settled at Matopo, receiving a piece of ancestral land through her father-in-law's intercession. During the War of Liberation they fled for Bulawayo, along with many others, but at war's end in 1980 "we came back and settled here," presumably on a more choice place than they had had before the war. She repeats Heta's claim that "people could settle anywhere and there was no one who could say anything," not even the mission authorities, who "took the issue to the Police"; it was a legal right that she expresses with the phrase "you know some things when you are educated."

And yet in other respects the culture of living on the land remained unaltered, and old forms of male privilege seemed to prevail. Eldah tells the bitter story of feeling compelled to take in her terminally ill husband, who had deserted her for a fast-paced city life twenty-three years earlier. Christian tenets on forgiveness mattered, but so did the traditional Ndebele teachings on death and burial. She recalls her son, Sicelo, entreating her to "please take care of our father so that we can bury him decently" and not to let him die "in an unknown place [where] we never get to see his bones." Ultimately this peace-making was linked to male-led social security within rural Ndebele kinship culture. Through marriage Eldah's was one of five Gumpo households, an intricately interwoven social organism headed by one of the eldest of her brothers-in-law. He was the "one with the cattle that helped all of us; he took care of all the family business like accepting and payment of 'bride price' for the daughters and sons respectively." Because of him, "all then became well." Her life had been shaped by hopes for racial equality, but ultimately a semblance of security came with old Ndebele teachings and patriarchy-based kinship ties.

This clan-based security, however, has been threatened by postcolonial Zimbabwe's shaky economy. Modern men may be weeding fields alongside women, but the failure of the country's economy has weakened rural society and ended certain pathways of female agency. She says that the "law of any homestead . . . when we were growing up [was] that [the] first thing in the morning, before a woman did [anything], she had to bathe, then she started sweeping the yard." And then she cleaned the other buildings of the homestead: the kitchen, the toilet, the sleeping chambers, and the sitting place. After the cleaning, she headed for the fields to plant or weed. Her only reprieve from this hectic work schedule was the marriage of a son, and with it "a new daughter-in-law; we would smile, the elders would be happy and say 'the old lady is now going to rest.'" But since Zimbabwe's failing economy, the young men have been lured to the cities, and there have been no daughters-in-law.

Other women decry this loss of postwar social cohesion for other reasons. Josephine Siziba, born in 1945 and married in 1968, complains that as a widow she has farmed alone and so has needed to be responsible even for the fences and cattle herding; indeed, if "[your] cows eat in other people's fields, they will beat you."[18] She has no grown children or nieces and nephews to help her; they have all fled to the cities. Priscilla Sibindi, a younger woman who married in 1994, also speaks of women planting, weeding, and herding, just as they did "even long back." But one change reflects the recent decay of rural society, that is, an increase in individualism: "Ah, each person takes care of their own needs," except "if there is an *iLima*," that is, communal work.[19] But *iLima* occurs less often than it did "long back," and women increasingly find themselves alone on the farm.

One indication of women's historical difficulty on the land is the way they imagine the land. They suggest a hard and troubled relationship with it and one rooted in history. It's a perspective especially apparent when their views are compared with those of the men. The seven Ndebele men interviewed for this book tended to speak casually of the soil as "'holy' because it was made by God" or because "God put us on it" or because "it belongs to God."[20] Two of the men spoke of how soil had been made dirty by human action but implied that it was still inherently good.[21] The women were more direct, even unequivocal. They spoke of how human oppression in the past had made the soil "evil" and how "sin" troubled the environment.

Eldah Gumpo, for example, explains Matabeleland's prolonged drought using an example from the Old Testament. She says that the drought is a result of "sin," and there is but one hope: when "we repent of our sins, the rain will come."[22] Eldah Siziba echoes this sentiment without elaborating on what the particular sin is; she wonders "if the reason we are having less rain is due to our iniquities; the clouds just gather but it never rains."[23] Effie Moyo offers a similar analysis, but now with reference to animistic ideas: "Long ago, when God had worked very hard during a rain storm, He would rest on a tree, and not [weigh in] on a person. But now you hear that someone was struck by lightning. It is because of our sins."[24] Only Zandile Nyandeni offers that the "sins" affecting nature relate to the colonial land policies of the past. She speaks of the government forcing the Ndebele to leave the Matopo Hills and then violating the arable land with such invasive techniques as land contouring. It may seem unfair, but she declares that the ancestors are withholding the rains because they are still angry about the time the colonial government forced the Ndebele to suffer.[25] Zandile concludes that "the land is holy," but only if you come to it with "a meek spirit," and in time, if you "surrender" everything to God, the land "will prosper."

The history of the Ndebele women on the land comprises numerous intersections, but especially one of gender and race. Power based on race has brought instability to this land. Stability for these women has come in the form of local kinship ties and a personal faith that links environmental upheaval with divine displeasure over historical inequality.

Siberia: Narratives of Suffering and Edenic Salvation

The story of Mennonite farm women becomes even more complex when one visits the Siberian settlement of Apollonovka (formerly Waldheim). The available texts here are two memoirs; neither memoirist highlights the environment nor agricultural history per se, but both write these themes into their narratives of religious and familial oppression. Mennonite settlers, along with millions of Soviet citizens during these years, suffered severe repression as despised capitalist *kulaks*, pathetic religious devotees, enemy ethnic Germans, or all three. The local reality here is that farmland was taken away from the Mennonites by Communist authorities. The farmers became alienated from the very land they had cleared as settlers.

Two memoirs by Mennonite women born early in the twentieth century, Elizaveta Bekker and Maria Janzen, tell a story of how sectarian and ethnic women within the Soviet state lacked much agency to begin with. They didn't stand to benefit from the liberation the Soviets promised peasant women on the collectives, nor did they actively resist collectivization through political engagement. For these Mennonite women, Soviet agricultural policy had all the elements of repressive "harvests of sorrow," to reference Robert Conquest's book title, and little of the resistance culture of the *bab'i bunty*, the Russian women who subversively protested collectivization.[26] Rather Bekker and Janzen reflected the worlds of "women without men," as described by Marlene Epp in her study of migrant women on a trek out of wartime Soviet Union. Bekker and Janzen were farm women without men, but their trek was not so much a geographic relocation across national borders as a virtual trek of oppression, from one moment of suffering to another, and all within the Soviet Union.[27] And yet, like the refugees in Epp's account, these women found their dignity in surviving oppression. They especially sought to turn an often difficult natural environment into a tool of survival and hope. Farmland, for example, is rendered in their memoirs not only as a site of suffering, where women whose men have been arrested on trumped-up charges are expected to do double the work, but it is also a site of hope. Both writers are alienated from their own farms through forced collectivization and ethnic cleansing, but both also find agency by working within the Soviet system. There they find their own microcosm of environmental connectedness, in the gardens and the forest, all made possible by their own agricultural knowledge.

The memoir of Maria Janzen, nee Derksen, born in 1916, tells the story of her family from the time of their settlement in Waldheim in 1911 to her son's marriage and her own bittersweet silver wedding anniversary in 1960. The memoir, written in 1991, when Janzen was 75, is a testimonial of religious faith and begins with the words "My life story, with the help of the Lord." But it is also a story framed by nature and agriculture. Janzen begins her narrative in positive enough terms: as she puts it, despite the aftermath of the Bolshevik Revolution, with famine and epidemic, and both her mother's and her stepmother's untimely deaths, "life was good in Waldheim" in the 1920s. But then things changed dramatically with collectivization in the early 1930s as "the

farms were all taken and *kolkhozes* organized . . . and a difficult time began."
Life became even more difficult with the myriad arrests during Stalin's Great
Purge of the mid-1930s. When Janzen's young husband, Gerhard, was arrested
in 1934 for being the stepson of a well-to-do *kulak* farmer and cheese factory
owner, the farm and what was left of it became Janzen's only hope. As she puts
it, she was "well provided for, the granary was full of wheat and we had two
cows in the barn."[28] She even had enough produce to take in an unexpected
visitor, a terrorized and destitute young mother with small children, one Sara
Janzen, who would later repay her debt when their fortunes were reversed.

Horrific times ensued as Maria Janzen's brother and father were also arrested by the black-car-driving NKVD, the Soviet secret police. Now the harsh
realities of the Soviet collective began to set in. True, Janzen was elated when,
quite unexpectedly, her husband was released in 1937, and within the year
twins were born to the couple. But then in June 1941 Germany invaded the
Soviet Union and her husband was once again forced from home. Janzen con-
veys this rupture with reference to an innocent enough moment of work on
the *kolkhoz* fields. The collective workers were tending a field of sunflowers
"when a wagon came bearing the news that war had broken out and a notice
that all the men were to be taken; my Gerhard was operating the tractor and
he was taken from the field, with the tractor, along with others." It was a reminder that the farmland had lost its function as a shelter from the outside
world. Now, in Gerhard's absence, it also became a place of almost unbearable
suffering. With her husband's sudden departure, "things were difficult at the
kolkhoz; every day more men were taken away, and the women had to toil from
early till late." Work on the collective did have one consoling aspect to it,
writes Janzen: while working in the fields, the women could talk about their
men without being overheard. They were treasured outdoor moments, despite generating uncontrollable and bitter tears.[29]

In time the *kolkhoz* became an even greater curse. Because Janzen was
deemed a German, she was drafted into the so-called *trudarmiia*, the infamous forced-labor army in the Soviet Union, which meant that she was exiled from the collective and from her children. She implicates a lust-filled
kolkhoz brigadier in the story of her arrest, implying that he offered her freedom for sexual favor; in her words, referring to the famed Old Testament
story, she suffered the "sin against Joseph in Potiphar's house," paying a price
for refusing to be sexually assaulted but, like Joseph, saving her soul and her
dignity. The *kolkhoz* took on an even more heinous form after Janzen learned
from afar that because of food shortages, her sister Gerta could no longer care

Mennonite milking brigade at the Kolkhoz Apollonovka in 1935, just four years after collectivization. Photo courtesy of Peter Epp, Isul'kul, Siberia.

for her children, the twins. Frantic, Janzen was able to secure a ten-day pass and returned to a joyful reunion with her children but was incensed to learn the ethnic reason for the food shortages at the *kolkhoz*: "The Russians had [food] enough to live; their men were at the front, but they didn't have to pay; we Germans had to surrender everything." Janzen had only one option: hitching her cow to a sled, and with a single pail of potatoes, she took her twins and sneaked out of Apollonovka one night and headed to Sosnovka. There Sara Janzen, whom she had assisted a long decade earlier, now worked as a factory brigadier and offered her a job. This, coupled with her own farm knowledge, allowed Janzen to survive. She found an abandoned house, which she heated with the manure from her cow, a fuel with which she also cooked her potatoes. She even had the cow rebred, producing a source of fresh milk. She concludes, "So I had a good life with my children."[30]

The old *kolkhoz*, however, continued to haunt Janzen. An official she refers to as the "very bad *predsedate*," whom she names as Sergeevich Vakhushev, informed on her when he discovered her whereabouts, and consequently she was rearrested. In hysterical defiance of the police, she clung to her children even as she was sent to a squalid prison. In her crisis, she "promised the Lord that should I ever get out of here, I would serve Him." And then she encountered a miracle: without explanation, she was granted the position of prison gardener. With her twins by her side, she now embarked on a life tending a field of cucumbers, which allowed her children fresh air and all the strawberries they wished from an adjacent garden. For Janzen and her children, the garden was an Edenic lifeline.

After the war Janzen's life improved incrementally, although there were new moments of repression. She was elated to be released and hastened to return to Apollonovka to await Gerhard's return. But she discovered that her old house had been destroyed, and the "bad *presedate*" put her to work alongside other women, all "working 'terribly' hard as the men still could not come home." But then things looked up once again. Quite unannounced, Gerhard returned one day in 1947, "during seeding time," and at once was put to work as a tractor driver, working "night and day." But then he was arrested only a month later for violating a ten-day military pass and did not return until nine months later, having been tortured and physically reduced to "skin and bone." Janzen now, feeling "great love for him in her soul," became fiercely protective of him.[31] She demanded that her husband be spared work on the *kolkhoz*, relenting only when he was made the night watchman for the apple orchard and vegetable gardens. Gerhard too had won an Edenic reprieve. Amid the best apple trees imaginable, writes Janzen, he found peace, and his health soon returned. In the 1950s Maria and Gerhard rediscovered their old faith, spoke of being "revived," and were baptized. They lived a simple life on the *kolkhoz*, finding peace in evening strolls through the garden, the private, outdoor nexus of nature allowed each household. In 1960 the twins married fellow believers, and another generation of rural life in the Soviet Union took root. Though the environment itself was not a central feature in Janzen's memoir, it appeared at crucial junctures in this narrative of a Mennonite farm woman crafting her life alone, mostly without her husband, within the world of Soviet agriculture.

The extensive, 350-page handwritten memoir of Elizaveta Bekker, nee Dik, covering the years from 1911 to 1965 at Waldheim and its environs tells a similar yet somewhat distinctive version of the "woman without man" story on a Soviet collective. It too is interlaced with references to agriculture and even more often to nature. Indeed, emotion and nature intersect in Bekker's childhood even before the Russian Revolution. She describes her joy-filled conversion to the faith in 1912 at age thirteen in deep snow by a "half-fallen down *semlin,*" the sod house from the settlement years, feeling a "forgiveness of sin" and being baptized in the Mennonite Brethren church. She reports on a horrific accidental fire in 1916 that destroyed her family's stately "full farm" with reference to the wider environment: the event occurred because "we were all in the fields" or picking "stone-berries" by the brook.[32] Hard work outdoors added innocence to her world and heightened the tragedy of her well-to-do farm family's loss.

The Bolshevik Revolution and the subsequent collectivization changed the very nature of Bekker's challenges. Like Janzen, Bekker seems to link environmental extremes with other difficulties. She writes of the summer of 1919, when the forces of the White and Red Armies battled each another, overrunning Waldheim and terrorizing the Mennonite farms with competing demands. But nature, that is, "the rainy summer of 1919, a difficult year," added to the terror. The ample precipitation resulted in a heavy crop, but the wet conditions made it "almost impossible to bring in" the harvest. The environment is rendered an even more hostile subject after collectivization in 1931. Because she had married Ludwig Bekker, the eldest son at the nearby Bekkerchutor estate—covering a thousand hectares and housing sixty-five milch cows—she was declared a *kulak* along with the other Bekkers and subjected to severe punishment. Several of the Bekkers, as well as the young couple, were exiled into the wilderness, a three-week journey northward. Upon their arrival in the north they faced impossible privation, accentuated by intense "rain, day and night, for an entire month, so much so that we saw no sun and it seemed that both people and God had abandoned us."[33] Later, as Bekker lay sick from typhoid on an earthen floor in tight, rough quarters, winter set in, with the wind blowing through the walls and their thin clothes.

In 1933 the families found their freedom and at once made their treacherous way back to Waldheim, where they presented themselves to surprised but thankful family members. Unlike the *kolkhoz* in Janzen's rendition, the collective in Bekker's story is a place of hope in the midst of chaos; here it is the NKVD that epitomizes evil. When, upon their return they were declared to be *stimloss, personae non gratae* with no papers, they could only look with envy upon those allowed to work the *kolkhoz*. Indeed, Bekker, her husband, Ludwig, and their five children were forced to live secretly among their relatives, always ready to relocate in a hurry; thankfully, they were able to plant a garden and eat the fruits of their own labor. After two years of odd jobs, often for nothing more than a pound of butter and a pail of potatoes, Ludwig broke down and registered as a *kulak* with the local NKVD. The couple now faced

new and onerous restrictions, but at least they had their own house and garden and were surrounded by extended family, and eventually Ludwig and
Elizaveta were also allowed to work. Bekker writes that "we were completely
at peace in our poverty," with much singing and playing of instruments. They
were also relieved when they no longer lived under the stigma of *kulak* and
especially that "my beloved husband could work . . . as he had in 1931." As it
was, she was required to "labor terribly hard," as were all the Waldheim
women, and she had to learn to meet the demands of an abusively critical brigadier.[34] And yet, she declares, life was good.

But it was a short-lived reprieve. A brief few months later, in August 1937,
Ludwig was separated from her. He was one of the first of some twenty men
arrested at the collective over the course of the year. Even before the NKVD
took him away on August 7 in broad daylight, a protracted and cruel debacle
in front of his five children, she had lived in abject fear. As her household stood
vulnerable, she regularly sought divine reassurance in the birch forests around
the village: "Oh, if the forest could talk, how much I prayed and screamed to
God." But it was to no avail, and she was forced to fashion a life without her
husband. Life on the collective was hard, indeed doubly unjust, for as a
"woman without a man, I had to work especially hard on the *kolkhoz*." And yet
rural Soviet life had an almost healing rhythm to it. In 1938 the *kolkhoz* workers began seeding wheat on April 17, individual families began planting their
own potatoes on May 8, and Bekker and her little family acquired a treasured
weanling, their promised larder for winter, on June 2. Ironically, when the
little piglet escaped from their backyard one day and a desperate Bekker and
two young daughters found it just before nightfall in the forest, she seemed
to forget her terrible spiritual anguish at Ludwig's arrest, writing that "once
again we have been blessed. Oh yes, the Lord answers prayers."[35]

Even when the war broke out and her three beloved sons were drafted on
the same day, life was bearable. True, the women worked ever harder: her own
daughter Lies, for example, turned the mix mill by hand "for an entire week,
from early till late," and where "two men had once turned it, she now turned
it alone." Pride and dignity are written into this very rendition. Then too, the
kolkhoz allowed Bekker enough time off to further develop her trading skills.
"We had actually become rich on the *kolkhoz*, as we had so much: . . . a cow
that gave us 10 liters a day, we had 11 sheep, 3 pigs, 1 big ox, and added to that,
a granary full of grain . . . and 60 liters of sunflower oil. Yes, we lived as one of
the privileged. We had no need." She writes that if only Ludwig had been
there, everyone would have been happy. Her spirited approach to life showed

itself on the occasion of her own horrific arrest in February 1945, followed by an interrogation in which she was made to "endure terribly much." She writes that she put up such a fierce resistance to the NKVD officers that they acquiesced, declaring that "such a spirited woman belongs on the *kolkhoz*." In an agonizing aftermath, she writes, "I cried so bitterly I could not see . . . and worked on the *kolkhoz* like an ox."[36]

After the war, life slowly became more settled. The first of the boys, Ludwig Jr., returned from the military in 1945, and in 1949 Bekker married a widower, a father of four young children. In the 1950s church services began again, with a "harvest service" in October 1955, the first such service in twenty years, drawing some three hundred people, all now concerned only about "preparing for eternity." In July 1958 she rues that "all the villages are to be amalgamated" into the Sovkhoz Medvezhinskii and blasts a new "horrible brigadier" for confiscating lumber she and her husband had procured for a new house. But now when bad weather arrived, it came laced with hope. She describes the "very difficult" fall of 1958, for example, as "this winter came so early" with a heavy snowfall on September 19, but Bekker demurs, "It looked so pretty, the green trees and so much snow, that people could drive by sleigh."[37] At every turn, Bekker's life and her emotions ride the contours of the weather and of nature, presented to her as forest, garden, or fields of the *kolkhoz*.

The memoirs by Maria Janzen and Elizaveta Bekker tell the story of the first decades of the Soviet state from the perspective of Mennonite women. They are also testaments to suffering, accentuated with references to the environment. The two women express differing views of life on the *kolkhoz*, but both write of fighting for their children and grieving their arrested, tortured husbands. These "women without men" assert themselves with their inherited agricultural knowledge. Drawing deeply on their religious faith and surrounded by fellow German speakers, they survived both the oppressive collective and the difficult Siberian winter. In their own narratives, old ethnic identities intersect with gendered worlds, and both shape the women's interaction with nature. Sometimes the physical environment was their foe, but more often it was their refuge.

Iowa: Degrees of Gendered Social Integration

The story of the Iowa Mennonite women represents the polar opposite of the account of the women in Soviet Siberia. By comparison, life in Iowa was peaceful, almost idyllic, shaped by a set of incrementally commercialized farm households. Indeed this account reflects broader developments recorded by historians of the North American plain. For the first part of the century the typical sociological model, such as that proposed by Sonya Salamon, for agrarian, midwestern household economies held sway and revealed mutually dependent gendered work roles, especially apparent in ethnoreligious communities of German origin.[38] In the middle decades, the midwestern story as outlined by Deborah Fink and others suggests a far-reaching transformation, a rural disjuncture, as old household economies yielded to male-driven commodified and specialized production that effectively ushered women from the farmyard into middle-class homes.[39] But then in the last decades of the century, as Jenny Barker Devine has argued, Iowa farm women in particular adapted to the new economy as "agrarian feminists" and struggled publicly to keep the family farm economically healthy, especially during the troubled 1980s.[40] This scholarly trajectory outlines an evolution of women's status ultimately linked to a farm's history within the broad theme of modernization. This story is revealed in the writings of Mennonite women in Washington County and its environs. It is told by three documents: a set of early-century letters; a midcentury farm diary; and a late-century collection of minutes from a farm women's organization. Together these texts illustrate the slow shift toward increasingly commercialized farms in southeastern Iowa and women's changing roles in this local context.

The first of the three sources is the early-century letter. Two such letters, one from 1911 and another from 1913 (8 and 14 pages in length, respectively) were penned by a teenaged Mennonite girl, Nettie Wittrig (later Preheim), from a farm near Noble and mailed to a married sister living out of state.[41] They provide little direct information about Nettie, but a family history indicates that she was the youngest of eleven children—ten girls and one boy— most of whom were married by 1911 and residing nearby.[42] Despite her relative youth, Wittrig writes with a sense of authority, fully identifying with both the local rural community and her farm household's economy, and its dependence on nature's cycles.

The August 2011 letter, written when Nettie was fifteen, begins as most farm letters do, with a report on the weather: "We have been having plenty of rain, a little more hart [hard] than people wanted, [that is, for] those that hadn't thrashed yet. We had a heavy rain Sunday night and the wind blew awful." Then she turns to the farm community, expressing an intricate relationship to it. She writes about having just turned down a state music scholarship because it required attending school in the town of Washington, eleven miles distant, and "mamma don't want me to go, she is afraid she'll get too lonesome." Having reaffirmed her place in the community, she proceeds with her report: they as a family have visited the home of Uncle Pete; her parents and sister Fannie have been to the doctor in Wayland; the church's mission program is progressing; "little Judy" has begun to walk; her married sister, Susie, has a new "fire-less" oven. Then Wittrig segues to the farmyard: mother and the girls have canned no less than "14 quarts of crab apples" from the small apple tree, which "was just loaded." From there she proceeds to the fields, where "we thrashed this afternoon" and "we got 205 bushels," the *we* signifying a bounty for the entire household. She is also her own person as she announces in the very next line that "I got a new shambra dress; it is light blue in color. . . . It looks dandy." In the remainder of the letter she moves in and out of the house, to the garden, the chicken coup, her extended family (her generous brother-in-law Roy, her Aunt Margret Shantz), and the plastering bee at the local schoolhouse.[43]

The 1913 letter, written when she was seventeen, echoes the one from 1911. It is addressed to "dear sister and all," a message from one household to another. It too begins with the weather: "the first snow last night, the first snow worth speaking of this winter." It turns to the rural community: the health of a neighbor, the church election with eight of the thirteen Sunday school and missions officers women, then births and weddings and reports of trips. It moves to the farm: she reports on "our" 125 chickens, all "doing fine," still producing six to thirteen eggs each day, fetching a price of twenty cents a dozen. Then she proceeds to hog butchering, and then back to the community with a mention of "Jake Shantz who hasn't drawn a sober breath in three months." She speaks about a wolf, the "largest ever seen in this part of the country," then returns to the church business meeting a third time before signing off with a joke, saying that if she keeps writing she will have to pay parcel rate and advising her sister to perhaps read the letter in two sittings.[44] Teenaged Wittrig is enmeshed in her community and at the center of the farm household.

The second source for this study, the 1942 diary of Gertrude Mae (nee Yoder) Brenneman, from near Wellman in Washington County, describes a farm woman's world a generation later, when the technology was more complex and farm products were more specialized.[45] Like Wittrig's, Brenneman's world is gendered, with a specific female section of the farm household but one in which women have status, exercise power, and fully identify with its economic structure. And here too is a story of local community, with the family linked to a Mennonite congregation, the Lower Deer Creek Church. A family history describes the Brennemans as a typical rural family: Brenneman was a dark-haired, cheerful woman of thirty-eight in 1942, married for eighteen years to Ora Roswell Brenneman and mother to four grown children—Nita, Audrey, Lowell, and Clarie—all still at home. But this same history describes them as living in a different world than that of the early-century Wittrigs. Significantly, the 1930s had left the Brennemans relatively poor for the times; although they owned a Model A Ford car, they had no electricity and needed to rent a tractor for fieldwork.[46]

The diary, unlike the family history, describes Gertrude Mae Brenneman's daily work routine as a farm householder in a world indelibly shaped by seasons. The notes for the month of May offer a springtime version of Brenneman's story. On May 1, a Friday, gender lines seem clear: she and her daughters worked in the house, while her husband, Ora, was outdoors. But significantly, while Brenneman writes that the women and girls made plum butter, baked cookies, mended, and cleaned, she also writes that Ora "cut out some hedges and harrowed some" and oversaw their neighbor Schwartzentruber as he custom plowed the first eleven acres for corn planting. Of course, Brenneman has a stake in the amount of land plowed; her comment that the day was "just awful windy" and "warm" is implicitly a statement that this was not a good day for the plowing of any soil. Her entry for Wednesday, May 13, clarifies her gendered world, one not merely rooted in the house but extending directly to the farmyard and, once again, vicariously, through her husband, out to the fields. She writes that after ironing and painting, she and the girls planted "the truck patch" with vegetables meant for a local market and then hoed the garden. Then that evening she delivered eggs from her henhouse to the neighboring Wenger family and dropped in on "the folks" on the way back. In a single evening she had connected the household economy, the local community, and her extended family.[47]

The gendered lines that intertwined the family with the environment remained intact for autumn but shifted to reflect new imperatives when winter approached. September 1 fell on a Tuesday, a warm and uneventful day on which Brenneman and the girls finished the ironing and then canned corn before heading outdoors, where they beautified the yard, mostly by mowing the grass. Meanwhile, Ora and the boys worked in "the timber" in the morning and then took "slop and water to the hogs" out in the field in the afternoon. But just as in the month of May, September brought a widening and narrowing of gender division from day to day. On September 9 the division was pronounced, as it was Ora who drove to the city, presumably with the family car and trailer, to sell some hogs, while Brenneman and the girls canned plums, washed clothes, and sewed. But fall also brought female and male together for the corn harvest and apple pickings, and even favorite outdoor social events. Thus, on Thursday the seventeenth Brenneman picked corn and beans, while "the girls shucked the pig's corn" and "Ora picked up some apples." Then, on this "warm and windy" day the family gathered at the "Noahs'," where, along with the "Lloyds, Homers and Rays," they surprised "grandma" with a wiener roast. Despite the common midwestern patronymic references to the rural household, Brenneman is hardly the demure housewife. With an eye to the weather and everyday work of all members of the household, she is at its center. In her lifeworld the farm was of upmost importance: her reference for September 27—"Had a killing frost. Nice day"—marked a major event for the farm household for the year.[48]

A third document describes the worlds of Iowa Mennonite women during the last decades of the century, a time when new technologies and scales of economy recast the way farm women engaged with the environment. The minutes of the Greene Township Farm Bureau Women's Club over a thirty-five-year period, from 1959 to 1994, report on the activities of a women's association in an Iowa County township, just north of Washington County. Judging by their surnames—Brenneman, Erb, Groff, Hess, Miller, Stoltzfus, Stutzman, Swartzendruber, and Yoder, among others—this club's members were mostly Mennonites. Clearly, most of the activities were meant to enhance the women's worlds as effective homemakers but now also as helpmeets to their farming husbands, as enlightened citizens of modern America. At the all-day annual organizational meeting held in October 1959, for example, the "14 ladies present" heard the visiting county home economist Alverda James explain the mechanics and culture of the sewing machine, after which the women sat down to a noontime "potluck dinner." But if these activities invoked images of homemaking, the agenda for the afternoon took women into a new and broader arena. They heard the club's chair, "Mrs. Chas Eckhardt," outline the year's goals, which included understanding "trends in agriculture" and learning about "civil defense, mobilization, citizenship and hospitality." Moreover, their planned field trips for the year included visits to the new 1,200-foot-wide hydroelectric lock and dam at Keokeek, the Shaeffer Pen Company's paper mill at Fort Shaeffer, and then the more distant Mormon settlement at Nauvoo, Illinois.[49] The Mennonite farm women were in the process of leaving their little rural worlds in Green Township, physically and culturally, and learning about human interaction with the environment in the broader world.

Over the decades, this pattern of reinforcing a world of female domestic-
ity while introducing the women to a new national and regional perspective
on environmental issues continued. Their awareness of their place in history
is demonstrated by their September 1962 meeting, at which each woman was
asked to talk about an artifact from her family history before hearing a pre-
sentation by Mary Stutzman "about 'ole times with Penn[sylvania]. Mennonites," one that solicited a hearty discussion on how "they used to do
things."[50] The understanding was clear: the Green Township women were not
Mennonites in any old-order sense of the word. Indeed, their meetings were
decidedly focused on modern-day America. At a November 1963 meeting, for
example, the women were introduced to new trends in hog marketing, in particular the "'Shift' plan and hog buying stations."[51]

The way the women considered environmental issues also suggests new
roles for them, now as consumers, citizens, and even tourists. Thus in June 1961
they attended an "outdoors cooking" lesson from the barbeque expert Ed
Schwartzentruber, the husband of one of the members, and in March 1962
they engaged in new ways of thinking about land use in the county, specifically "county zoning" and the work of the regional Conservation Board.[52]
Then in the 1970s and 1980s the women were presented with new ways of
thinking about the environment itself, approaches reflecting national cultural changes. In April 1977 the women toured the agrarian Amana community, twenty-five miles distant, where they admired a religiously driven simplicity, but as tourists and not as descendants of onetime plain people.[53]
Then in November 1978 they brainstormed on ways to "save energy" and
made a list that reflected more the suburban bungalow than a traditional
farmhouse; the list included "turning off TV when not watching" and "turning down the thermostat."[54] It was also as modern, engaged citizens that in
May 1994 the women toured the American Eco Systems factory to see how it
manufactured the biodegradable Pink Magic "multi-purpose cleaner concentrate," which doubled as a "septic treatment compost accelerator and Pit and
Lagoon treatment."[55] These women demonstrated a broad ecological engagement, but not necessarily one related to agriculture. The worlds that concerned these farm women had come to embrace a broader regional and even
national discourse.

The Mennonite women of Washington County and neighboring counties
shared a particular history. During the first half of the twentieth century it was
characterized by a strong sense of the local, rural place, defined by church and
kin but especially by the self-sufficient farm household, requiring the labor of

both men and women. During the second half of the century they engaged a wider world altered by mechanization, enjoyed easy access to town, and embraced specialization of farm commodities. By century's end, a middle-class culture of civic engagement and professionalization had begun to define the farm woman's place in an advanced economy guided by an interventionist state common in the Global North. In none of the other places examined in this chapter was incremental modernization, intersected by a gradual shift in gender relationships, as apparent as in this Iowa community.

Bolivia: Immigrant Imperatives and Generational Difference

The Old Colony Mennonite women's story of Riva Palacio, Bolivia, is sharply at odds with the history of the Mennonite women of Washington County, Iowa. Indeed it is similar to a group of Anabaptists, the Amish of Washington County, who would not have recognized themselves in the account above. The Old Colonists of Bolivia and the Amish of Iowa share a common commitment to tradition and simplicity that has fundamentally shaped women's lives. Sociologies of Amish women have long emphasized their high degrees of status and even relative power within their rural households and communities. Despite living within patriarchal societies, women had an unspoken standing because of their central role in their respective groups' commitment to self-sufficiency, strict social boundaries, and a religiously ordered simplicity. Indeed, it would seem that the Amish community was utterly dependent on women's work in achieving goals linked to its very raison d'être, ensuring that "within the family they [were] . . . independent and assertive."[56] More historically oriented work by Katherine Jellison and Steven Reschly, for example, demonstrates similar findings. They write that during stress-filled times such as the 1930s Amish women experienced less "harsh" forms of patriarchalism than did other US women, a reality still apparent in the 1990s, when Amish women still "assert[ed] their sense of participation in the family enterprise."[57]

This equation, linking religious teaching, economic activity, and status, also explains the relationship between women and men in the overtly patriarchal horse-and-buggy Old Colony Mennonite community at Riva Palacio. Bent on a strict separation from the wider world, lived in local contexts remarkably distinct from life in neighboring communities, women here have appeared quiet and subordinate in public, but in private they have been assertive, fully engaged farm householders. Indeed, insofar as the Old Colony Church in Bolivia was committed to changelessness, women's roles were crucial. They were inscribed in communal effort, rooted in farming and the self-sufficient community. This narrative is apparent in the testimonials of Old Colony women in different social stations: a middle-aged widow, an elderly matriarch, and two middle-aged married householders. A comparison of a diary kept by the widow (both as an active middle-aged householder and later as an elderly member of her daughter's household) and the oral histories of the two well-established married women tells this common story. They are all women who derive their status from their farm households and, with a particular concern for generational succession, the satisfaction of seeing their own children thrive on farms.

The diary of Katherina Friesen (nee Wiebe), with examples from 1980 and 2002, underscores this identity from two points in her life. In 1980 Friesen was a 52-year-old recently widowed farm householder living in Chihuahua, Mexico.[58] The main events in Friesen's life for that year were her husband's illness and then his death in February. She writes in starkly emotional terms, bitterly struggling with his painful illness and then untimely death, a struggle that also intersects with an intense spiritual wrestling. But significantly, she is still the farm householder in this crisis. In her January 29 entry she fervently prays, "Dear Savior Jesus, help me, forgive me all my sins," but then she adds that "it is a little windy; today we made a new floor in the pig barn and worked a bit in the garden." Even after her husband dies and she grieves bitterly, she references daily farm life and its link to the climate and the biotic. In one entry in April, she writes that "this week we pruned our trees"; in another from May, that the water in the well is very low; and in one from June, that they have received "a little rain," but such a small amount that they still have had to purchase water for $2.50 a ton. It is evident that farming without her husband is hard: on June 19 she notes that "it is still dry and hot, and currently we are plowing, but the boys really need father, and so do I; it is very difficult." But she is also buoyed by favorable climatic events and their effect on the farm. On September 6 she reports a long, slow rain and describes the day as "nice and wet," and she is exuberant that the well once again is producing water, adding, "the loving God be praised many times!"

In 2002 Friesen's life changed significantly as at age 74 she migrated from Mexico to Bolivia to be with her married children at Riva Palacio.[59] As an older woman and a new immigrant, she was more removed from the productive center of the household. Her diary, for example, is a testament of a deeply emotional, spiritual questing, as well as a detailed social record telling who visited, who died or was born, what tragedy befell whom. And yet, as Friesen becomes integrated at Riva Palacio, she becomes the commentator on the success of all aspects of the community, including its agriculture, and she identifies with her extended family's farming operation. Her comments on the weather reflect a farmer's imperative: on September 15 she notes "a nice rain today," while December 10 marks "a beautiful day," although she is aware that "it all needs rain," especially the freshly planted soybeans. In time, the weather is linked even more directly to farming potential at Riva Palacio: on January 20, 2002, she notes that her children are thinking of returning to Mexico because "at the present it is very dry. . . . The grain is beginning to dry up; it seems very dismal to me, how are people supposed to continue living here if a harvest does not occur?" Other references directly connect the physical drought to spiritual life. On November 22 she writes: "It seems very difficult to me; what we will do if the loving God will not let it rain; oh Heavenly Father, do not dwell on our many sins, be gracious to us again." But in her very next entry, on December 22, she is buoyant, anticipating the approaching Christmas and grateful for rainfall, so much rain that the "Abram Wiebes," the household of her daughter and son-in-law, had to temporarily halt their planting because of wet fields.

This sense of belonging to the farm household is expressed even more explicitly by the oral history of two middle-aged Old Colony women at Riva Palacio; they tell the story of mothers and farm householders. Indeed, these women speak as authorities on farming practices, outlining strategies for working within nature as women, but also as economic partners with their husbands. The account of Nettie Blatz, born in Mexico in 1955, the third youngest of fourteen children and married to Dietrich, is intertwined with agricultural history rooted in Bolivian soil.[60] She begins her narrative by describing her move to Bolivia as a teenaged girl. She indirectly refers to farming in Mexico, saying the push to relocate to Bolivia began with the introduction of cars, trucks, and rubber tired tractors. As she puts it, her family moved in April 1969 "well, because my parents didn't want to go along with everything, the way it was going. . . . They hadn't been taught that they could use cars themselves." Farming for the Old Colonists was to be a means of protecting their communitarian ways from the individualizing technology of the outside world, and the rubber-tired tractor threatened that fundamental life goal. As Nettie puts it, the move was about an obedience to the church's *Ordnung*, which prohibited modern technology.

But in her society, where religious commitment has been expressed by material signposts and a distinct lifestyle, Nettie is less comfortable talking about religion than she is describing the Old Colonists' engagement with the

Bolivian environment. She recalls that life in the early years in the bush marked a joyful time for the youth, for both the girls and the boys. After a day of hard work clearing trees and shrubs, she would meet up in the evening with "a lot of boys and girls that laughed a lot," and when it rained they "walked by foot through the village . . . in bare feet." When Nettie recalls life on her parents' farm as a young, unmarried woman, she describes work in the house and the garden, but she is never far from the fields. Her parents' first priority after clearing some forest was to begin planting as soon as possible; they learned from their neighbors who had arrived earlier, who in turn had just "asked the natives." After the first year of picking the roots and branches from the freshly deforested fields, Nettie and her siblings became the sorghum and corn harvesters, cutting everything by hand. Nettie also recalls that the farm began to use fungicides, which tropical Bolivia seemed to require, describing how on an "already quite big field of corn . . . we walked and sprayed everything." The freshly broken forest land was remarkably fertile, yielding plentiful potatoes and carrots in the garden but also corn crops, and all without synthetic fertilizers. She recalls their quick adaptation to a new environment: "We didn't know how to grow greens in the beginning, but we seeded that also," and they did so counterintuitively in the winter months of July and August, for during South America's summer "it's much too warm." Nettie is eager to talk about harnessing nature in Bolivia; as she recalls life as a young woman she easily writes herself into this settler narrative.

Nettie is not only the pioneer in this chronicle; she is also the more established farm householder, connecting the domestic sphere to both the barn and the fields. She describes the farm's experimentation with peanuts and chia seed, which she gives a decidedly culinary spin by saying that chia "looks like pig feed . . . but it's very good for people. . . . I used it for a while . . . to mix in our food. . . . One spoonful before breakfast and then pour hot water over it." Then she moves back to a commodity-centered perspective: "Some have raised this [chia]. . . . It goes up to $5000 per tonne." She also offers that chia is difficult to grow, its pods near harvest time prone to shattering in the wind and its seeds easily blown away. Nettie once again returns to the domestic: "It doesn't taste like anything. . . . Now peanuts, they taste like something," especially if they are mixed with chocolate. But then she reconnects to the commercial side of agriculture: "Oh, but the native [Bolivians] have very many big fields, such big fields and it looks like it would give a lot." She highlights again the vulnerability of growing chia, how stormy weather can easily shatter the pods, and says that their neighbors have protected their chia by cutting it early; she adds that their crop "looked very nice."

Nettie's description of the farm may be from a gendered, domestic center, but her description of the farm's overall aims are gender-neutral. As a couple, she and Dietrich have done well; when each of their children marries, they "receive twenty-five hectares, and then we give them some cattle, and some household goods and furniture." Nettie begins her farm narrative by talking about learning agriculture as a young girl from her parents; she ends with reference to the ultimate aim of providing both daughters and sons with land of their own.

The oral narrative of the second of the two middle-aged women interviewed, that of Anne Peters, is similar to Nettie's.[61] Anne's story also begins with the trip to Bolivia as a youth; it was 1967, and she was ten. She vividly recalls the chartered flight to La Paz with her parents and nine siblings, the truck ride to Riva Palacio in the bush, and early life in the village of Rosenthal. She introduces a gendered division in the process: she recalls that at first the women and girls of their migrating group lived in a hotel in Santa Cruz while the men built the houses at the settlement. Then, when they finally moved into their "temporary house" on the colony, mother and the girls reassembled their domestic world with bedding, clothing, and a few dishes brought from Mexico. Like Nettie's account, Anne's seamlessly links the house to the wider farm. She recalls that in the beginning the available food was strange to them. But then the family purchased a few cows, in short order obtaining milk as they had in Mexico, and soon began to produce the familiar commodities of butter and cheese. She then moves decidedly to the fields: "The first year we [planted] . . . just so, a bit so, with a hand planter . . . then we were able to clear land . . . [and] raise crops." In time they hired local Bolivians, who cleared more land.

After her marriage, Anne and her husband started their own farm. They had each received a cow from their parents, and "he had already bought himself one cow, so we had three cows to begin with." During their first decade of married life they slowly built up the farm and began their own large family. In 1996 they constructed their first substantial house, and later they had the deep satisfaction of seeing each of their twelve children married and all but one living in Riva Palacio. In the end, as Anne returns to an earlier decade when the children were still at home and unmarried, her story becomes overtly gendered again: "Oh, the girls, I have just taught them how I learned it . . . making food and baking, cleaning up and everything, and milking, and the boys, they work outside like many do, like grinding grain, feeding cows." And yet without elaboration she mentions that both her daughters and her sons have helped milk the farm's twenty-eight cows. Ultimately the farm household is a social nexus of two genders.

Anne's story at age 57, like Nettie's at 59, is at the center of a farm narrative. And like Katherina Friesen at age 52 and later at 74, both Nettie and Anne understand that their worlds are firmly set within their rural community, shaped indelibly by season, crops, animals, and climate. They share the imperative to survive in nature, thus fulfilling a religious calling that took them to Bolivia in the first instance.

Conclusion

The history of Mennonite farm women from places in both the Global South and the Global North has been intrinsically linked to nature and religion. In each of the communities described above, women practiced a religion of the everyday, saw the hand of the divine in nature, conferred a certain sacredness on the vocation of agrarian householder, and met neighboring farm women at local churches. But the precise nature of their worlds was affected by a variety of identities that intersected with the religious: class, race, ethnic, civic, and generational.

The first variable, class, seen most prominently in Java, placed well-to-do and poor women into distinct social categories within the village, whose class lines were ameliorated somewhat through common GITJ membership and the very idea of female landownership. In Matabeleland, a second factor, race, was evident in the BICC missionary Hannah Francis Davidson's early-century diary and indirectly in the postindependence hopes of the Ndebele women a hundred years later. Siberia posed a third query, how gender was constructed within a centrally dictated economy and overtly repressive state, a difficult world made more difficult by ethnic identity. In Iowa, women's lives projected a particular trajectory of modernization and civil society in the United States, an evolving social equation that reimagined farm womens' status over the course of a century. In the fifth community, Riva Palacio in the Bolivian lowland, the question of age and generational location within the household, highlighted the relative power of well-established middle-aged women having achieved the ultimate of goal of seeing a new generation of faithful rural householders taking root.

A comparative analysis of Mennonite farm women in five communities from around the world challenges any notion of a homogeneous experience of agriculture and gender, even in communities that share a common religious base. Certainly, ideas of self-sufficiency, tight-knit community, religious consciousness, and a commitment to remaining on the land fashioned the lives of all these women. But the women in each of the five places were also affected by localized issues, including particular civic cultures, distinctive climates, relative degrees of modernization, and divergent government agendas. Ecofeminism, raising issues of power and resistance, helps frame the historical inquiry into these women's lives. Intersecting concerns from their various locales also suggest how their worlds in the natural environment reflected difference, fluidity, and change. The role of women in securing the central aims of their agricultural households within their respective communities defies a singular story of gender and nature.

CHAPTER 6

Farm Subjects and State Biopower

Seven Degrees of Separation

In 1925 the elderly Manitoba settler Peter Elias sat down to write his memoir, reflecting on his fifty years in western Canada. His story of the migration from Russia to Canada included all the details of valiant settlers sacrificially leaving well-established "old-world" farms for the unbroken frontier of the "new world," where they turned the plow to the prairie. But at the heart of Elias's narrative was neither pioneer ingenuity nor sectarian effort; rather his account hinged on actions by two governments, both bent on creating a modern state and society. Elias emphasized rumors from 1871 that Russia's new laws regarding universal military service were ending old privileges and forcing Mennonites "to perform military duty"; the rumors were threatening enough to spark the emigration fever of the 1870s. And then it was a different government that came to the Mennonites' rescue: "God then led events in such a way that at this time a man came from America [i.e., Canada], William Hespeler, sent by his government to recruit immigrants. He promised them the freedom to practice their faith. Was this not a measure of relief to those troubled souls?"[1] Based on this invitation from Canada, a delegation of twelve land scouts put forth in 1873 "to inspect that land that was offered."[2] Four of the twelve delegates would accept the Canadian offer, which included military exemption and rural block settlement, allowing the Mennonites to rebuild their cohesive agricultural community. Significantly, they did so by means of an implicit contract: the government would provide land and religious freedom, and the Mennonites would quietly work the fields, collaborating in the Canadian colonial project of building a modern commodity-export economy from sea to sea.

According to a second memoirist, the Manitoba farmer Jacob Fehr, the Manitoba-bound Mennonites were warned by coreligionists who chose to settle in Kansas to avoid western Canada, with its "cold, long, raw winter." Fehr, like Elias, also placed government at a pivotal place in the narrative of transplantation. Indeed, the reason why seven thousand of the seventeen thousand Mennonite settler migrants to the North American grassland stuck with Manitoba was their sense of trust in and obligation to the state. "Our leadership remained firm," wrote Fehr, because "it had once negotiated our freedoms with the Queen. The contract was in writing. . . . If we would hold on to our teachings [on humility and nonresistance], no law could touch us. Our leaders and our Elder did not want to break this solemn promise."[3] The sectarian farmer's commitment to the state was that fields would be plowed and a rural economy built. The state's pledge to the Mennonite farmer was that it would grant sufficient land procured by treaty with local Indigenous peoples to build a rural society on the Manitoba prairie. Indeed, while the Mennonites would farm their land in peace, the government would attend to their needs as a paternal, benevolent guide.

The government's interest in driving the push into "frontiers" especially in the world's temporal grassland regions, all helping to "replenish the earth" for settler colonies, is well established. James Belich, for example, emphasizes the singular importance of a "reasonably strong central government and a workable system of public finance" in settling the US interior.[4] In many other places in the Anglo world, settler society also went hand in hand with other state apparatuses: in Nova Scotia it was a naval base, in Australia a penal system, and in South Africa a nascent state-funded advertising scheme that facilitated the settlement process.[5]

The global story of the state's role in settlement, however, is more complex than simple government leadership. James C. Scott argues in his 1998 *Seeing Like a State*, for example, that the modern state has advanced a certain "high modernity," in which science and technology, and experts within this impulse, are deemed fundamental to rural social well-being and, by design, serve also to boost state power. Scott is deeply critical of this historical phenomenon, especially when "an imperial or hegemonic planning mentality . . . excludes the necessary role of local knowledge and local know-how."[6] Of particular importance for this study are sections of Scott's book devoted to state intervention in agriculture, a form of "domestication" that seeks to integrate farmers and their communities into the domain of the state, a process in which agriculture is standardized, even simplified.[7] Scott sees such social engineering most clearly in Soviet collectivization, its failure evidenced in its litany of environmental degradation, but he also sees it in certain heavy-handed postcolonial projects in Tanzania and environmental overreach in the United States.

More recently, environmental historians have argued that the adverse environmental effect of state intervention should not be assumed. The Canadian historian Tina Loo has posited that "high modernist" government action has led to both instances of environmental desecration and a safe guarding of nature with nonexploitative "ontological" and "improvisational" knowledge.[8] Within rurality, the state has both undermined and empowered farmers' agency; its policies have alienated farmers from the land in some instances but also directed a sustainable agriculture in others. A similar argument is made by the German historian Joachin Radkau. In his *Nature and Power*, Radkau writes that "environmental history is . . . always the history of political power—and the more it moves away from practical problems on the ground and into the sphere of high-level politics, the more that is the case." And yet, as his richly documented work reveals, those state policies invariably affected agriculture at "the ground" level and did so in contradictory ways. Sometimes state power provided the ruling classes with access to resources and limited farmers' latitude on the land. But the state could as easily side with local and communitarian interests against the wider and global projects even as it bolstered its own *biopouvoir*, its biopower, in this instance derived from regulating the biotic.[9]

This chapter explores the historical relationship between the state and Mennonite farmers in the seven communities. It takes as its model Scott's comparative analysis but also Radkau's call for comparative case studies of numerous places, working to make sense of the complexity of state action on the environment. Indeed, as Radkau insists, among "a lot of zig zag developments" there is "no single master narrative," and he argues that "specific insights . . . can be gained most clearly through a wide transnational and comparative analysis."[10] The relationship as it pertains to Mennonite farmers certainly defies a single master narrative. In this respect even overarching dichotomies—the Global North versus the Global South, grasslands versus tropics, settler versus colonized—are inadequate to the task at hand. However, a "transnational and comparative analysis," as Radkau puts the challenge, can indeed shed light on the role of the state in the life of Mennonite farmers, especially with regard to agricultural expansion and soil stewardship. It can also illuminate the ways in which farmers frame this relationship, as, clearly, some communities have felt freer to criticize government policies than have others, given disparate civic cultures around the world.

In order to understand the full range of this relationship, this chapter takes two approaches. In the first part the focus is on what seems to be an especially complementary relationship, that between the Mennonite farmers and the government in Manitoba. This relationship evolved over time, and thus a longitudinal study, over the course of 125 years, from the inception of the community to the end of the twentieth century, is revealing. Three phases of state intervention, all aimed at an incremental agricultural expansion, but with a sensitivity to local farm cultures, seem especially important: first, the managing of settler society; second, "improving" the land; and third, promoting agricultural expansion.

The second part of the chapter assumes a lateral study, one based on oral history, and looks at the ways in which farmers have cast the history of their relations with the state in the recent past in all seven communities. Oral history clarifies the farmers' own perspective on the question of the state's involvement in agriculture. It also enables a direct comparison across seven far-flung communities and overcomes a significant unevenness in textual resources. This approach reveals a particular continuum of increasingly tenuous and oppressive relationships, complicating the very idea of modernity and its effect on farm-state relationships. The fact is that government land-management programs were not equally amenable to the interests of sectarian farmers and to the goal of sustainable agriculture, or to either.[11] Such a continuum crisscrosses the globe, zigzagging in and out of settler and postcolonial places, in the North and the South, in the tropical and the temperate zones. Thus, we first visit the farmers of Manitoba and then those of Java, Bolivia, Friesland, Iowa, Zimbabwe, and Siberia, each in turn. The order reflects the happenstances of place, the particular intersection of national politics with local cultures and the imperatives of farming in specific physical environments.

Manitoba: Documenting a Century of Complementarity
Of the seven settlements studied here, it was in Manitoba that the government developed an especially stable and close relationship with its farmers. Indeed, the iconic idea of metropolitanism within Canadian history, depicting the federal government as directing settler history, even in the remotest of rural districts, is at the foundation of Manitoba Mennonite history. Historians of

western Canada—variously invoking phrases like *fashioning farmers, imperial plots,* or *powering up Canada*—have offered numerous ways in which the state imposed its will on nature.[12] Other historians have shown the state's specific involvement in ordering environmental transformation, whether by transforming the biotone of the forest prairie edge in central Saskatchewan into "efficient" mixed farms or by turning a wet prairie in southern Manitoba, a veritable "large soup bowl," into well-drained monocrop wheat farms.[13] The Canadian federal government may have the appearance of social engineering, but as it regarded the Mennonite farmers in southern Manitoba it understood that its final goal required a sensitivity to local autonomy.

Mennonite writers have noted the irony of this approach. In a 1972 essay, the sociologist Leo Driedger describes the enabling role the state played in privileging Mennonite settlers over Indigenous inhabitants. "Mennonites, although historically suspicious of governments," writes Driedger, were "able to settle in large concentrations on the Canadian prairies only because the Canadian government had cleared away the Indians and Métis in the West." Perhaps it was a bewildering benefit, given the Mennonites' own commitment to nonviolence, but it is one of which scholars are increasingly aware.[14] In their 2017 studies on the Mennonite settler story in Manitoba, including the Rural Municipality of Rhineland, Susie Fisher and Joseph Wiebe write variously of the effects of Canadian state action and Mennonite settler complicity: Fisher refers to a forceful "reconstructing" of Indigenous "terrain," while Wiebe speaks of it as a form of "political violence."[15] But having cleared the prairie for settlers like the Mennonites, the government stepped back from its position of power, granting them military exemption, parochial schools, and vast land blocks on which to build their farms as they saw fit. Indeed, as the Mennonite scholars Adolf Ens, William Janzen, and others have argued, the Mennonites eagerly registered the "free" 160-acre quarter sections and plowed the requisite acres. And when they ignored the official survey lines, the government acquiesced with a special "Hamlet Privilege" clause to the Dominion Lands Act of 1872. It allowed the Mennonites to live in their premodern farm villages, the *Strassendörfer,* superimposed on the homestead system.[16] It also signaled a willingness on the part of the state to respect local cultures.

This overture by the state to settler society in Manitoba is also apparent in the ways the federal and provincial governments dealt with Indigenous inhabitants. Thus not only did Ottawa pursue Treaty 1 with the Anishinaabe in 1871, extinguishing their claim to southern Manitoba lands, but government surveyors hastened to draw the lines of the Mennonite West Reserve in 1875, well before the 1887 staking out of the adjacent Anishinaabe people's Roseau River Reserve. Moreover, Ottawa's experimental farms to advance prairie agriculture—one in Indian Head and another at Brandon—sought to produce "reliable information for the best methods and practices for local conditions," but without foisting those ways onto the Mennonites. Indeed, the Mennonites' own records suggest that they had already revered "black summerfallow" and that they learned the benefits of fall plowing in cold climates on their own.[17]

The government of Manitoba, in the meantime, updated its County Municipality Act in 1877, but in such a manner, according to the historian W. L. Morton, that minority communities like "the French parishes and the Mennonite village were incorporated smoothly" into the province's market economy, without disrupting their religious or even their educational programs.[18] The government even allowed the Mennonite settlers their own governance scheme. In a 1876 newspaper report, the West Reserve farmer Peter Wiens, for example, described the Mennonites' transplanted bicameral system: church leaders "manage . . . the affairs of the church," while "for the management of their temporal affairs, roads, bridges, etc., to keep them in order, the colony has a *Besirksamt*," a colony authority, also referred to as the *Gebietsamt*, based on a system of local governance they had known in Russia. The jurisdiction was headed by an elected *Obervorsteher*, a colony mayor, while each village had its own *Vorsteher* or *Schult*, a village mayor.[19]

Even when in 1880 the provincial government imposed the municipal system on the West Reserve, ultimately producing the Rural Municipality of Rhineland, it strove to accommodate Mennonite sensibilities. True, the 1880 act resulted in a decade of immense strife, emanating especially from the western half of the West Reserve, home to the Old Colony Mennonites. But eventually the provincial government reached an unofficial understanding with the Old Colony church leadership that the *Gebietsamt* and the official municipality could work side by side. The eastern half of the West Reserve, by 1890 synonymous with Rhineland Municipality, however, was home to the more accommodating Bergthaler Mennonites, who, according to the historian Gerhard Ens, "welcomed municipal government." In fact, writes Ens, "transfer from *Gebietsamt* to Municipal council was smooth with a good deal of continuity," including the carryover of the Bergthaler *Waisenamt*, overseeing the Mennonites' distinctive partible, bilateral inheritance system. Ironically, at this very time the old open-field system collapsed in Rhineland, and farmers adjusted to the homestead system based on the quarter section, that 160-acre (64-hectare) perfectly square parcel of land meant for each farm household. Farmers now worked the land to which they had legal title, meaning that they could mortgage their properties and no longer needed to consult with other farmers on the question of crop selection.[20] The state apparatus had created an orderly settlement process and then began to sway farmers to abandon their old communitarian approaches to land management.

Over time, the close cooperation between the Rhineland farmers and the state continued, but in particular on issues related to environmental management. No document makes this point more clearly than does a short 1944 history of Rhineland by the long-serving Rhineland general secretary H. H. Hamm, a Mennonite. He begins his book by lauding the benevolent role of the government as trustee of the natural environment: Mennonite settlers, writes Hamm, had been drawn "to the land of the 'free' upon the invitation of the Dominion government" and found in present-day Rhineland a land of "no railroads, no roads, only buffalo grass, swamps and mosquitos." The government and the farmers together transformed it. True, there were tensions, most importantly in education; the 1916 province's Public Schools Attendance Act, wrote Hamm, brought "radical change" and caused such "deep concern" among the traditionalist Sommerfelder Mennonites within Rhineland that many emigrated to Latin America in the 1920s and 1940s. But their leaving also made room for more "liberal Mennonites," those with a "more progressive outlook."[21] They would welcome state intervention to propel "progress," certainly one of Hamm's favorite terms.

At the top of the list of achievements spelling progress for Hamm was the province's long-term drainage program. Begun in 1902, it turned the northern sections of Rhineland, "waste lands, flat as a pancake," into soils that were well "drained and brought under cultivation." But Hamm especially highlighted a crescendo of state-sponsored activities in the 1930s, a time of severe crises in wheat growing Rhineland. In 1931 the Rhineland Agricultural Society was created in close cooperation with the provincial government and began the process of replacing old monocrop wheat farming with a modern agriculture of speciality crops such as sugar beets and sunflowers. Then in 1935 came a series of drought-fighting measures, including dugouts for cattle and soil-protecting shelter belts planted under the federal Prairie Farm Rehabilitation Act. From Hamm's perspective these developments were all rather serendipitous.[22] They clearly indicated a new, more intrusive role for a benevolent state within the natural environment.

Records from the second half of the century reveal an even more integral phase of this relationship. They record the state's long reach in directing farmers to increase their production through soil stewardship but also through new technologies, especially those linked to the chemical. Again, the local booster newspaper, the *Altona Echo*, begun in 1941 and expanded in 1955 under the banner *Red River Echo*, is the most accessible text in telling this story. Indeed, the weekly seemed eager to report on any story related to the modernization of agriculture in Rhineland, including the government's role in the process.[23] Regular news releases from both the federal and provincial departments of agriculture sent a trusted message to farmers: keep your soil healthy by remaining rooted in locally tested, reliable, old organic ways but judiciously employ new scientific and chemicalized methods to increase your yields.

The Mennonite farmers, thus entered the realm of modern agriculture accepting the outreached hands of an intrusive state. A typical report, this one from June 1957, describes federal Department of Agriculture field trials in "the Morden-Altona district of Manitoba," in which "considerable damage to sunflowers downwind has been experienced from 2,4-D drift." The report offered advice from none other than the federal herbicide expert J. E. R. Greenshields: to avoid such chemical drift, use the more stable amine solution rather than the volatile ester compound.[24] These words of caution on chemical use were, of course, also instructions on how to apply them successfully. And mixed with such counsel were reports of visits to Rhineland from a wide variety of agricultural experts, usually possessing respected British Canadian ethnic surnames—Greenshields, Robertson, Wallace, Wood—all offering professional advice on how to build a modern agriculture. Among dozens of such visitors making headlines, each potentially advancing a governmental *biopouvoir* in the postwar decade, were the "Manitoba Weed Commissioner" in 1948, the "Provincial Potato Specialist" in 1952, a "Water-Resources Expert" from the Department of Industry and Commerce in 1957, and the "Manitoba Soil Specialist" in 1959.[25] It was a stellar display of a benevolent state promising to advance the best interests of the farmer.

This increasingly paternalistic approach by the state was most vividly exhibited in the weekly reports from the local agricultural representative, the "Ag Rep." An employee of the Manitoba Department of Agriculture, the ag rep was often a trained agronomist offering free advice to local farmers. As such, the ag rep had status and authority. In a typical March 1959 missive from Rhineland's ag rep E. T. Howe, another British Canadian, the topic was manure spreading. Under the headline "Congratulations Husbandmen," Howe spoke of driving through the municipality and being impressed at seeing manure piles on snowy fields and pleased that "so many farmers [had spread their manure] during the winter." He also counseled farmers to rely on "green manures," clovers, alfalfa, and peas, which were often "a more effective means of increasing this priceless element of organic matter in the earth of our fields" than using "bags of chemical fertilizer." The latter, said Howe, did nothing to improve internal drainage in clay soils or to neutralize "undesired . . . alkaline patches." He was not adverse to "artificial fertilizers"; he merely wanted farmers to know that "a high level of organic material" made the soils "more effective."[26]

Throughout the 1960s Howe maintained this balanced and trusted approach to agricultural counsel: in 1964 he lambasted black summerfallow, in 1966 he castigated straw burning, in 1967 he heralded an eight-year crop rotation.[27] When Howe retired in 1978 and was replaced in Rhineland by Ahmed H. Khan, the balanced approach continued. Indeed, even though Khan was a Muslim immigrant from Pakistan and clearly not British Canadian, he nonetheless spoke with authority as a university-trained agronomist. And like Howe, he worked to create bonds of trust with local farmers. He voiced respect for the Mennonites' cultivation traditions but coaxed them to consider the chemical. In one of his first reports, for example, Khan advised using herbicides on black summerfallowed land, the old practice of tilling the soil for an entire season that Mennonite farmers had first practiced in Russia. But Khan's idea was also innovative, outlining a "combined tillage-chemical method" that could "save 3 to 5 tillage operations" in southern Manitoba conditions. He even suggested a specific routine: summerfallowed fields should be "sprayed in early spring with a broad leafed herbicide, followed with tillage operations at progressively later dates."[28]

Government representatives presented themselves as reasonable arbiters of the farmer's place in the environment. Simultaneously, the state promoted an expanding agriculture. Over a 125-year period, government intervened in the worlds of the Mennonite farmers in a variety of ways, facilitating their settlement, "improving" the land, and boosting new technologies. Usually these interventions conveyed a certain respect for local knowledge, making for a particularly complementary relationship.

Manitoba: An Oral History of Remembered Change

The written texts—memoir, local history, and newspaper columns—tell a story of departments of agriculture or their equivalent seeking farmers' trust, nudging them to greater food production. Oral history interviews with Rhineland farmers corresponds rather neatly with the text-based sources. Indeed, most often farmer-state relations from the past are cast in overtly positive terms, with but an occasional rather subjective complaint of state overreach.

Stories told to these farmers by their parents place this affinity as far back as the 1930s. Ray Siemens, the son of the local agricultural reformer J. J. Siemens, for example, speaks of his father's experience with the government in the interwar and postwar periods as characterized by excellent working relations. As Ray sees it, J. J. Siemens certainly was recalcitrant in his own ways, demanding perfection from his sons and workers and easily at odds with the local Bergthaler church leadership, who opposed "the things dad tried to do in the community," especially in founding a secular-based agricultural society, which the church considered socially destabilizing. But in creating the Rhineland Agricultural Institute, the elder Siemens found support from the government for his own love of experimenting with and demonstrating new crops. Indeed, he worked closely with the Department of Agriculture in Ottawa, "trying new things like soybeans," and before the soybeans, the sunflower, the important crop from the 1940s. According to Ray, the sunflower was the direct result of the "work and contact" his father "had in Ottawa."[29]

Other farmers, recalling more recent experiences with the government, applaud the very structure of the state. The organic farmer Joe Braun portrays his encounters with a variety of farm support systems in a positive light. He cites as examples the 1970s-era program of tax-free, dyed or "purple gas" for farm vehicles, then in the 1980s the Goods and Services Tax (GST) exemptions, the "claw backs on farm related inputs," and finally, in the 1990s, the "government-subsidized crop insurance programs." Perhaps these federal and provincial initiatives paled in comparison with those for European and American farmers, placing Canadian counterparts "at the bottom of the rung in terms of government support," but Joe expresses gratitude for what has been received. He even applauds Canada's "Supply Management" system, with its quotas and guaranteed profits for chicken, egg, and turkey production; perhaps they were restricted to only a select group of farmers, but they arose from "our Canadian identity" and commitment to "support each other."

Joe also recalls positive encounters with specific government officials and reserves complaints only for the incompetent. He describes working closely with the agronomics branch of the Manitoba government to establish a provincial Farmers Market Association in 1988 and then a local farmers' market in the town of Altona in 1989. He still has fond memories of "two great guys" from the Manitoba government: "very helpful; that first winter, they came to every meeting: . . . tremendous civil servants!" Joe also approves of the various ag reps he has encountered, especially Ahmed Khan, even though he was an outsider, "from Pakistan; he was Muslim." Khan had no experience with grain or greenhouse farming, but he "always had a phone number" for an expert, and "he always said, 'go right to the top, right to the top, don't bother with those underlings.'" Joe's only complaint is with the provincial crop-insurance adjusters, with whom he has had "more than one fight." He recalls the time his peas suffered a minus 7° frost: "The guy from the Crop Insurance came out and I said, 'this crop is shot,' and he says 'no it isn't,' and I says 'yeah it is,' and he says 'no'"; the argument went on, and finally "we had to go with his word." It was irksome, as the peas were clearly frozen to the ground. But this was the extent of Joe's unpleasant experience with the state.[30]

One subject of significant debate among Rhineland farmers has been the Canadian Wheat Board, the Depression-era government-run wheat-marketing agency, which had a monopoly on wheat exports from 1935 to 2012 and pooled profits for farmers. The retired farmer Grant Nickel supported the Wheat Board, as it took pressure off the farmer to find the best prices. He compares the egalitarian old way with the open markets, which have benefited only the large farmers and the grain merchants. When "you sell it to the big guys in Winnipeg . . . that live on [luxurious] Wellington Crescent," says Grant, they profit from selling globally, but "they pocket the money" for themselves; the old idea of pooling profits has disappeared. He predicts that his neighbors who farm large acreages and market most of their grain in the United States would have opposed the Board.[31] But even Melvin Penner, the farmer of thirty thousand acres, who came to oppose the Wheat Board, acknowledges that back in the 1930s the board "had its place." Indeed, it became irrelevant only because "things changed": wheat production plummeted, and canola and other specialty crops became prominent. Melvin even asserts that historical government subsidies that were spent locally served to bolster the rural economy; hence, no one complained that the farmer was "sucking on the government's purse."[32]

In recent decades, government intervention in the name of environmental stewardship has been controversial. Benno Loewen focuses on provincial regulations on manure disposal and is nostalgic for the time when farmers piled up manure in the fields and spread it as time allowed over the course of the winter. "Nowadays, you've got the manure, you bring it out, you have to put it, incorporate it . . . within a few hours" because officials are so concerned "with runoff, like by drainage or creeks, you have to stay away, so and so far.

There was none of that when we started." Benno has grown weary of government regulations. But Mary Loewen, his wife, interjects and reminds him of the "benefits that we've seen" from government programs "like the Wheat Board, or . . . Supply Management; that has really helped: many, many farmers." Benno concedes that the Canadian Wheat Board "was good because I am not a salesperson"; it was better to let government experts sell the wheat.[33] A consensus of farmers concedes the point: government intervention has had its benefits. As the farmer Melvin Penner puts it, "somebody's got to monitor what we're doing."[34] Evidently the government watchdog has been able to guard both the small and large farmers' interests.

The history of state-farmer relations in Manitoba has been a long one, shaped and reshaped by the colonial project of refashioning the prairie in the interest of the metropolis-based economic and political power. The prevailing narrative here has been that the state has helpfully fought both deluge and drought with programs supported widely by everyone except an occasional myopic Mennonite sectarian leader. Indeed, a survey of this Manitoba history over the course of a century and a half demonstrates an especially reciprocal relationship between government and the Mennonite farmer. This story, however, has been experienced in distinct ways in the six other places under study here.

Java: Joining the Green Revolution

The Javanese farmers at Margorejo who were interviewed for this project speak about the government's historical role in agriculture almost as positively as do the Manitoba farmers. Their memories of course do not cover the first half of the twentieth century, when the Dutch colonial government dictated the terms of Indonesian agriculture in broad strokes, which were then applied more precisely by white farm managers. But it seems that when the local farmers of this mission-run village embraced the August 1945 declaration of Indonesian independence, they put themselves in line to take advantage of the postcolonial landholding and agricultural policies of President Sukarno's government, in power from 1945 to 1967. They certainly embraced his land-redistribution policies and later the Green Revolution goals of President Suharto, the second president of Indonesia, from 1967 to 1998.[35]

The Javanese Mennonite pastor and historian Danang Kristiawan writes about this postcolonial relationship in mostly positive terms. He describes a two-step land-distribution process that favored Margorejo's farmers. The first step occurred in 1957, when the seventy-five-year-old land-lease agreement between the Mennonite Church and the state expired and the Indonesian government took back the land. Yet the government saw to it that "the people were still allowed to cultivate the land as usual, even though they could not own it."[36] Then in 1960 the government passed the Basic Agrarian Law, ordering the equal distribution of agricultural land, meaning that the farmers at Margorejo were now able to own the land even if many farms were only one-third to one-half of a hectare in size.[37] It is true that some farmers resisted Sukarno's introduction of elements of the Green Revolution in the 1950s, including chemical fertilizers, hybrid seeds, insecticides, and standardized planting systems. Others blamed the government when a severe drought resulted in food shortages at Margorejo in the early 1960s, and hyperinflation tested the local economy in the mid-1960s. Nevertheless, they warmed to the Green Revolution in the 1970s and 1980s.

And although the regime of Suharto was ruthless and devastated myriad farm communities sympathetic to Communist ideology, Margorejo, as a village, saw mostly benefits from the state. During Suharto's first five-year plan in the early 1970s, for example, local agriculture was transformed as new shorter-stalk, faster-growing rice varieties, now dependent on chemical fertilizers and pesticides, were introduced and promoted by government extension services. Kristiawan also notes that simultaneous government spending on infrastructure, including credit for mechanization from the Bank Rakyat Indonesia (BRI) and the creation of local offices of the National Logistic Institution to purchase the farmers' rice, helped to modernize Margorejo. He agrees with the assessment that rice-production goals had resulted in "decreased farm income due to high dependency on chemical inputs, the damage to the environment, and increased gaps between the rich and poor" by 1984. Nevertheless, he speaks of the Green Revolution as a "seeming success story."[38]

Kristiawan's relatively positive observation is bolstered by the oral history he conducted among Margorejo's farmers. They credit the government with helping them establish their farms in the first instance. Adi Retno, a farmer and director of the local farmers' union, operates a one-hectare farm that he inherited from his parents, land that he says was originally part of the "allotment from the Dutch Mission." But he emphasizes that the farm's real foundation came from government land policy; indeed, his family received its land soon after the land at Margorejo "was returned to the government" in 1957. Farmers at that time were given an opportunity to make a "submission to government," requesting their own land, as "every family got half of one *bau*," about three-quarters of a hectare, and each farmer also received a small "land allotment for their house."[39] Later the government also opened up its national forestland to the poorer farmers wishing to plant cassava among the trees. The farmer Sugiman explains that he had his beginning in agriculture only after working first in distant Sumatra to earn "enough for our family." But then his real boost came from the government: "As a Christian I must be grateful to God, praise God . . . that . . . the government started to allow people to plant secondary crops in the Teak forest," in the foothills of Mount Muria, near Margorejo.[40] Sugiman sold his cow and concentrated on growing cassava, pea-nuts, and corn on a half-hectare patch of forestland rented from the govern-ment. This move allowed him to make a good living.

Other farmers laud the government for imparting scientific knowledge. The farmer Kliwon, who converted to Christianity from Islam when he moved to Margorejo, tells of learning how to use fertilizers from the local farmers' union, which had obtained them from the government; the farmers' union showed "what farmers need, like [plant] medicine, seeds, fertilizer." He knows that the policy came into effect after 1970, because that was "the year when short stalk rice variety was introduced," the high-producing variety requiring synthetic fertilizer.[41] Sugiman even hints that the government has enabled farmers to contest multinational companies, which try to sell more chemicals to farmers than are actually needed. He cites a government recommendation when he declares that "manure is very good for cassava, also urea and organic fertilizer," and recites the recommendation to mix organic fertilizers and urea in a 1:2 ratio. He interjects that from his own experience this ratio applies only for rice in the wetlands; in the drylands, "urea and NPK is the best [as] organic fertilizer is slower."[42] But the fact that he has his own ideas on fertilizer doesn't negate his respect for the government research.

True, government intervention has not been without controversy. Indeed, the government's push for farmers to modernize their farming techniques has been somewhat contentious. Retno explains the Green Revolution's effect on farming with reference to new varieties of rice: in "the past, rice grew for 6 months, now rice grows for 3 months."[43] The changes came with the introduc-tion of a new rice seed known locally simply as PB and developed by Indone-sia's Federal Rice Research Institute. But with the new varieties came the ne-cessity to use fertilizers. Retno acknowledges that in the 1970s, soon after fertilizer was introduced, the government provided a subsidy and farmers be-gan to "really depend . . . on fertilizer . . . urea, calcium, phosphorous." With the increased usage of fertilizer, farmers also moved from planting in irregular lines to planting in straight rows, bringing order to the rice paddy, allowing for space between plants and hence more air circulation between them. This standardized spacing also allowed the farmer to work their rice fields more precisely with tiller-tractors and to apply fertilizer more efficiently. Retno, though, has some doubts about the new system: he thinks the soil has become less fertile and notes that farmers in 2014 were using five times as much fertilizer (five quintals of fertilizer for one hectare) as in the 1970s. And yet, even Retno is impressed with the state: "I think the role of the government is positive. As I said before, the government or Agricultural Department provides a lot information for the farmer."[44]

The GITJ peasants in Margorejo almost uniformly portray the government as having played an active role in securing their well-being. They point out that it assisted them first with the land-distribution policies in the 1950s and 1960s and then with the full reach of the Green Revolution in the 1970s and 1980s. While some farmers express worry about their dependence on synthetic fertilizers, herbicides, and miracle rice seed, they do not question the motivations of the state itself.

Bolivia: White Sectarians and a Postcolonial Food Policy

Ironically, the horse-and-buggy Old Colony farmers of Riva Palacio in Bolivia, having settled the country's lowlands without desiring to become integrated into the nation-state and sometimes at odds with government agencies on environmental issues, speak about a certain closeness to the state. This Mennonite group's old communitarian aims of separation from the wider world seem to have comfortably aligned with the postcolonial state's bid for national food security. Then too, as the 1997 "Retrospective and Answer from the Reinlaender Mennonite Church" declared to Bolivia's government, traditionalist Mennonites have deferred to the state as a matter of belief. The farmer-preachers who authored the "Retrospective and Answer" were unequivocal in this stance: "We believe and profess that the government of the land is established by God and that a land without a government cannot survive." They committed their community to obedience to the national government as long as its laws did not contradict the Mennonites' religious teachings on nonresistance and agrarian simplicity. And thus even though they wrote that Mennonites "cannot take any revenge . . . cannot take any part in war," they also understood that the government was there to "protect the innocent."[45] And like Elias and Fehr in Manitoba in the 1870s, they spoke about an implicit contract: religious freedom in exchange for a quiet tilling of the land, which ultimately helped the state achieve national food security. The 1997 document at once created a social boundary that isolated the Mennonites from the wider Bolivian society, and it set the stage for a workable relationship with the state.

In fact, in the interviews undertaken for this study, several Riva Palacio farmers spoke of the state as the Mennonites' historical champion. They offer as an example the Bolivian government's help with the 1967 land acquisition. The farmer Johann Boldt, aged 80, of Campo 11, is succinct in his analysis: "I think the government was always in favor of us." He recounts the migration from Mexico to Bolivia with a basic equation: "Bolivia had a lot of land for farmers. . . . And we went there to look at the land, spoke with the government and they said come on over here" and help turn the lowlands into farmland.[46] Cornelius Froese, 64, of Campo 21, has an even more benevolent view of the government. He recalls that "when we bought this colony, the government of the time didn't even want us to pay anything; they wanted to give it to us."[47] It was the Mennonites who thought it better that they pay for it in the event of any potential claims on the legality of their ownership. The farmer Peter Klassen's account historicizes the event: he points to the settlement of Canadian Mennonites in the Paraguayan Chaco in 1927 and says that their arrival alerted the Bolivian government to the benefit of attracting Mennonites to Bolivia. In fact, the Bolivian government "was very interested in all the work that the Mennonite" settlers had accomplished in Paraguay, and "for this reason they were looking, and wondering, 'we have a lot of bush and need workers'"; perhaps the Mennonites might be interested in developing the Bolivian Oriente.[48] The situation proved to be mutually beneficial.

Although most Riva Palacio farmers laud Bolivia for letting them run their own German-language schools and local affairs, they also acknowledge the state's role in managing the environment to the ultimate benefit of the farmers. Johann Fehr, of Campo 27, recalls that at first the government was too poor to provide any help but that in time the state did become involved.[49] Isaak Peters, 57, of the same village, recalls state Forestal agents coming to tell the Mennonites to plant shelterbelts and that these government officials were taken seriously: "To leave windbreaks? Yes, it was [a directive] from the government forestry department, the Forestal, but now, we ourselves know this; we see this, oooh. . . . No one wants to take down the wind breaks."[50] Mennonites eventually saw the advantage of stopping the wind. The shop owner Cornelius Peters, 59, of Campo 31, corroborates this concern, saying that when the government came to the colony to talk about planting shelterbelts, they "commanded, or obligated us to plant trees to block the wind." He regrets that the Mennonites didn't know better, as "it would have been advisable to leave a stretch of [forest] land between the villages, with [native] trees from the land; this would have been better."[51]

The trust in government forestry officials, however, has not been blind. The teacher Jakob Buhler, 57, of Campo 17, says that according to colony protocol any agricultural initiative of the Bolivian government needs to be filtered through the colony *Vorstehers*, the colony mayors. It checks overbearing government agronomists, who "come at times to explain how one has to plant so that you have a better harvest" but can be somewhat shortsighted. He explains that the officials' agronomy stems from a research station in the tropical El Beni department, to the north of Santa Cruz. But even if their plan "doesn't work in this place," the much dryer south, he thinks the officials are well intentioned.[52] At times when the Mennonites have received bad advice, they respectfully have shown the government what worked and what did not. Gerhard Martens, 32, of Campo 35, says that when forestry officials have inspected the colony's trees, to "see how it is," they have indeed listened to the Mennonites. They were able to show that the government reforestation program floundered on the colony's south side, where "the land is 'tired' and there is no force for [the trees] to grow," but functioned "a lot better on the north side of the colony," which receives more rain and is more fertile.[53] Young Gerhard's analysis of the limits of state knowledge is more subtle than that of some of the older farmers. Abram Reimer, 69, of Campo 32, for example, insists that shelterbelts failed because government-issued saplings were simply unsuitable "for this climate and for the wind." He also laments that the Forestal was not present in 1967 with its request to "leave strips of natural forest"; the natural strips would have outperformed any shelterbelts. Still, the Forestal has never yet fined a Mennonite farmer, even though, as Abram confesses, not all farmers have complied with its directives.[54]

The Mennonites know that their presence in the Bolivian lowlands has not been without controversy. After all, they are German-speaking, pacifist white farmers, often without Bolivian citizenship, and politicians have threatened them during elections. Yet, the farmers say the politicians have been easily swayed by the argument that Mennonite farmers are invaluable to the country. The machine-shop owner Peter Friesen, 77, of Campo 1, recalls the time in the 1970s when President Hugo Banzer threatened to revoke the Mennonites' religious privileges because they refused to teach their children Spanish. The Mennonite leaders simply invited the president to visit them and see their agricultural progress; and when "he came to the colony he was happy."[55] Other farmers have their own version of Banzer's change of heart. Johann Bolt implies that it occurred because Banzer knew "that Bolivia needed to produce things to live and . . . they were building a factory for [vegetable] oil, and Mennonites had started farming soybeans." Eventually, asserts Johann, the Mennonites would produce 77 percent of Bolivia's soybeans, proving themselves indispensable to Bolivia's economy.[56]

Ultimately, the oldest farmers return to the land purchase of 1967. Abram Thiessen, 71, also of Campo 1, references it when he insists that "generally all the governments have been good for us." He insists that even Evo Morelles, the first Indigenous president of Bolivia, elected in 2006, respected the original land deal as legal and transparent. At first, says Abram, "the *campesinos* were saying that Evo would take away the Mennonites' land, [but as for] me, I was always a little calm about this; these lands were bought directly from the government." Any talk of the Mennonites losing their land, repeats Abram, is merely electoral chatter, and in the end even the champions of Bolivia's dispossessed have not threatened the white settlers.[57]

The postcolonial government and the sectarian farmers share a common interest: growing food in Bolivia's lowlands. The Mennonites' relationship with the state has not been without tension, but overall the farmers insist that the working relationship has been good since the start. The basis of that relationship is the national goal of food security on soils protected from degradation, and the mutual desire that Mennonites participate in the global soybean market.

Friesland: "A World Going Crazy"

The relationship between the Friesland Mennonite farmers and the state is a complicated one. It is not that these farmers oppose regulations per se, but they have generally been wary of government intrusion, expressing a form of regional alienation common in rural postwar Europe, thus making Friesland the fourth closest to the state in our continuum. Certainly, given Friesland's low-lying topography, farmers in the area have known civic engagement for centuries. Indeed, a strong local government was required to build and maintain the dikes, the very foundation of agriculture in Friesland. When today's farmers speak of the government, they applaud policies of the wartime and postwar periods that shaped agriculture to meet the national food-security needs of each decade. Of particular importance in this history were policies from the 1940s, when, in the historian Jan Bieleman's words, "the government was faced with the almost impossible task of bending a surplus economy into an economy of scarcity," and also those from the 1960s, when the

State Extension Service pushed a policy of agricultural "scale enlargement," mostly through mechanization.[58]

If these factors aligned farmers with the state, other developments since the 1960s have alienated farmers. The famed Mansholt Plan of 1968, the Common Market's first comprehensive agricultural policy, seeking farm modernization and land-parcel consolidation, was welcomed in some respects. But with the Common Market also came a plethora of regulation. And then the rise of the Dutch environmental movement in the 1970s brought even more restrictive laws. Of particular concern to farmers was the replacement of the Land Consolidation Act of 1954 with the 1985 Land Reconstruction Act, giving new priority to natural scenery based on the ideal of environmental purity.[59] It was a time when the average Dutch farmer, in Bieleman's words, began to see "a world going crazy," with activists on every dike.[60]

Dutch Mennonite farmers express an ambivalence: they have historically been open to civic engagement, but recently they have become apprehensive about government overreach. Some farmers tell stories from generations ago lauding some forms of agricultural regulation. Yme Jan Buiteveld, of Poppenwier, the young farmer from a three-generation farm, says that state intervention has been necessary for agriculture, especially for modernizing it. He contextualizes the benevolent state as one arising from post–Second World War needs and seems to link it to earlier times of regional regulation. He recalls stories of the implementation of the official regional Fries Rundvee Stamboek (literally, "Frisian breeding book"), a program of pedigree recording dating back to 1879, and the founding of the dairy-registering body, the Melkcontrolle (Dutch association for milk control), in 1914. The difference is that back in the day it was about "judging cows by the beautiful hide and all that stuff, but now it's about the functional properties, the height, the milk" quality.[61] In his view the focus on increased production is a positive one, and certainly regional authority has received farmers' support.

Other farmers recall useful state interventions during the interwar and postwar periods. Some had to do with land justice. Jan Buiteveld, a retired farmer from Nes, says that it was a time when the landowning nobility was extracting impossible rents from the disadvantaged farmers, who were required to meet in cafés to bid against one another for the rented land. Then, "the rental law came [and] you couldn't do that anymore. That protected the farmer . . . to some extent . . . because the nobility really was plucking those farmers."[62] Other farmers applaud the state for reclaiming the sea. Klaas Hanje, of Offenwier, on land once owned by the local Mennonite church, recalls how his farm increased in size with the reallotment of the Sneek-erhoutpad, fifty thousand new hectares available to farmers, and then later in 1954 also the Haskerveenpolder, another massive new source of land. He gives credit to the government for initiating this reallotment of land and creating special land banks, with fifteen-hectare parcels that farmers could lease and thus enlarge their farms. Then in 1973 the Hanjes were able to afford the purchase of the church land they had been renting because, as Klaas notes, at that time "our government, The Hague, still had a guaranteed price for milk," and farming was profitable.[63] When the state helped improve dairy-cow genetics, guarantee markets, reclaim land from the sea, or ensure equitable distribution of land, it was considered a crucially important force in the lives of Dutch Mennonite farmers.

Yet when the farmers reflect on more recent state activities, they are more critical. The dairy farmer Margriet Faber, of Tzum, has a pragmatic answer: "Yeah, well, the Netherlands . . . cannot do without rules. It's just . . . every little stem matters, every, square meter is, cultivated. . . . I think it's natural that there are rules." But, she says the Common Market's arrival is of concern. She recounts how difficult it has been for the Netherlands to fit in with the rest of Europe, having to face such a plethora of regulations that it drives some in the dairy industry "insane."[64] Even farmers who grew up valuing civic engagement have their doubts about recent regulations. Douke Odinga from Pingjum is more philosophical, but he is also apprehensive about the intrusive state. He says that farming is "unique" because farmers own their means of production, the ground on which they work, producing commodities, and sustainably as possible. But "public opinion and regulations" have made things difficult for farmers, limiting what they can do with the manure, the number of cows they can keep, how much milk they can produce, even how much the Netherlands itself can produce. To his mind, if farmers know how to manage growth well, they should be left alone. Or if farmers know how to "process animal manure so that it becomes a kind of fertilizer then . . . that's only good, then you need less phosphate and calcium . . . so, yeah, why isn't that allowed, but ok, it's not allowed."[65]

The independent-minded Frisian farmers are most ambivalent about the role of the state in agriculture when it comes to managing the environment. Many farmers applaud the state playing some role in protecting both the soil and the animals. Jelke Hanje, of Wildehornsteringel, applauds the environmental laws from the 1970s that limit the number of cattle: "The government wants to do that, and I think that's good because otherwise it gets out of balance."⁶⁶ But other farmers think the state has gone too far. Yme, from Poppenwier, is concerned that the state has given itself too much power on this issue. True, he agrees that there needs to be a check on agriculture's impact on the environment and is critical of the cruel American system, in which dairy farmers have congregated thousands of cows in one confined facility. But matters in Friesland have gotten out of hand. He recalls an unpleasant visit from the Antibiotics Administration, which gave him a passing grade on barn cleanliness but cited him for bad farming practices because that day his cows were inside the barn and not outside in the pasture. He was incredulous: "Yeah at that moment all the cows were inside because it was warm weather and yeah those cows think, 'God, uh . . . [I'm] not . . . sit[ting] there baking,' so they go inside. So she [the inspector] says, 'Look I can't give a positive evaluation to your pasturing,' 'uh no?' I said, 'hello, I put the cow first and not the consumer.'" Yme is also incensed at the manure-disposal regulations: he explains that "if we're going towards so many people on the planet, after 2020, then I say we need all the manure we can get."⁶⁷

Some farmers think the state bureaucracy is simply out of touch with how agriculture actually works in the everyday. Hendrik Bosscha, a retired farmer near Akkrum, puts it rather bluntly: the state bureaucracies "never ever ever ever want to listen to people who are in nature and know how to work with nature"; how, he asks, are farmers "to take any advice from them there in The Hague?" He thinks state officials are out of touch with farmers, and he tells the story of the year his milk quota was cut back "by people who sit up front there" because his herd contacted infectious bovine rhinotracheitis, IBR, and production on his farm fell. The bureaucrats in The Hague simply did not understand agriculture and had no idea that of course IBR would cause milk production to go "wacky," but only temporarily.⁶⁸ Hendrik knew his farm's milk production would increase again; there was no need to adjust his quota.

Most of the Doopsgezind farmers in Friesland interviewed for this project say that they applaud the historical role of the state in simultaneously boosting farm production and protecting the environment. But many also insist that in the last decades the state has indeed overreached in a "world going crazy," with activists who don't care about food production and so-called experts who don't understand nature at the local level.

Iowa: Midwest Independence versus Big Brother Insurance

Of the seven places in this study, Washington County, Iowa, has received more government aid and input than most, even as large swaths of the wider nation seemed to revere libertarian ideals. The Depression, for example, produced federal conservation initiatives, and the Cold War era led to national food-security programs; both were regarded as inherently "American" in nature and still resulted in an increasingly intrusive government. The historian Douglas Helms points out that "the Depression awoke the nation to the inter-related problems of poverty and poor land use," and after viewing classic films such as *The Plow That Broke the Prairie* even the well-watered midwest-ern Corn Belt voiced fears of soil erosion and supported the Soil Conserva-tion Act of 1935.[69] Then, during the 1950s, as Kendra Smith-Howard has argued, the Midwest in particular was "touched profoundly by the interna-tionalized political economy" of the Cold War, resulting in large increases in sugar beet and soybean production; together, these crops fundamentally altered the Midwest's farm landscape. She concludes that farmers, despite the frontier myth of the independent yeoman, got caught up in government programs, subsidies, incentives, and insurance schemes.[70] And as the envi-ronmental movement took off in the United States in the 1970s, more agen-cies intervened to direct farmers' interaction with nature.

Like many US farmers, Iowa's Mennonites and Amish share an ambiva-lence toward these government programs. They recognize the history of at least some legitimate state involvement in agriculture, especially in ensuring that good soils remain a national asset, but they also express an almost liber-tarian concern about these intrusions into their lives. Indeed, Mennonite farmers across Washington and neighboring counties and across the ideolog-ical divide—including the progressive Mennonite Church USA adherents and the somewhat more traditionalist Beachy Amish and Conservative Mennonites—recall state programs with skepticism.

Some of this reserve is rooted in Iowa Mennonite and Amish farmers' his-torical Anabaptist teaching on simplicity, self-sufficiency, and equality. The farmer and seed dealer Bob Miller, near Kalona, a member of the progressive Mennonite Church USA congregation in Iowa City, suggests that not only have state programs been unwelcome but they have been a force for outright injustice. "I think government has a positive role," he says, "but there's a prob-lem with how we're allocating our resources." He argues that the federal crop-insurance program is skewed in favor of large and well-to-do farmers. He applauds the conservation programs begun in the 1930s as well as the full deployment of the Environmental Protection Agency programs in the 1970s. In fact, he is overtly critical of the state's governor, Terry Brandstad, for cut-ting back on conservation programs, exclaiming, "Earth to Terry! Where are you coming from? The rest of the world is counting on us not to abuse our water. . . . That's where we should be spending our money, instead of gifting based on a previous crop history," thus benefitting the wealthiest farmers. But he also doubts the effectiveness of old programs, including those that established county or extension agents in the earlier part of the century. Perhaps they were beneficial at the time, but their services have become redundant in the wake of web-based sources since the mid-1990s; sardonically he credits the county agent for having been wonderfully "good in terms of the 4-H program" for children.[71] For local farmers, government has become ineffectual at best, a tool of injustice for small operators at worst.

Other farmers are even more outspoken in their critique of government, in particular its insurance schemes. Calvin Yoder, another Mennonite Church USA member and an organic farmer from just outside Wellman, says he has opposed anything other than "a type of catastrophic insurance." He believes that crop insurance has merely driven up land prices. "My cows are my crop insurance," he says, and he thinks Mennonites need to return to their dependence on God and community. Like Bob Miller, Calvin is even skeptical of the usefulness of extension workers like the county agent. In recent decades it has been universities such as Iowa State and their agricultural programs that have helped young farmers understand both the biotic and operational sides of agriculture. But he is most critical of the federal insurance schemes, which he believes have been a scam, pumping millions of dollars into the coffers of rich farmers without benefiting the local community. He applauds the Amish, who have historically declined state insurance money, and spurns talk that on their 80- to 100-acre farms the Amish have not been efficient: Is the farmer "efficient if he farms 1000 acres and has to have government money to survive? Or is our Amish [farmer] efficient on 100 acres and feeding his family, with no government money? Now who's efficient?"[72] Clearly it's a rhetorical question.

The conservative Beachy Amish as well as the Conservative Mennonites, both of whom are car-driving "plain people," take a distinctive view of government programs, expressing overtly libertarian ideas based on religious principle. Delmar Bontrager, a Beachy Amish farmer from just outside Kalona, says his opposition to the federal Payment-in-Kind (PIK) program, introduced in 1983 with the aim of reducing corn acreages by a third, was based on principles he learned as a child. He asserts that "we were brought up . . . that you don't take government assistance, unless you are almost dying. . . . No, we never took government assistance. I know there were guys who [did] during

Amish farmers in Kalona, Iowa, inspecting straw at Kalona Barn Sales, a family-run auction business founded in 1947. In recent times the Amish have also participated in workshops on sustainable agriculture funded by the government (i.e., the USDA). Photo by John Eicher, Altoona, PA.

some of those years [and thus] kept their operations going. For us to keep going, we got side jobs."[73] He longs for the time when Mennonites and Amish were left alone.

Perry Miller, whose parents were Amish and who is a lay minister in a local Conservative Mennonite Fellowship congregation, is even more critical of the state. He says that "the government just doesn't know how to handle it; I think sometimes we have guys in church, with the wisdom they have, who are much more qualified to take care of those things." He likens agricultural programs to welfare schemes, which might be good in some respect but are often misused. Perry combines an old self-sufficiency with a sense of injustice. He too is aghast at lucrative payouts allowing farmers to drive brand-new equipment and pickup trucks. In fact, he thinks government programs have merely allowed farms to expand beyond their capacity to survive. Government, says Perry, has upset an equilibrium that is based on efficiency and good farm management, and thus most Conservative Mennonites have not participated in government-support programs. Ultimately, he says, the Bible "calls us to be obedient to those that are in government as much as possible and I think we'd do well to do that."[74] But the way things have evolved with intrusive state programs, this biblical injunction has become hard to heed.

A majority of Iowa Mennonite and Amish farmers interviewed for this project think that the government, and particularly the federal government, should have stayed out of agricultural programs and that Mennonites should not have accepted government aid. As the "quiet in the land," they have paid their taxes but remained skeptical of the effect of the government in securing a sustainable agriculture. In this respect they have combined an Anabaptist sense of justice with the classical US adage that "government is best that governs least."

Zimbabwe: The Hostile ZANU, the Friendly ZAPU, and the BICC Farmer

The Ndebele people of the Matopo Hills, in southwestern Zimbabwe, have had a particularly complicated relationship with the government. Indeed, they have been twice victimized. First, they suffered the indignity of Ian Smith's apartheid government before the 1980 liberation, and then came the post-1980 massacre of twenty thousand of their fellow Ndebele at the hands of the Shona-dominated Zimbabwe African National Union (ZANU) government. The black BICC antipathy to white colonial government is well established. Scotch Malinga Ndlovu, the native son of Matopo, is overtly critical of both the white government and the white BICC missionaries who managed the Matopo Mission farmland until the 1960s. He highlights the Land Apportionment Act of 1930, which divided the country, with five thousand whites obtaining 19.9 million hectares and the majority blacks given a mere 3 million hectares for purchase, with another 8.8 million being put into black reserves. And the BICC, argues Ndlovu, "lacked both the will and desire to challenge the inhumane forced mass movement of people from their home areas." He cites the historian Terrance Ranger's finding that it was not until the 1950s that BICC members developed a political consciousness. This concern with social justice developed only after the Zimbabwe Africa People's Union (ZAPU) leader Joshua Nkomo's well-known 1953 visit to the holy Dula shrine in Matopo Hills, where he received a divine message that "the country would be freed but only after a bloody struggle." Inspired by Nkomo, the BICC youth became politically engaged, with an increased "antipathy to [white-run] state interference with [black] agriculture and cattle raising."[75]

Matopo farmers placed a great deal of hope in government following the victorious War of Liberation, but they were mostly disappointed with the 1980 electoral victory of the ZANU and the loss by their own ZAPU party. Then they were terrorized by the subsequent violent attacks on its members by the ZANU government. Some farmers do have good memories of the government-dispatched workers of the Department of Agricultural, Technical and Extension Services (AGRITEX), created in 1980. Sikhanyisiwe Masuku, for example, credits "the government program in which we were getting seeds for free," and she gives special recognition to Mrs. Ngwenya, an AGRITEX officer, who helped her become a certified "master farmer," excelling in crops, poultry, and cattle production. As a master farmer, Sikhanyisiwe learned to be especially observant and keep meticulous farm records. Perhaps she made mistakes in her fields, but she learned to keep notes to assist her in changing her methods when necessary. She has, for example, carefully monitored planting: "If I put my seeds in October, I wrote it down so that I could check on progress and to analyze and see the best planting time for the next farming season."[76]

The farmer John Masuku also applauds the government extension workers who appeared after the war. They introduced "new systems of breeding" and animal care for a wide range of livestock, including cattle, pigs, sheep, goats, and donkeys, as well as chickens. This knowledge, he says, helped him engage the market economy and earn his grandchildren's school tuition. But John dismisses the overall effect of the AGRITEX officers and joins other farmers in critiquing their effectiveness and questioning their political neutrality. He has gained more knowledge and strength from deeply rooted, local cattle culture. "In our culture if I want to slaughter a cow," he says, "I call my relatives and we will have a feast," and while feasting they share farming experiences. Understandably, he has no love for the pre-1980s white regime, a period when the white man claimed as much land as a galloping horse could cover before tiring out, in one instance stretching all the way to Esigodini, thirty kilometers to the northeast. But he is also critical of the post-1980 ZANU government. He complains that although it took the land from the whites and shared it among the black majority, its tribal-based food policies ensured that the land "hasn't been producing ever since."[77]

The surprising inference is that the land was producing before 1980 and that the AGRITEX officials have been ineffective in increasing agricultural production. Eldah Siziba has her own reservations about AGRITEX agents: true, they provided useful advice on new seeds and fertilizer application, but she still wonders whether synthetic fertilizers are inferior to manure, indeed whether "fertilizer destroyed the soil." Like John, Eldah indirectly criticizes AGRITEX by offering an unusual shout-out to white farmers. She says that historically whites produced better harvests than the blacks and speculates that it is "since the whites used to respect the soil; that had an effect."[78] Alfred Gumpo, who has seen AGRITEX officers as somewhat disengaged political appointees, shares her perspective. He is thankful that the AGRITEX folks stayed away as much as they did. As Alfred sees it, at Matopo the "land belongs to the church and the government doesn't interfere much, as it knows that it belongs to the Lord. . . . The government knows that Brethren [in Christ] are respectful people and it respects us as well." While Alfred has an antipathy to white repression, he is nostalgic for the stable economy he recalls from pre-1980 days. He grants that "olden officers," extension workers from the era of Ian Smith, offered useful advice: they "used to insist that we put drains so that the soil doesn't erode" and they would "organize a field day" heralding farmers who "did the right thing, who had the better harvest." Alfred even expresses nostalgia for the time when farmers of the previous generation supplied white-owned shops in Bulawayo, saying that today "there is no business." He wishes the government would assist agriculture by creating a farmers' market in town.[79]

Government-appointed extension workers, to the minds of a number of Matopo farmers, have not had the folk-based knowledge resources that John Masuku recalls from his clan feasts. Nor have they extended the benefit of modern technology and chemical-based agriculture that was allowed to the privileged white farmers before independence. These circumstances have resulted in lost faith in the Shona-dominated government.

Siberia: Steppes of Treachery and Incompetence

Not surprisingly, among the seven places studied, it is in Siberia that the farmers describe the most tenuous of relationships with the state. After arriving in Waldheim in 1911 in full compliance with Russian settlement laws, the Mennonites quickly fell out of favor with the government following the Bolshevik Revolution in 1917.[80] For more than seventy years their relationship to the state was at best tension-filled and at worst treacherous. Numerous accounts describe Mennonites facing civil war, famine, collectivization, and then postwar slave labor camps. The historians Colin Neufeldt, Peter Letkemann, and others have explained the effect on Mennonites of the terror of collectivization, the stigma of being labeled *kulak* (capitalist farmer), the wholesale exile of ministers, and the almost impossible agricultural quotas set by Stalin.[81] Fewer historians have written in English about Mennonite life on the *kolkhoz*, the early collective, and then the later *sovkhoz*, the massive state farm. Yet over a sixty-year period Mennonites were asked to participate in the centrally planned economy of the Soviet Union, accepting the ubiquitous tractor, the new electricity-powered barn technologies, and eventually the use of chemical fertilizers and herbicides. When asked about their experience with the state, former *kolkhoz* and *sovkhoz* workers hardly celebrate the tractor or electric-powered milking parlors; rather, they reflect mostly on the loss of freedom and the inefficiency of a state-run agricultural apparatus.

The local historian Peter (Piotr) Epp, born in 1960 in Apollonovka, raised in a family of nine children and baptized in its church of Mennonite Brethren roots, for example, is critical of the heavy workload on the *sovkhoz*. "My memory of my father when I was a small boy is that he was gone from early in the morning till late at night," six days a week, and during harvest for weeks on end, home only when it rained.[82] But he reserves his most pointed disdain for the state-sanctioned authorities who were sent to the farms, whose primary aim was to please the central planners: "When Moscow said combine, we combined, no matter that we just endured five inches of rain" and that it meant that tractors were required to pull the combines through mud. The Mennonites' only response was to maximize the allowances they were granted; for example, because the same system permitted each household to raise but one pig per year, farm families invariably turned it into an animal of immense weight, five hundred pounds and more.[83] He contrasts the system to the freedom in the post-Communist era, when "all they [local farmers] can think about is, if they can but build a barn and then raise pigs; if one has 100 or more pigs, you can make a good living, with 200 pigs, you can make a very good living."[84]

Other Mennonite state farmworkers speak of their open contempt for the Soviet system and link this disdain to their religious faith. Ivan Ivanovich Peters, born in 1955, worked as a cattle herder, or in his own words, a "cowboy," enunciated in English, on the state farm that encompassed Apollonovka. His narrative of life on the *sovkhoz* is a story of opposition to the Soviet system. He begins with the religious, recounting that "throughout my childhood I believed that God existed" and that in February 1978, at the height of Communism, "I consciously gave my life to God, that I would serve him." His was a subversive faith that saw *sovkhoz* society and culture as a sham, but his story moves quickly beyond religion. The state farm was run by so-called "learned people . . . wanting to teach us to do things that were completely irresponsible. We would say 'why would you do that, it would be terrible for the cows,' and they would say, 'No, do what we say.'" He recalls the time when he encountered an especially inept city-issued veterinarian. As a cattle worker Ivan cared for dairy cows in the barn in winter, but in the summer he and his neighbor Peter took the animals to a distant pasture, some twelve kilometers from the village. One day the veterinarian came to the field to inspect the herd, arriving on horseback. At the *sovkhoz* headquarters, officials had given him "a horse, a draught horse, to ride to . . . the field to see if the cattle were healthy or not." Not used to galloping, which the veterinarian demanded of it, the horse was dead tired and very hot upon arrival. And then to Ivan's dismay, "this learned person . . . wanted the horse to eat right away. We said 'what are you doing? He has just arrived, he shouldn't eat, he will die!'" Ivan scolded the veterinarian, telling him unequivocally that the horse "needs to rest an hour and a half, so that the heart is normal." He had no respect for this Soviet official and blames the "socialist system for not advancing" agriculture. From Ivan's perspective, "people were afraid about this and didn't talk; but we saw everything."[85]

Similar accounts describe the experiences of two tractor operators, Vladimir Friesen, born in 1939, and Abram Dirksen, in 1936. Vladimir and Abram share stories of Soviet terror, of fathers sent off to forced labor camps during the war and mothers toiling in oppressive conditions on the *kolkhoz* to feed their families. True, both Vladimir and Abram recall positive stories from the collective and state farms, the only rural society they ever knew. They remember the time of their youth, working in pure air, speaking Low German, the everyday language of Kolkhoz Kirov, and their pride in being tractor drivers on the larger *sovkhoz*. In part because neither Vladimir nor Abram was a "believer" at the time, they had an easy time gaining the coveted role. Vladimir was even selected for tractor-driving school, while Abram became a driver in 1959, after returning from work as a mining expert in Vorkuta, in the far north. Both men adapted to the Soviet system of professional subcategories, with Vladimir driving a track tractor for thirty-three years and Abram driving a wheeled tractor for twenty-five.

But this is the extent of their positive memories. In fact, Abram returned from the far north to be close to his mother, who at 60 was still being forced to work extremely hard on the newly expanded *sovkhoz*. And although he himself was spared hard physical work as a tractor driver, Abram describes a mind-numbing industrial model of agriculture: "The caterpillar would work day and night. Then there were always two tractor drivers: one all day, one all night. Come to refuel and more and more. They didn't rest."[86] Vladimir, in particular, laments the spirit of the time, working endless shifts, rarely seeing

Tractor drivers Abram Tevs (Toews), Jacob Klippenstein, Viktor Friesen, Ivan Shiling, and Isaak Wall taking a break from cultivating sunflowers in front of a typical birch forest with three Soviet-built tractors at the Sovkhoz Medvezhenskii in Siberia, ca. 1975. Photo courtesy of Peter Epp, Isul'kul, Siberia.

the children except late in the evening or early in the morning, before 6:00 a.m. Only in the cold, short winter days was there more time for them.[87] For the most part, the life that Abram and Vladimir experienced on the state farm was culturally at odds with their religious and family values.

But like Peter Epp, both Vladimir and Abram reserve their sharpest criticism for the inept Soviet system of central planning. Abram recalls the Communist ideology, learned in school. where "we studied only Stalin; he was in every book . . . everything was our dear Joseph Vissarionovich Stalin and [ironically] he considered us all enemies and spies." This ideology was reinforced on the *sovkhoz*, where members of the Communist Party had all the power, directing the school, running the village, "all the important positions."[88] And they carried out instructions from the regional office in nearby Isil'kul or even from far-off Moscow, an exercise of folly in the eyes of the tractor drivers. Vladimir says that their *sovkhoz* never attained high yields and that central planning made no sense: "There was a plan, for example to seed . . . early," but without realizing that "if you plant grain in cold soil you won't win; you need warm soil."[89]

Abram recalls one especially wet spring when they were told to seed no matter the mud: one tractor couldn't pull the seeder through the sticky soil, so "one of the others hitched up with a rope and two tractors dragged three seeders." When mud clogged the seeding spouts, the drivers just let the seed fall on top of the ground, thus meeting the official quota for the day but knowing that a poor and weedy crop would follow. He also recalls a year when the combining was so delayed by innumerable repairs that they were overtaken by winter: "Everything was under snow [so we] took out the pitch forks, all of the *sovkhoz* workers, went out to the field [and] we pulled [the un-threshed grain] out of the snow." He adds that sometimes "there were good years," but they seem to have been exceptions to the rule.[90]

Sadly, the central planners never understood why their unrealistic quotas were not reached. Abram recalls the mediocre harvest of 1957, the year the smaller *kolkhoz*, Kirov, was folded into the massive *sovkhoz*, Medvezhinskii: "There were always laws or plans from above, in Moscow, 'this year you will achieve 13 quintal from each hectare' [and they] didn't even know what kind of land we had. They didn't know anything. They just wanted . . . 13 quintal in all of Russia" and promised that if this target was met, "we would have lots of bread. But we only got 8, 9; it was in general poor agriculture." Abram concedes that things improved in 1973, when Medvezhinskii joined Novorozhdestvenskii, because now regular deliveries of specially blended synthetic fertilizers began to arrive, and consequently the farms reached their desired quotas of 12 and even 14 quintal.[91] But Abram still has little positive to say about the Communist system.

In fact he has little praise for any kind of large-scale agriculture. Under a capitalist system, no one is ever satisfied: "Now 18 [quintal] or 20 isn't enough. But it is never enough, always it needs to be more, without ending."[92] He only sees a dichotomy of Communism versus capitalism. The small, independent farms of Waldheim from before the Russian Revolution are nowhere in sight.

Conclusion

The accounts of how state power influenced farming practices vary among the seven places studied in this book. A century-long portrayal of government's involvement with sectarian farmers in Manitoba offers a congenial story of complementary aims: early laws encouraged a transplanted agriculture, later there were programs of land "improvement," and finally came a discourse-based approach to expanding agriculture with modern technologies. In each era the state's actions allowed for Mennonite sensibilities against too close a relationship with the outside world or a "high modernity" driven by a singular obsession with technological innovation.

This story differed by degree in each of the other places. Accepting Radkau's challenge of comparative analysis, this study of seven communities reports on the disparate results of multiple ideas from former times and from around the world. As the farmers spoke to the question of the role of the state in their lives, they gave voice to a wide range of experiences, from the reciprocal in Manitoba and Java, to the somewhat more tenuous and critical in Bolivia, Friesland, and Iowa, to the patently hostile in Matabeleland and Siberia. Clearly, these experiences point to Radkau's own observation that in environmental history "epochal turning points arise from a change in . . . structures of power, and from the alterations this brings about in everyday habits."[93] In each of the places, Mennonite farmers found ways to practice their everyday faith, but clearly some government agricultural policies were more amenable to those practices than others. While the state in Bolivia, Friesland, and Iowa, for example, could sometimes appear overreaching in agricultural policies, those in Matabeleland and Siberia dramatically curtailed the agricultural aspirations of Mennonite farmers. Clearly, any story of the state's agricultural and environmental policies is also an intensely local story.

CHAPTER 7

Vernaculars of Climate Change

Southern Concern, Northern Complacency

Mary Dube moved to Matopo as a young woman in 1958 when she married, and there she became the Brethren in Christ Church mission school's librarian. But she also farmed with her husband, and after his death in 2011 she farmed alone. For Mary, the natural environment is a text through which God speaks to people: "Because God has time in the rainy season . . . he will be communicating with the people and also be giving them food so that they do not die of hunger." Mary easily invokes the divine when she describes the farm's bounty. But when she talks about environmental hardship, the debilitating effect of climate change, for example, Mary becomes clinical and scientific and talks about the effects of the Indian Ocean's El Nino. She recalls the year it rained too much and also the times of prolonged drought. "The weather has really changed from what it was," she says; "long back, the rain would fall in its rightful time," ordering the farmer's work on the land. Now even ancient signs of impending rain have changed. In the past, when the sun would have "lines" across it or be encompassed by a rainbow the old people would declare, "'This year there is plenty of rain'; they call it *izinga*," a big rain. Or they would know that if the wind came "from this side [of the mountain], there would be plenty of rain." She adds, "These are the observations of grandparents as they were teaching us."[1]

Climate change has ended these benevolent signs from nature; in fact, it has devastated them. Tough weather has caused the fruit bushes to disappear and threatened certain trees—the *umtshekisane*, the *umsuma*, and the *umpumbulu*—with extinction. It's the same story with the animals. Long ago Mary saw a lot of wildlife: the antelope, the duiker, the ostriches, and the amalanda, which flew in distinctive V formations just before a rain. They have all become scarce. Unfortunately, when she tells the youth of her village about extinct fruit bushes—the *izadoli* and the *izadenda*—they don't even know what she is talking about. They can't remember the rains or the season when the bushes would flower. Even more unfortunate, people have stopped respecting nature and now they only take from it. Looking for meat, they "kill animals willy nilly"; searching for firewood, they "cut all the trees." Mary believes that climate change is real, and she implies that human complacency, seen in the youth who don't know or care, and human greed, which led to killing the duikers and ostriches for their meat, are somehow complicit in this malignant alteration.[2]

Many Mennonite farmers from other parts of the globe don't share Mary's perspective. Indeed, like farmers from countries with some of the highest educational standards in the world, Mennonites have often questioned whether the climate has indeed changed. In an important 2016 essay, "Big is a Thing of the Past," the science historian Deborah R. Coen suggests that the reason only two-thirds of Americans accept climate change as real is that the very phenomenon "exceeds our mental capacities," for "only at the scale of the planet does the catastrophic irreversibility of the greenhouse effect become evident."[3] Climatologists, world historians, and environmental scholars, especially since the dawn of the space age, it would seem, are best positioned to grasp the magnitude of this event. Yet in her comprehensive *Global Population* Alison Bashford argues that "the vision of a singular planet, the spaceship Earth, is better understood as a new rendition of a planetary imagination that was already many generations old." If it was easy to imagine the entire planet as a natural organism in the time of Malthus, who spoke of "taking the whole earth instead of this island," England, his home, as a unit of analysis in the late eighteenth century, why was it a problem for two-thirds of Americans in 2016?[4] One of the reasons for the resistance to the basic idea of climate change, writes Joyeeta Gupta in *The History of Global Climate Governance*, is that the "essentials" of this "problem" are more political than scientific, more about human rights and the guarantees of equal access to the earth's resources than about a measurable historical event.[5] Indeed, the debate on climate change often pits the developing Global South, suffering directly from the effects of climate change, against the highly developed Global North, which is most culpable in it.[6]

These disparate ideas on climate change are also apparent in the ways farmers from the seven Mennonite communities speak about this issue and in the vernaculars they employ in describing their experiences with it. The farmers from the Global South seem to be much more stirred by it than do farmers from the Global North. Both sets of farmers speak about severe weather in the recent past, even a change in climate since the time they can remember, but their interpretations often differ dramatically. And those understandings seem especially affected by issues of relative wealth and vulnerability, and most specifically by the level of access to advanced technologies able to mitigate the worst effects of climate change.

To focus more singularly on this issue, and to move expeditiously between the grand and the everyday, this chapter is based solely on interviews with farmers who were asked how they have experienced climate change, and then only with farmers from five of the seven communities, three in the Global North and two in the Global South. Arranged in order of the intensity with which the farmers say they have experienced climate change, the chapter begins with Matabeleland and Bolivia in the South and then moves to Friesland, Iowa, and Manitoba in the North. Ultimately, this story is even more complex than a dichotomy between the North and the South, as local variables—environmental zones, politics, ethnicity, and religious sensibilities—ensure that the five locales have their particular versions of climate history and its effects on them.

Matabeleland: The Malice of Human Greed

The farmers at places like Matopo Mission in Matabeleland readily speak of changing weather patterns, and they invariably link them to destructive human behavior. Farmers here insist that humans—sometimes national or corporate leaders but just as often people in local circumstances—have had a role in climate change. Historically these farmers are known for their spiritual view of the land, blurring the line between the creator and creation, at odds with a dichotomous Northern viewpoint. In this community of the ox-drawn plow and the donkey-pulled wagon, farmers have also placed their hope in the divine more often than in the technological. This spiritual linkage to the land has led farmers to conclude that the soil and the climate have been under siege from humans' overreach.

At Matopo, talk of climate change comes only after utterances of gratitude for the environment. Mary Dube's prayer—"God, take care of our fields"—is echoed by numerous farmers. For example, John Masuku, born in the same year as Mary and the onetime regional manager for Dunlop Corporation in the city of Bulawayo and father to children in London making good, enjoys his urban ties. But he revels in being solitary on his farm. Happy to be back home on his native Matopo farm, he speaks of his love of "being alone" in the fields, where he has had time to ponder, to be introspective, and to observe the intricacies of nature. For John, farming has a sacredness to it. "If you look at nature," he says, "you will realise that it is the great hand of God" that has created it; "these things do not happen on their own. . . . When a certain tree grows and blooms and puts on fruits, it is the hand of God . . . providing food for his people, and the animals and insects as well." When John invokes the divine he is not being pious: he is outlining the environment as it should be, measured by seasons and sustenance, and connecting one species to another.[7]

Like Mary Dube, though, John easily moves from worship and gratitude to bitter talk of the history of weather extremes and the science of climate. He recounts a recent flood in the Matopo Hills that was followed by a severe frost: "Heavy rains and a big snow, that was too much snow, I have never seen such. Doves died, you would find birds seated, having been frozen. Many people who had grown up by that time will never forget such." He says it was two months before the snow thawed and that this was not an isolated event. Indeed, John has seen an increasingly volatile climate. He knows it from experience and even from scientific records: "We used to know that if you hear the rains falling like this, or if you see clouds like that, it means it's going to rain heavily or not. But now clouds just gather and nothing happens." In the past, at Easter or around April 18, Independence Day, Matopo always had rain, but

Brethren in Christ Church farmer Alfred Gumpo and helper hoeing the fields at Matopo Mission, Matabeleland, 2015. Photo by Belinda Ncube, Bulawayo.

now all that has changed, and if it rains at all the rains come late and stop early.[8]

Farmers are bewildered, and even the scientists are still unsure, says John. But then he offers an analysis; he blames modern ways and links climate change to the history of environmental degradation. He begins with the soil, explaining that "long back" farmers used only manure, which "revives the soil and gives it food and adds to its fertility, while fertilizer washes away the food in the soil." Synthetic fertilizer, invented by humans wanting quick results without having to mess with manure, produces a weakened soil and then stunted maize, which has to be harvested early and consumed quickly, lest it spoil. John also links other forms of human greed to weaknesses in the natural world. For example, people have cut the trees without realizing that "the rain is . . . drawn by the trees, hence we have to keep them growing"; sadly, people have lost touch with the cycle of life.

The subsequent multiyear drought in turn has caused birds and insects looking for green foliage to move to other regions. John lists the disappearances: the wild fruit *amakhemeswane*, the *uxakuxaku* and *imikhiwa* trees, and the animals—the giraffes, the baboons, and the wild geese. Some of this wildlife simply has vanished, and human recklessness is linked to it all. John offers as an example Matopo's trees: "Trees used to purify water but now the trees have been destroyed. God intentionally put the roots underground so that they could wash the water, but there are no trees now. There is a lot of stagnant water here and it has turned green." As John sees it, humans have lost sight of the divine in nature, and as a result they have stopped nurturing it.[9]

This need to nurture nature is part of the narrative of all Matopo farmers, especially the elderly. Zandile Nyandeni, a former Anglican schoolteacher born in 1936, recalls the terrible drought and famine of 1947 as "very bad, it was really . . . very bad," so horrific that people mixed water and anthill soil, declaring "that the soil is good for your body, even if I have not eaten, but if I drink this I would feel better." And yet 1947 wasn't much worse than the drought of recent times; what is worse now is the human carelessness with the environment. Zandile speaks of multiple ways in which soil and nature have been degraded. Soil erosion is caused by water flowing in strange ways not seen in the past. Soil exploitation occurs without good crop rotations to maintain the soil's health. Animals are destroyed, drowned for sport or killed without good reason: the anteater, the *mantswane*, and the *mahelane* are all gone. Meanwhile the trees are being cut without regard to the future: "Long back, people did not cut the trunk, but only the branches, knowing the trees will sprout again, but not . . . the entire tree." Zandile even laments the loss of marauding grasshoppers and recalls an especially horrific invasion in 1949, when the trees were bright red with the insects. The weakened state of current vegetation can't even sustain an old-fashioned grasshopper plague. Zandile doesn't want an end to all tree cutting; she just wants an old balance to be restored. The reason is simple: "We get the air we breathe from the trees, when the breeze blows, it also comes into us, it is good air." But human activity has destroyed the good soil, the pure air, and a predictable climate.[10]

Most of Zandile's older neighbors share a version of her story, many even more unequivocally angry at the human myopia at the root of changing weather. Another child of the 1930s, Aaron Tshuma, says it most explicitly: "The climate has changed and . . . it is caused by people's failure to respect it." He adds that in the distant past his grandparents and even older people still alive in recent years shared a profound respect for the environment. He explains that because the BICC has taught members to respect one another and also the environment, the culprit must lie outside the community. Without explaining, Aaron links environmental problems specifically to the fact that "the country is not in good hands; those in authority are committing crimes to the environment and it's now different from what we had long back."[11]

Other farmers see greed closer to home. Moses Tshuma, born in 1937 and a resident of Matopo since 1959, voices a common refrain: "This is not what we used to see when we were growing up. . . . Long back, it was nice. . . . During spring, trees would bloom and by October people would be saying . . . the *insewula* [first rain], will come." And then it would rain for at least a week. Today greed colors everything, even when God sends the rain: "People hurry to collect and hide it," hoarding it for their own farms. In Moses's account, climate change has not only been the result of greed, it has also produced it.[12] Another elderly farmer, Elliot Gumpo, who has faced dislocation and poverty, sees greed in brazen behavior in the hills around Matopo. He sees the loss of ancient ties to the land as being at the root of it all: "Now people just burn the bush and in the process [this] affects animals and vegetation. . . . Long ago there were specific periods for doing this. Now people want to hunt lots of animals and forget about tomorrow."[13] He refers to the ways of yesteryear, when each clan hunted specific animals on specific days named after the clan, a rudimentary but effective system of conservation. The Ndebele culture was based on sharing and sustainability.

For the farmers of Matopo there is nothing incomprehensible about climate change. A somewhat younger farmer, Timothy Sibanda, born in 1952, grateful that independence gave him a chance to cultivate land once owned by a white man, is unequivocal on matters of the environment. On the one hand he believes that the air is dirtier than in the past, even though "we can't really say it is, as we cannot see it"; it's a belief based on being "told that we mustn't cut trees until they are old so as to serve the purpose of cleaning the air." But just because pollution in Matabeleland seems intangible doesn't mean that Timothy disputes the idea of climate change. His own experience substantiates the science. He points to the cyclone in 2000 that produced intense rainfall and compares it with the gentle rain of yesteryear: "Long back, the rain wasn't something scary, we would play in the rain when we were young boys herding cattle. Nowadays when it starts raining you have to ensure that all the children are indoors."[14]

Other farmers offer stories of climate change with reference to personal hardship. Sibonginkosi Dube, who moved to Matopo in 1967 and raised three children there, has her own way of measuring the drought. In earlier decades, she says, "we struggled to use the 'bushy' roads during rainy season, because of the swamps; but now, there is no rain" and the bushy roads are dry. The immediate result of the drought is possible hunger; certainly she has seen her own farm animals—the donkeys, goats, and cows—suffer from lack of feed.[15] Her experience is specific to her farm, but her analysis parallels that of her neighbors. Mary Dube, with her firm belief in the science of climate change, hints at struggle and possible suffering. She says that decreased yield on her farm, only fifty bags of maize, sweet potatoes, and other vegetables, is worrying. But she insists that to date there has been no hunger per se, and although it is a limited amount of produce, "still we do not starve."[16] And yet a hint of desperation colors her dismissal; climate change could still bring about hunger.

Climate change for the Ndebele BICC farmers can't be countered with new technologies, either mechanical or biological. Erratic weather means the possibility of hunger, and the nostalgia for a past that cannot be reconstructed is no basis for hope. Modern mentalities have robbed the Matopo farmers of their forests and fauna, but they are without the modern machines to address the injustices.

Bolivia: Regret for Breaking the Bushland

Climate change also has had far-ranging effects on Mennonite farmers in Bolivia. But here the climate has been repaired to an extent by an increased sensitivity to nature, albeit without a strong base in scientific knowledge.

When the Old Colony Mennonites of Riva Palacio, in the Bolivian lowlands. talk about environmental change, they begin with stories of how they have experienced it in personal terms, but they base their stories on little scientific detail. As one of the Old Colonists' stated objectives in Bolivia was to protect their rudimentary parochial schools, which focused only on those subjects required to replicate agrarian simplicity and religious faithfulness, they have taught no sciences.[17] They didn't hear the terms *climate change, global warming, climatology,* or *biosphere* in school. But they have all been taught by their parents about nature, and all have experienced it in the everyday.

Indeed, as an agrarian people the Old Colonists know well the intricacies of nature, weather, and climate. They readily tell stories about weather, debate how to meet its capricious demands, and recount its life-giving qualities. For example, the Bolivian lowlands have a higher precipitation rate than the old homeland in northern Mexico did, typically with abundant spring and summer rains during the months from November to February. But farmers from across Riva Palacio colony speak of both very dry years and severe rainstorms. The teacher Jakob Buhler, of Campo 17, recalls especially the long drought of the late 1980s and 1990s, "the worst" the Old Colonists had experienced. It was a time when many farmers had to sell their cattle and even the well-to-do had to spend their bank savings. Jakob recalls how disappointing this change of fortune was. At first, in the late 1960s, Bolivia had seemed to be the land of bounty; indeed, the only challenge was to clear the forest and adapt to the country's strange seasons, especially its hot and humid summers. But then, after they had cut the trees and adjusted to the weather of the Southern Hemisphere, the weather became erratic. The drought of one year easily led to a deluge the next, or a drought in one part of the colony hampered crops, while too much rain in another ruined them. Jakob says that when he first lived in Campo 31 it would rain so little that it didn't pay to harvest, but then he moved five kilometers away to Campo 17, and there were bountiful harvests and perfectly timed rainfalls. And now this year "has been a year, we've never seen this before, with all the rain."[18]

Jakob and his neighbors understand the reasons for some of these weather patterns. Riva Palacio, for example, sits on the ecotone between the Amazon and the Chaco, meaning that even within their colony the northern sections have more rainfall than the southern reaches. But they also have an explanation for weather changes themselves, especially the drought and dust storms. They recount that upon arriving in the bushland in the late 1960s they relentlessly cut trees to plant their crops, but because they longed for the refreshing dry winds and open vistas of northern Mexico, they had no compunction in clear-cutting. Johann Fehr, aged 57, offers a narrative that combines elements of both natural beauty and human bravado. He recalls that "when we got here, it was all forest . . . very beautiful . . . very good land." The environment allowed quick economic advancement, and within two years of arriving the settlers were growing profitable crops of corn, sorghum, and even soybeans. But that advancement came with destruction. Johann recalls his role as a bulldozer operator and the satisfaction of completing the two-decade process: "I remember when we finished clearing. It was '85. We had finished everything. In '67 we had entered [the bush] and by '85 we had finished [cutting it down]." But then the settlers reaped the consequences of their shortsightedness. He recalls the sandstorms of the 1980s: the "wind was really strong, and we didn't really know the climate very well. We didn't know what was going to happen. We had cleared everything; we didn't even leave a single windbreak."[19] The immediate challenge was survival itself in their self-made dust bowl; most of the settlers had little money, and government programs were nonexistent. Adaptation of some kind was required.

Farmers were often forced to make desperate moves. Farmer Abram Enns, of Campo 21, recalls how 178 days of strong wind in a single year brought high sand dunes right up to his house and robbed him of his desire to work. He remembers 1989 in particular, the driest of years, when no family harvested a viable crop. At one point he was even driven to borrow money from the cheese factory for a sack of flour. And the only way to feed their cattle and horses in the drought was to import *caña*, sugarcane, which was abundantly available from the wetter and tropical northern regions of the country.[20]

Other farmers even began to criticize some of the sacred practices of the Old Colony. The steel wheels on their tractors, required by the Old Colony church as a measure of simplicity and humility, were cited by some farmers for their negative effects on the land. The shopowner Peter Friesen, of Campo 1, explains that the prolonged droughts made the land less fertile, with less organic material. But he says that the steel-wheeled tractors also had an adverse impact, compacting the soil and decreasing its ability to absorb excess moisture during unexpected downpours. When they first started cultivating the land in the 1970s, he says, "the land was . . . softer, and it held the rain better. Now it is harder. When you prepare the land [with steel-wheeled tractors] it compacts it." And this means that water pools and floods the lands more easily. At least some farmers have taken drastic measures, leaving the colony and its steel wheels, willingly withstanding excommunication for accepting rubber tires.[21] The brothers Abram and Frank Rempel, onetime residents of Riva Palacio, in an act of rebellion founded the car-driving and rubber-tired-tractor Chihuahua Colony in 1989. Franz recalls the confrontation with the Old Colony church leaders: "At that time they were really strict, they wouldn't let us work with the machines we needed. They wouldn't let us work with a harvester that had its own motor. Always, they wanted you to work with a tractor with steel wheels." Franz simply "didn't understand why it was a sin." This rule about steel wheels was a central reason why they decided to establish their own colony in 1989, during the height of the drought.[22] The Rempel brothers saw the old steel wheel as keeping them from pursing an efficient agriculture, but other farmers also came to view the steel wheel as harming the soil through undue compaction.

Most farmers found refuge from the drought not through technological change but by turning to new cropping strategies. For example, the elderly Johann Boldt, of Campo 11, began to farm cattle, specifically the hardy Zebu strain. He purchased calves and raised them for slaughter, and, in the process he found a low-risk venture that complemented his dairy herd and soybean fields. Cattle raising, he said, "is much easier, because with farming, a month of drought, ruins the harvest. And for cattle, a dry month is not too bad, when it starts raining again [they improve]."[23]

Other farmers have a more complex narrative of farm diversification. The middle-aged Jakob Giesbrecht, of Campo 27, begins his story simply enough: the Giesbrechts first planted corn, then soybeans, and finally, with the drought, they turned to cattle. But then he adds layers to this story. In the 1970s "on the new fields," he says, "you would always plant corn because of all the trunks and roots," and "watermelon also would grow well in new fields." But then came the major breakthrough with soybeans, the low-profile miracle crop he references to the year of his wedding, 1986: "At this time it was all soybeans." However, soybeans suffered in the drought of the 1990s, and farmers turned to more drought-resistant crops. One day a Bolivian seed importer came around with a sample of sesame seeds, soon followed by another, coaxing the Mennonites to try producing a commodity they had never heard of. Sesame was a low-technology, labor-intensive solution, nicely suited to Mennonite families with many children, but in addition "it grew well and almost without rain." Sesame took off at Riva Palacio: ninety tonnes of product from the colony alone during the first year, a thousand tonnes the next year. But sesame soon lost out to crop diseases, and, as Jakob tells it, the Mennonites were ignorant of agrochemicals that might have countered the pathogens. So, the

Giesbrechts switched yet again, this time to drought-resistant peanuts. In the first year they planted only four hectares. The work was hard—they weeded by hand and fought funguses with fungicides—but they earned more than a thousand dollars per hectare. The next year Jakob expanded the peanut field to nine and a half hectares, which included a sandy dune on which "nothing really grew. . . . So I said to my kids, let's plant this 1-½ hectare [dune] with peanuts [as well]. And this small section produced three thousand dollars per hectare."[24] They had discovered the latest of a set of promising crops and reaffirmed their basic approach to the land: adapt to the soil and take good care of it.

Despite their variations, all these approaches by Old Colonist farmers acknowledged climate change and displayed a bullish commitment to adapting to its vagaries. How humans might alter climate change with a range of new technologies available to farmers in the Global North—land tiling, irrigation, rubber traction—has not been central to the formula of countering climate change at Riva Palacio.

Friesland: Fixing Erratic Weather

In contrast to the farmers in the Global South, those in the North express less concern about climate change. When they do talk about it, though, they more quickly speak of technological adaptation that has addressed its ill effects. Some farmers even claim that climate change has been but an abstraction perpetrated by the media or so-called experts, all exaggerating the effects of global warming. Often the farmers speak of their commitment to a healthy environment, but they don't necessarily see the link between a concern about climate change and sustainable agriculture.

The Doopsgezind farmers of Friesland recall a time in the past when weather seemed more stable, certainly when winters could be expected to be cold. The farmer Jelke Hanje, from just outside Joure, for example, recalls a time when the winters were always cold, an era when even Frisian farmers were among nationally acclaimed Dutch skaters. Jelke tells of his father being a professional skater, touring the canals of Amsterdam and those across the Netherlands, thus earning his living. Only when he married did he leave skating and become a farmer.[25] Jelke's brother Klaas Hanje, of Offenwier, places their father's passion in a time "when there was tough ice during the winters, well in that case you could go ice skating . . . a fantastic time for us. In my opinion there was a lot more winter back then as well." Cold winters, of course, made farming difficult, but from a boy's perspective the ice was nothing less than a vibrant neighborhood playground. He recalls the outdoor activity as exhausting for young children, but nothing that could not be remedied by a spoonful of cod-liver oil and a warm bed.[26] Winters since then, says Jelke, have become warmer, and rarely do the canals freeze over for any length of time.[27]

Years ago, not only were the seasons more predictable but so was the weather itself. Margriet Faber, of Tzum, who has run her farm since she was 19, says that even as a small girl growing up on the farm, she knew about the weather and how to predict it. She points to the horizon: "Very spacious, eh, a line of trees every now and then, but you can see the horizon, you can see the skies, you can see it change."[28] This knowledge has allowed her a certain meditative peace within nature. As a farmer she has always known "that winter is coming by seeing flights of geese coming, or noticing that it's going to snow or that there's a storm coming because there are no longer birds in the sky, well then you know for sure it's going to storm. That sort of thing influences your whole being." Margriet also recalls her late father, a quiet, reflective man, accepting any weather that came his way. She tells of a terrifying thunderstorm with multiple lightning strikes when she was a young girl. She and her father, along with a farm maid, were out in the fields haying with two horses hitched to two hay carts when the storm hit. Her father ordered the maid to grab little Margriet and hurry home; "he, himself grabbed the horses and carts, and just ran home, straight over the land, with the horses in hand; that's an image I'll never forget." Even in that frightening moment, she recalls, he was an inherently peaceful man, calming his animals, cows and horses alike. This way of living in nature, though, has changed over time, and farmers no longer know how to predict the weather. To explain, Margriet talks of the dog days, the period from mid-July to mid-August, when Sirius, the Dog Star, rises along with the sun. She recalls that these days could be treacherous, for if it rained at all, it rained a lot, with intense heat and humidity, but "if the dog days were bright and sunny, expect a fruitful year." But no one talks about the dog days anymore.

The increasingly erratic weather has made folk-based weather prediction a lost art. But another reason for this cultural shift is that farmers have been able to mask bad weather with technology—mechanical, chemical, or genetic— and crop-insurance schemes. The farmer Douke Odinga, from just outside Pingjum, has taken a decidedly modern approach to any climate change. He talks about having to adapt constantly: "It might be too wet or too dry. Do I have to irrigate, do I have to grow different varieties? There are always new varieties coming out of course, in wheat, beets, and potatoes. Do I choose a crop, that I have to spray a lot, or do I choose a variety which might have slightly less yield, but which is very resistant?"[29] The technology available to Douke has not been only mechanical in nature but also chemical and genetic. All Frisian farmers have had insurance schemes to cover the effects of bad weather. Indeed, to turn down insurance and rely on nature to mend the effects of bad weather has been unusual. When the "eco-touch" organic farmer Jan Idsardi, who has traveled globally but returned home to take over his parents' North Sea potato farm at Holwerd and become a full member of the Doopsgezind community, even serving as his church's organist, faced a deluge one year, he was determined to try to work with nature. As an organic farmer, he has emphasized "clean soil" as a means of combating disease in the crops, but he acknowledges the fact of climate change, including deluges in which rain came down "all at once." He says that farmers need to trust nature a bit more and recalls the time when the family's seventy hectares of freshly planted potatoes almost drowned from too much rain. But then, after filing an insurance claim, he experimented and asked for permission to leave the damaged potato plants in place on a few hectares to see how they would fair naturally. "Well they ended up growing nicely," and all he did was drain the field "as quickly as possible."[30] Of course, even then he had the tractor and access to the mechanical world to maintain the drainage ditches.

Older farmers do talk about machines that have countered inclement weather, but they acknowledge their limits. Hendrik Bosscha is a retired farmer and identifies as a minister at the Doopsgezind church near Akkrum. He and his wife, Baukje, have relied on technology to fight the effects of extreme weather. Some years ago, a period of extremely hot temperatures caused their hay to ignite spontaneously. Their response was to purchase a hay blower. Hendrik describes how it worked: "There has to be air, you blow air underneath [a floor], and then there was in the middle . . . a large barrel, it went along straight upwards. . . . Then you had to close it off [at the top], so . . . that way no air would escape. . . . And then the air had to . . . pass through the hay . . . and that is how it dried." Of course the machine had its limits, because the outside air had to be dry and there had to be no actual water accumulation under the hay. Hendrik has even more pointed stories of the limits of machines. He recalls the year a custom hay baler failed to come at an appointed time, indeed arriving just as it began to rain. He remembers telling the baler to keep working despite the rain; the plan worked for a while, but then when it continued raining and unbaled hay stood in water, "'Well' I said 'I do not have to deal with that now.'" He fired the tardy hay baler, who was "pretty riled up at me," and then patiently waited for the sun to come out and dry the hay naturally. Hendrik thinks that the dependence on modern technology has blinded farmers to nature's innate power. As he sees it, younger farmers especially have begun relying too much on machines: "The older farmers, those from . . . before the war . . . if they now see how the young farmers are doing things . . . [they say] maybe this [reliance on machines] is really going too far." He doesn't elaborate; in any case it is more of a sentiment than a reasoning.[31]

If the older farmers lack faith in machines, they are also resigned to a certain viciousness in nature. They disagree with urban environmentalists who imagine some idyllic, bucolic return to the wild. The fact is that nature has been capricious; sometimes crops have been hurt by severe weather, and sometimes animals have had to suffer through storms or in the heat of the sun. Jelke Hanje, for one, thinks the environmental crusaders who have imposed themselves on agriculture and more broadly on nature in the Netherlands have been unrealistic. He recalls seeing a film about the nature reserve Oostvaardersplassen, just south of the inland Markermeer. He was appalled that in the movie "wild animals are lying there perishing, that's all allowed because it's nature," but anyone seeing a farm animal lying outside in a pasture in the rain can get the farmer "arrested right away because it's animal abuse."[32]

Jelke's disdain for the environmental movement is matched by his brother Klaas's acquiescent approach to the environment. Farmers simply must accept bad weather and attribute it to God. And they need to help one another overcome its ill effects. He offers as an example one year when the Netherlands enjoyed a lovely late summer while in Italy people were drowning in rain. The aberrant weather has directed Klaas to a religious expression uncommon in rational, modern Friesland, but he is unequivocal: ultimately all "we can say, yes, Our Dear Lord . . . it's something He is in control of." He laments that younger farmers don't share this humble yieldedness or the sense of communitarian interdependence that he links to this outlook. Back in the war, says Klaas, "people united, people tried, to get things back, resistance started. But people needed each other, yes and that's all gone now. . . . We have become individualistic and everybody takes care of themselves."[33] Ultimately the best approach to combatting climate change, as Jelke and Klaas see it, is to return to old standards of religious faith and community cohesiveness.

The farmers in Friesland have taken a variety of approaches to climate change. Most agree that weather has become more volatile and unpredictable, but most also speak of the need to adapt to it. A majority of the younger farmers seem to have placed their hope in technology—irrigation systems, hay dryers, new genetics—while the older farmers emphasize how faith and community used to suffice to combat the effects of bad weather. But they all seem to agree that environmentalists from the city haven't understood the farmers and their fields.

Iowa: Ingenuity and Incredulity in the Midwest

The Mennonite farmers of Washington County, Iowa, and its environs have a similar viewpoint to that of their counterparts in the Netherlands. But given the wide spectrum of ideologies and theologies within the Anabaptist community here—ranging from the progressive Mennonite Church USA to the Old Order Amish—the discussion of climate change is also more complex in Iowa. Some farmers point to the historical evidence of it, while others deny that the climate has changed at all, explaining that extreme weather has always been a challenge for farmers. Each of these subsets of farmers, though, agrees that technologies have addressed the negative effects of adverse weather.

Some farmers recall times of bad weather without linking it to climate change. When Gerald Yoder, a member of East Union Mennonite Church, and his son Brent, of Wellman, for instance, reflect on the history of Washington County weather, it's not climate change but the tornado of 1984 that comes to mind. Gerald recalls that "it was really hard on this farm here . . . no it was bad! It was bad!" He remembers precisely how it came up from Missouri at 5:30 late one afternoon, hitting the Yoder farm at 9:45 that night before heading up to Wisconsin, where it killed fourteen people. He tells that of all their buildings, only the farm's two residences were left standing; several sheds and the grove of eighty evergreens were flattened, the farm equipment was badly dented, two dozen of their three hundred cattle were injured, and the pig barn full of fattened hogs was left roofless. Brent has his own memories of the evening of the tornado; they were racing home from a baseball game, "almost hydroplaning . . . there was so much water on the road, it was raining so hard." When they arrived home they all sat down around the kitchen table, relieved to be safe, and then his father came in from the barn "and all of a sudden . . . Dad said, 'go to the basement!' and we went. Nobody argued. . . . I remember . . . it felt like the house was so light it could go up at any time."[34] Fortunately the house stood firm and no one was hurt. It is this specific weather aberration that has stuck in the minds of Gerald and Brent, not the amorphous global phenomenon of climate change.

Yet some Iowa Mennonite farmers readily talk about climate change and applaud measures to address it. Bob Miller, the farmer and seed dealer, and a member of the urban First Mennonite Church in Iowa City, is satisfied that technology has addressed some environmental problems. He mentions clean diesel technologies as an example but also heralds no-till cropping practices. Though it has meant more herbicide use, this no-till practice has significantly reduced farmers' reliance on fossil fuels and promised a regenerated topsoil with less compaction and even added microbial life. And yet Bob worries about the sudden downpours the community has endured in recent years. He recalls the harmful seven-inch deluge his farm experienced in the late spring of 2013, just before the crops had established a good root system. Serious nutrient leaching resulted, or as he phrases it sardonically, "We put a lot of nitrates into the water, a higher level than what we've seen for a while." He believes that such rainfalls have been the direct result of human-induced climate change, as have the droughts between the very wet years. He is insistent: "Some of it's just nature [but] . . . I'm a firm believer that man does play a role in climate change. . . . We've had more extreme events in the last three to four years than we've ever had in my . . . memory." Bob sees genetics as a way to adapt to extreme weather. He speaks of his company's drought-resistant hybrid and GMO seeds that have produced acceptable yields on only three inches of summertime rain. And he defends this approach: "I look at it like God's given us a lot of tools to use in [plant] breeding, and one of the things that we do in our testing program is that we tend to stress the plants." It's all an effort to try to discover the best genetics to counter extreme weather.[35]

This approach has broad support in Washington County. Gilbert Gingerich, of Parnell, in Iowa County, also a member of a Mennonite Church USA congregation, says that as an older farmer he has seen significant climate change over a period of fifty years. He says that "someone [who's] been watching for 10 years wouldn't realize it so much." He also highlights the 1984 tornado, which tore up his farm's trees and damaged buildings and equipment, but to his mind that tornado signaled a larger pattern of difficult weather: in the "1980s we set lots of weather records, for coldest, hottest, wettest, driest, you know months and years. So . . . on Christmas Eve of 1983, we had a terrible, the worst, the coldest [day in] . . . a century," with temperatures reaching minus 30° and winds up to sixty miles an hour. But earlier that year, in June, a horrific heat burned up the corn crop, and then the very next June they experienced the historic tornado. Four years later, in 1988, they had the worst drought since 1934, so devastating that they had a corn crop of only twenty bushels an acre.[36]

Perhaps younger farmers haven't experienced the same changes, but Gilbert has seen both a general warming trend and more weather extremes. A longer growing season is a simple fact of history; the earth is warming up. He says he doesn't understand farmers who deny climate change: why is it that only "4–5 percent of farmers believe in global warming, but it's something like 95 percent of scientists believe there is scientific evidence" for it? Like the other farmers in the Global North who accept the fact of climate change, he thinks answers lie in new technologies, especially in renewable energies. And although his wife, Sandy, indicates a concern with GMO crops and their effect on human health, Gilbert applauds this particular science: "If you are able to control corn bores without using chemicals, just by splicing the genes, I think it is to everyone's advantage. Because it increases yield [and] reduces chemicals. The reality is that without GMO, we are unable to feed the world."[37] Then too, with GMO varieties farmers have been able to reduce their reliance on both chemicals and fossil fuels.

If farmers like Bob and Gilbert have their answers to climate change, some of their neighbors from more conservative Anabaptist branches minimize its very existence. Gilbert observes that some of his Conservative Mennonite and Amish neighbors don't accept the idea of global warming, insisting that it has all happened before, even in ancient history.[38] Marlin Miller, a member of a Beachy Amish congregation who farms just north of Kalona, is one of these farmers. It's not that Marlin doesn't care about the environment; in fact he is unequivocal: "We should care for the land as if it was God's. Leave it better than it was when we got hold of it. And I think farmers as a whole, that's their aim, whether you're Amish, Mennonite, or whatever, or even non-believers." Yet when it comes to questions of climate change Marlin is equally clear: "The climate change thing, I'm not sure that I just buy all of that. . . . A hundred years ago they also had changes." He thinks that all the talk about being green, not polluting, and creating a sustainable environment diminishes the role of God in nature. In the end he rests in uncertainty and places matters in "God's hands."[39]

This approach comes with other issues, both theological and scientific. A fellow Beachy Amish farmer, the minister Delmar Bontrager, who also farms just southwest of Kalona, shares Marlin's deep appreciation for the environment and sees it as God's "handiwork." For him, this belief necessarily entails a rejection of evolution: "To think that there isn't a God that made all of this. . . . And they say this all happened with a big bang? You know? I just don't believe that." He says he treasures the soil and advocates for soil stewardship because it was created by God. But this belief doesn't mean that he is worried about climate change: erratic weather is just bad weather. "I've said for years already, in the middle of a drought year, there's going to be at least [one] extreme gully-wash or something. I don't care when . . . it's going to happen sometime." And it usually does, and so farmers have creatively countered such weather extremes. Delmar applauds the use of new genetics—BT corn and GMO Roundup-ready soybeans—which not only have helped farmers fight off weeds but also have addressed the ill effects of drought. He asserts that "probably the biggest unbelievable change in the industry has been in seed corn. . . . [Recently] we had a dry summer, it was dry, so it only made like 130 [bushels an acre]." This yield is a lot less than the two hundred bushels per acre that farmers have come to expect in a typical year but much more than the sixteen per acre in drought years in the past. He jokes that scientists have yet to develop a corn plant that can survive being under water.[40] Technology has its limits, according to farmers like Delmar, but its accomplishments in fighting bad weather are impressive.

Complicating matters is a uniquely third perspective on the issue of climate change. Vernon Yoder, who also lives near Kalona, is a member of the most traditionalist of the various branches of Anabaptism in Washington County, the Old Order Amish. An organic farmer, he is skeptical both of modern science that claims to be able to measure human impact on climate change and of modern technology that promises to fight extreme weather. Indeed, he is skeptical of most things modern, including large machinery, same-sex marriage, climate manipulation, chemical farming, the information age, and secularity. He has his own view of things, and it is decidedly apocalyptic. He has no problem accepting climate change and even sees it predicted in the Bible. He recalls the 1993 flood in the area, an event climatologists called "a hundred-year flood," and since that time "we've had two of them. . . . Now how old am I?" Of course, it's a rhetorical question. He continues from another angle and recounts how, back in the day, tornados were a Texas and Oklahoma phenomenon, but now Iowa, in the Midwest, also experiences them. And then his interpretation of climate change becomes decidedly antiscientific: "I think it's a sign of the times. Jesus said in the latter times there's going to be storms and earthquakes and I think that's what that is." He, for one, doesn't need science to prove climate change. But nor does he applaud scientists who seek to manipulate the environment, to "make clouds" and induce rainfall, or indeed their obsession with collecting data in this information age. On all counts, says Vernon, "we are probably getting out of our realm," giving in to pride; and he adds that "mankind is not quite doing what the Lord wants them to do."[41]

The debate over historical climate change in southeastern Iowa is a complex one, revealing the disparate nature of the Mennonite community here. Like the Doopsgezinden of Friesland, the Mennonites and Amish of Iowa in the Global North believe that farmers' agricultural practices should not be dictated to them by urban experts. And here, as in Friesland, farmers reassure themselves that technology can address the effects of extreme weather.

Manitoba: Making Hay While the Sun Shines

Of the five groups of farmers compared in this chapter, the Manitobans in the Rural Municipality of Rhineland are the least concerned about climate change. They say that while climate change with ill effect may be true in the wider, global community, it just isn't so in western Canada. Some echo the ideas that aberrant weather has always been a part of Rhineland's history; others accept the reality of climate change but quote scientists who say western Canada is one of the few places on earth that has benefitted from global warming. The Manitoba farmers' more pressing concerns are how to deal with pooling water on heavy clay land in wet years, how to stop potential erosion of the soil's fine particles in dry years, and how to farm in Manitoba's perennial short growing season "while the sun shines."

The farmers of Rhineland, from the small organic producer to the very largest farmer, say climate change simply has not been a matter of much local concern. Joe Braun, the former grain farmer turned organic vegetable producer, says he has not experienced climate change "in my little world." It's not that he doesn't care about the environment: in fact he still regrets using chemicals on the grain farm that allowed him to increase it to seven hundred acres. Herbicides were the worst: "Of all the farming jobs, spraying I hated the most. The stench!" He also detested the pressure to make sure no neighbor ever saw an unsprayed strip of weeds, meaning that farmers tended to "overlap" when spraying, even if it stunted the growth of grain in those strips. But even though he is an environmentally conscious farmer, Joe claims that climate change really has not happened in Rhineland: "No . . . no, not here." He does admit that the weather has been erratic, and he shares his records indicating that his farm received only four inches of rain by mid-June of 2017, compared with sixteen inches by the same time in 2016. This variation worries him. He has also seen a recent infestation of the small black sap beetle in raspberry fields, and he's watched the Richardson's ground squirrel destroy his green pepper field. But he says it's all cyclical: he remembers the beetles being a problem in the 1990s, and as for the squirrels, "we've had a lot of people tell me that as kids they would go out on pastures and shoot these things, by the dozens." His rhetorical question is, "So, is this a cycle?"[42]

From quite another vantage point, that of one of Manitoba's larger conventional farmers, Melvin Penner, comes a surprisingly similar analysis. Melvin says that his H&M Farms is a rather typical operation on an old-fashioned foundation: "We have a mixed farm, I mean we raise pigs . . . we are a multiplier . . . so we have a mixed farm . . . we just added a 0 in the back. . . . We used to farm 3,000 acres, now we farm 30,000. Big hairy deal. We still do the same thing." But if his operation is in fact not a typical Rhineland farm, Melvin is similar to other Rhineland farmers in doubting that the climate has changed. He adds that if it were, he would welcome it, as "I really don't like minus 30 that much!" But he says weather is always changing: it was different in the 1950s than in the 1960s, and different again in the 1940s. "And then in the 2000s it's different than it was in 1980. I mean, in 1980 [everyone was saying that] we were going to have a drought forever. . . . Now, last year, we were pull-

Farmer Kevin Nickel harvesting weed-free wheat at his farm north of Altona, Manitoba, with his late-model John Deere combine, August 2016. Family friends Amy Loewen and Jude Loewen ride along in the air-conditioned, dust-free cab. Photo by Susie Fisher, Gretna, Manitoba.

ing combines [through mud], and all hell was breaking loose; today, the whole season has been beautiful." His sarcasm is as real as Joe's question is rhetorical: "The climate changed, from last year to this year!; the climate changed!"[43]

Old-timers equipped with stories of yesteryear can be even more dubious about climate change. Grant and Leona Nickel began farming as a young couple in 1964. Now, looking back, Grant is unequivocal on the question: "You know, people say that" climate change has happened, but "I can't say that it has. Oh, as soon as there's water on the field [everyone cries], 'climate change, climate change!'" He says that of course he can't remember back a century or more ago but that in his memory there has always been severe weather; the climate really hasn't changed. When he began farming in the early 1960s, the heavy clay land around the town of Rosenfeld was constantly under water and required annual interventions to improve drainage. The most rainfall he has ever seen was a seven-inch deluge on June 7, 1970, just after the peas were seeded; the sudden water washed the seeds out of the seedbed, and he had to replant everything. That year he didn't finish seeding till July, "but that's the way it is. That's what farming's about." He also sees weather vagaries in broad strokes: "1960 it was sort of dry, 1961 was very dry and then '62 came along and [it] was very wet in spring." It is all a matter of weather cycles. He believes people who talk of climate change in places around the world, but he doesn't believe it has happened in Manitoba. His wife, Leona, agrees: "On this farm we haven't really noticed it. You're seeding, like some years you got on early, some years you got on late, some years there were hail storms, some years we heard there were tornados." She offers that many farmers now are able to grow heat-dependant corn in Manitoba; she says that while it is a process that some might attribute to a warmer and longer summer, in her view it is as much a result of improved genetics as it is of changing weather. Grant offers that in "the big picture," the weather patterns that he and Leona have seen might not have occurred in previous centuries, in the year 1740, for example, but then again "who knows?"[44]

This is also the perspective of the very eldest of the Rhineland farmers interviewed. Peter Hildebrandt, married in the late 1940s, recalls the severest of snowstorms of the past with precise detail, but they are from 1941 and 1966, not from recent decades. About the one from March 1941 he says, "That was a very bad one. It hit so suddenly it was like an explosion. And various people in the community died." He recalls that the day of the storm dawned beautifully, with temperatures above zero and snow thawing on the roof. His father

knew from experience that "this type of day that we're having is very dangerous," and he secured all the barn doors before supper. And sure enough,
quite suddenly, that night the wind picked up and their large, stately house
began shaking: "Such a beautiful day, you know, and then all of a sudden, this!"
Peter also recalls the storm of March 1966. Returning from Winnipeg on a
Wednesday, he and his travel companions encountered the storm near the
town of Morris, halfway home: "Nobody said a word, we just kept on driving . . .
the snow was piling up. . . . We never made it home that night." They stayed at
the hotel in Gretna, just a few miles from home, for two days before they were
able to dig themselves out. He remembers these incidents with clarity, but regarding climate change he demurs and says he's never given it much thought.
For Peter, climate change is a nebulous phenomenon, difficult to grasp; legendary snowstorms, however, are real and empirically experienced.[45]

The problematic climatic factors that matter to these farmers are those peculiar to the Red River valley's heavy black clay soil. For Edna and Leonard
Wahl, who began to farm in the 1970s, it's always been a question of employing the best technologies in order to farm this land. They have rejected the old
idea that soil needs to be tilled and exposed to the sun, a process that led to
soil erosion; as Leonard puts it, "It was blowing dirt," and the basic answer to
this problem was that "we did a lot less tillage" and used more chemicals.[46]

Other, somewhat younger farmers see Manitoba's hundred-day or so frost-
free zone as the central challenge to farming there. But they think that agricultural efficiency, with a reduction in both chemical and fossil-fuel usage, is
the answer. The farmer Kevin Nickel says that Manitoba farmers have always
had to seize days of good weather, "these windows of opportunity," and they
have done so with ever-new technologies. He traces the time from the 1950s,
when the soils were precultivated and then seeded with a press drill, a process that led to unwanted soil compaction. Then he segues to the use in the
1960s and 1970s of the disker seeder, a machine that "revolutionized seeding
in heavy soil" as a single-pass implement, and then to the much more efficient,
much wider, higher-capacity "air seeder." But he emphasizes most the new
"air drill" of the late 1990s, which guaranteed "minimal soil disturbance" and
soil compaction with the help of glyphosate, marketed as Roundup. This controversial chemical, says Kevin, has both "greatly simplified and reduced our
chemical use" and allowed for less tillage, thus reducing fossil-fuel usage and
guarding against soil erosion.[47] For Manitobans, the main challenge is not climate change but the age-old problems linked to a humid continental climate
and heavy clay soils.

When Rhineland farmers do talk of climate change, they tend to emphasize its global rather than its local nature. True, Benno Loewen says he has seen Manitoba warming up, an increase in the number of "our frost free days," and later dates for the annual first "killing frost," now as late as November. And yet, Benno sees greater evidence of climate change globally than in western Canada. He recounts the disasters and intense flooding across the United States and indeed around the world. As a member of the Rhineland Municipal Council, he has attended conferences aimed at planning for the local consequences of climate change, and heard that the temperatures by 2040 will be at a dangerous 1.5 degrees higher than in the early 1900s. But he also remembers one speaker saying that Manitoba and Saskatchewan would be "the best area for climate change" globally because by 2040 their temperatures would resemble those of Nebraska and Kansas. Such a change, Benno agrees, would fundamentally reshape agriculture in Manitoba, indeed making it easier to farm.[48]

Manitoba farmers have not been worried about climate change in general. The older farmers recall severe weather in the past, and the younger operators see Manitoba's main weather features as a constant: a relatively short growing season with more than adequate precipitation. They welcome a longer growing season, and equipped with new technologies and chemicals, they have been able to grow soybeans and corn, crops traditionally not planted in western Canada. Many farmers in Manitoba welcome climate change, especially if it is accompanied by global warming.

Conclusion

The farmers from the five different agrarian communities discussed above have experienced climate change in dramatically different ways. But their farms—located variously in Matabeleland, Bolivia, Friesland, Iowa, and Manitoba—can also be grouped in a dichotomy pitting the Global South against the Global North. True, most farmers speak of the fragility of the natural world, which they see as a treasured creation in need of protection and nurture. Many also speak of increasingly erratic weather patterns, mostly in the form of drought and heat waves but also as deluges of rainfall and even as tornadoes and cyclones. And many suggest that farming is more difficult now than in the past and that significant adaptation has been required.

Despite some similarity in ideas on changing weather and climate, the differences between the North and the South are pronounced. When the farmers in the Global North speak of climate change, they more easily complain about the history of official overreach in combatting environmental problems or about a skittish, urban public, ill-informed on rural issues. At the same time, farmers in the North more often think of climate change as too complex and vague a problem to be understood and as problematic in other places on earth but not in their own. The farmers in the Global South, on the other hand, speak of changing weather not in terms of broad global issues but with reference to their very ability to survive in the everyday. They also more easily blame their own actions or the actions of their neighbors for changes in the environment. And although easily invoking God as the creator, they leave the divine out of their descriptions of climate change. Rather, they describe it in rudimentary physical terms, based especially on their own experience.

Commensurate with these disparate approaches are the ways climate change has been translated into an economic equation and a political problem. For the farmers in the North, climate vagaries have been addressed through new technologies. For farmers in the Global South, without plentiful financial resources, adaptation to new climates has been limited to crop selection and other rudimentary actions, such as tree and meadow planting. But climate change has mattered most to economically vulnerable farmers; left behind in the march of modernity, some have lashed out against an imagined sense of what might have been and offered a vision of what must be. The farmers of the Global South have a more intense personal experience with climate change, but they also more easily blame themselves. As Lydia Barnett writes in "The Theology of Climate Change," it is much easier to acknowledge environmental "sin" within the Anthropocene if it is apparent that it is you who has committed it than to be told by an outsider that you are wrong.[49] Whether climate change has been seen in economic or political terms or has been experienced firsthand, these five farm communities have each responded to weather and climate in their own way.

CHAPTER 8

Mennonite Farmers in "World Scale" History

Encountering the Wider Earth

In 1986 Johann Fehr, a horse-and-buggy Old Colony Mennonite from Riva Palacio, Bolivia, made his first of many trips to the United States. He was accompanied by his father-in-law, whom Johann admired: "Oh he was a very intelligent person. He spoke English well. He was born in Canada. . . . [And] he was very well informed in this business. And he did a lot of business."[1] Their mission was to search for secondhand farm equipment made redundant by increasingly large farms in the United States. Traveling by air and by bus, they stopped first in Seminole, Texas, where they connected with Old Colony Mennonite cousins who had migrated there from Mexico in 1976 hoping to farm the state's semiarid western plain. Johann and his father-in-law then headed north, by hired car or bus, into the Midwest, visiting equipment dealerships. They had a special interest in small diesel tractors and pull-type implements, including small combine harvesters, silage cutters, and soybean planters. They wanted inexpensive machinery suited to their modest, household-based farms in the tropics, where agricultural practices were ordered by particular religious teachings on simplicity.

Johann describes traveling north as if it were as straightforward as tilling a field of soybeans. After they purchased the pieces of farm equipment at various dealerships in the Midwest, "we brought them together" and shipped them to Houston by transport truck. There the men dismantled the machines, "put them in wooden boxes and from there . . . sent them by ship." On that first trip in 1986 they made their shipments via the Atlantic Ocean, down to Buenos Aires and then across the continent by rail to Bolivia. In later trips they transported the equipment via the Panama Canal, to the Pacific port of Arica, Chile, and then across the

Andes into Bolivia—to La Paz and Cochabamba by rail, then down to lowland Santa Cruz and Riva Palacio itself by truck. As Johann explains, the latter route "cost a little more, but it was closer and much faster, because the other way, it would take up to five months." After his father-in-law died, Johann undertook numerous similar trips with his brother-in-law, Wilhelm Buhler. Recently Johann has traveled less: "The trip is very long and it takes at least a month and I have a big family, I have nine children."[2]

The irony of this international linkage, of the most intentionally antimodern of the seven communities trading with the richest country on earth, points to an indisputable reality of modern agriculture. Farming in each of the seven communities has been shaped in some way by its interaction with the wider world. The local Mennonite farm may have been built on kinship-based social networks and inherited cultural codes, but it was also part of a global community. Indeed, global ties and worldwide phenomena often shaped the very way the farmers from the "seven points on earth" worked their land.

Farmers have encountered the global in numerous ways, of course. First, as in Johann Fehr's case, the farmers often had direct connections to other countries, mostly related to trade of some sort. Here, shaped by a "world systems" economy and a "center-periphery" continuum, poor places on the margins found the cast-off equipment of more wealthy regions, which allowed them to participate in a commodity export market with significant effects on local environments. Second, these farmers were also connected to the global through local encounters with the effects of global climate change, as seen in chapter 7. In I. G. Simmons's words, here were the marks of intense human intervention in nature, the result of a massive "ecological footprint . . . of humanity . . . transformed from the patter of tiny feet to a crushing boot of Pythonesque proportions."[3] The everyday activity of these farmers in their fields—whether from the use of synthetic fertilizers, insecticides, or fossil fuels—simply had a cumulative global effect that upended established climatic cycles and jeopardized the health of the soil. And in yet a third set of circumstances, farmers were party to a common set of developments that happened simultaneously in many places, or in Donald Worster's words, "bedrock historical issues shared by virtually every nation today and communities in every part of the planet," including the "professionalization of nature" and global markets that broke the "nexus between local people and local resources."[4] Everywhere, it seemed, old lore-based knowledges were being replaced by scientific study propelled by parties with vested interests in their propagation. And, fourth, farmers have also been affected by what scholars have variously dubbed "Transnationalism" or the "Global Turn." Here, as a voluminous scholarship argues, global communications—faster international transportation systems, digital information networks, and "time-space" compression—fundamentally altered ways of being.[5] Such a transformation has reshaped internationalized agricultural knowledge, with profound effects on the way weeds, fertility, and soil preservation have been managed.

Certainly the seven Mennonite places studied here were linked globally in each of these four ways and more. Those ways also included the participation in international food aid and rural-development programs, in the global diaspora of farm families, and in the heightened sense of a global environmental citizenship on a fragile earth. Then too, as Johann Fehr's story points out, not only have these global relationships taken many forms but they can be studied in many ways. At the forefront are empirical studies, with either a micro- or a macroanalysis and infused with quantitative data.[6] But these relationships can also be studied more personally, by asking farmers how they experienced the global turn in the everyday, phenomenologically.[7] How have they imagined a sustainable earth in global terms? How did they as farmers learn to exploit international connections to the benefit of their farms? How have they seen events in the wider world shaping the way they tilled their soil locally? How did those same events circumscribe their relationship to the land?

To obtain this subjective sense, this chapter is again based solely on oral history, a method that also allows for the consideration of the "softer concepts . . . of being human," including emotion—fear, loyalty, anxiety, excitement—with which the farmers tell stories of uncertainty in the wider world.[8] And as in all oral histories, this chapter moves easily between the past and the present, as farmers invariably conflate the past in some way with the present.[9] The chapter also claims less to represent the majority of farmers in any given locale than to report on a range of stories from each place. Of course oral history once again marks a common source, allowing for a comparative analysis of seven places with significantly differing historical texts.

This chapter begins by comparing the global engagement of the most ideologically exclusive of the seven farm communities, Riva Palacio in Bolivia, with one of the most acculturated of the communities, Washington County, Iowa. It considers the irony of the Bolivian Old Colonist engagement with the most powerful modern economy in the world, finding in it the tools to keep their small farms viable and their rural communities isolated from that very world. It asks how the Iowa farmers, holding dear the American nationalist refrain of "feeding the world," have used advanced technologies and produced university-trained agronomists to meet that goal. It reflects on the paradox that Bolivian Mennonites have visited Iowa to seek the tools to ensure agrarian simplicity, while Iowa Mennonites have sojourned in Bolivia to help its agriculture modernize.

The chapter then proceeds to the farmers from the remaining five points, from the most globally integrated, Friesland and Manitoba, to the least connected, Matabeleland, Java, and Siberia. Again ironies appear. Friesland, for example, was not only a rural society but one that was linked to a former imperial power, which meant that it was also linked to a global network. The Netherlands had exported commodities for centuries, sent out foreign missionaries, participated in a massive agricultural diaspora, fought to keep a former colony, and played a central role in the Common Market. But if such ties emboldened Friesland, they left places like Java disadvantaged, marred with a colonial past that undermined efforts for the former colony to enter the global community on equal footing. Similarly, the British Empire gave Manitoba farmers an early network through which to export their wheat and then, from a position of economic advantage, they could envisage the world as something transformable through church-based aid bodies. Matabeleland, also once a member of the British Empire, found that this association failed its aspirations for a just and stable society.

But what the seven places had in common was that the global was filtered by local culture—mythology, religious aims, social codes of cohesion. And in each place this dialectic shaped the farmers' relationship to the land, indeed the very way they imagined nature. Given the complexity of the twentieth century and the varied climates of the seven places, though, no basic pattern emerges. What becomes clear is that "world scale" events affected the way farmers imagined their place on the land and their right to exploit natural resources.

Bolivia: Importing Farm Implements to Secure Tradition

The story of the Bolivian engagement with the globe was in fact more complex than importing farm implements from distant shores. As members of the most recently founded settler society studied in this book, Riva Palacio farmers possessed a vivid collective memory as privileged white settler migrants. They recall negotiating international borders, adapting to the host country's strange environment, and playing to a new homeland's national food-security policies. Indeed, they were keenly aware that they were farmer immigrants in a particular geopolitical context. Leaving drought-ridden northern Mexico for the green Bolivian lowlands in 1967 and 1968, they answered a global call to help Bolivia attain self-sufficiency in food. They knew that they were global citizens without full cultural or even legal citizenship in any one country, but they were also aware that their productive labor on farmland served as their virtual passport. Johann Fehr recalls once being told by a native Bolivian that "you're not a *Boliviano* . . . you're a foreigner: that Bolivia didn't matter to me."[10] On the one hand, the accusation was correct: Johann had not even sought to become a citizen of Bolivia, citing cost and time. On the other hand, he did have a permanent visa, and more importantly, he felt that he had earned a right to live in Bolivia by farming its land for forty-six years. Indeed, the desire for land was crucial in this migration southward. True, most Old Colonists point to church disagreement in Mexico over rubber-tired tractors, trucks, and electricity in drawing them to Bolivia. But the migration narrative of many farmers, including the retired *Vorsteher* Abram Reimer, is all about the lack of land for the next generation. In addition, Abram says, Mexican *agrarios* were threatening to take land from the Mennonites, who, as one recent study indicates, considered them "foreigners despite being citizens."[11] By contrast, in Bolivia, asserts Abram, "everything was open and free."[12] Farming the bushland of its eastern lowland may have been physically difficult, but politically it was easy: Bolivia wanted the Mennonites.

Farmers who speak of the 1967 and 1968 migration from Mexico to Bolivia in religious terms also link it to the Mennonites' global search for farmland. It was simply necessary in their quest for agrarian simplicity. Many Old Colonists speak about it in ambiguous terms: Elisabeth Thiessen, for example, quips that "it did not work out for us in Mexico."[13] But Isaak Peters is more explicit: migration ensured that "we could stay within the religion," resisting ideas that "come in from outside," ideas at odds with the teachings "my father left . . . to me, and I to my sons, and [ideas] you have to remain with." Without theological pronouncement, he lists the rules of the church: only natural light, only steel-wheeled tractors, only horses for transportation. Isaak says that if you want it any other way, you simply have to leave the colony for the wider world and "take another religion."[14]

These farmers saw exotic Bolivia as a place with the land base to allow for the replication of their religiously informed agrarian simplicity. The bookstore owner Abram Redekop provides more detail: "There in Mexico, [some] people started to put rubber tires on their tractors, and because it wasn't permitted by the religion to have rubber tires, [our leaders] thought about coming to Bolivia. . . . So that's why [we] came."[15] He doesn't explain that the real problem with rubber-tired tractors is that their road gears reduce the travel time to town and the tractors' greater efficiency encourages ever-larger farms, thus undermining the communitarian nature of the community. A steel-wheeled tractor has kept the city at bay, the farm small, and the farmer humble. Of all the places the Old Colonists considered migrating to, the Bolivian Oriente had enough land to secure these religiously determined goals.

Migrating from one country to another also made the Old Colony Mennonites aware of the requirements of farming in disparate climates. Johann Fehr has vivid memories of farming in Mexico. He compares its fresh mountain air with Bolivia's lowland heat and says, "You never get used to sweating like you do here." But agriculture in the Sierra Madre was difficult: perhaps irrigation and synthetic fertilizers have made farming in Mexico profitable in recent years, but in the 1960s mountain-valley aridity simply stifled good

Farmer Wilhelm Buhler, of Riva Palacio, Bolivia, overseeing the peanut harvest, 2014. The US-sourced John Deere tractor is equipped with steel wheels according to Old Colony Mennonite teachings. Photo by Ben Nobbs-Thiessen, Winnipeg.

crops. He hastens to describe the necessary adaptations in the new homeland, itself a transnational story. Indeed, to that end, says Johann, the migrants to Bolivia looked about them in South America and found their answer among Canadian-descendant Mennonite settlers in the Paraguayan Chaco who had farmed there since 1927. The Paraguayans "suffered a lot in the first years," he says. And yet, even if their initial agriculture was primitive—traveling by oxen and high-wheeled carts—they were challenged agriculturally and forced to "work together in everything, with grain, cattle, milk," leading to their remarkable successful cooperatives.[16] In Paraguay too the Bolivian Mennonites found horses and dairy cows more suitable to the tropics. Abram Thiessen talks of visits to Paraguay, as well as to Argentina, in the late 1960s after discovering that Bolivia's creole cattle produced "very little milk." But even these forays were inadequate, for they found what he refers to as the Paraguayan Hollandes cows too "delicate" for their needs. Thus, they developed mestizo cattle, some of which produced an abundant thirty-five liters of milk per day.[17] These international ties led them to fine-tune the agriculture they employed in Bolivia.

The migrants in Bolivia also learned to grow a range of new crops, most notably soybeans, and again they did so with an international connection. Abram Klassen recalls farmers being coaxed into soybean production by the soy-exporting company Aceite Rico, owned by a Mr. Marinkovic, a Croatian national from Cochabamba, in the Andes. Remarkably, Marinkovic had the power to reshape Riva Palacio agriculture and did so twice; first, he signed farmers up for soybeans, and then, says Abram, he cut off their credit. This action forced many farmers into dire financial straits, which, coupled with the drought, took them back to the old practice of cheese making, for which they had become famous in Mexico. It was a mixed blessing: Marinkovic's charm had introduced the Mennonites to a new commodity, and his parsimony forced them to remember the value of a commodity from the old homeland.[18] Johann Boldt, 80, a cattle fattener and former *Waisenmann* (head of the colony estates committee), has his own version of the transition to soybeans and it too involves an international connection; it centers on an Argentine exporter linked to Santa Cruz's Fabrica de Aceite, which in about 1983 was looking for a farming community that could produce five thousand tons of soybeans annually. Johann says that Riva Palacio responded enthusiastically, and it was the Argentine connection that was important.[19] Mennonites who had migrated to Bolivia to escape the wider world found the global marketplace alive and well in the new land. The soy market cajoled them to cultivate the forest soils; a rediscovery of milk turned that very land into more sustainable meadows.

This was the context in which farmers traveled north to the United States to find the equipment necessary both to adapt to Bolivia's new climate and to grow products for global markets. Some farmers emphasize Bolivia's tropical climate as the catalyst for these international farm-equipment forays. Enrique Wiebe recalls that his first tractor in Bolivia was a gasoline-powered John Deere from Brazil that "didn't really work . . . in the heat" in Bolivia, and so farmers turned to cooler-running, diesel-powered tractors that could easily be obtained from the United States.[20] Other farmers cite Bolivia's fruitful climate as the reason behind the search for farm equipment in foreign lands. Cornelius Peters, an Old Colonist preacher, speaks of the difficulty of cutting heavy crops of sorghum by hand, a task that was just too hard for a single family. And hiring Bolivian help for this work was difficult too, as cultural differences undermined good working relationships. Thus Cornelius procured a mechanical harvester from the United States, and he, the preacher, created a fad. He recalls that the machines were but "little ones, like the John Deere 30, small machines, but when one person had one . . . they would call him a *ricacho* [rich man]; it was amazing!"[21] Soon everyone wanted to be *ricacho*, albeit humbly, with a secondhand US castoff.

Most recently, for products and machines Bolivian Old Colonists have moved far beyond Paraguay, or the United States for that matter, to China. Abram Thiessen, who has traveled to the United States many times, from Texas to the Dakotas with stops in Kansas and Oklahoma, predicts that China will continue to gain importance. It has produced inexpensive herbicides and increased Riva Palacio's dependence on pesticides, herbicides, and fungicides, with some farmers "spraying 6 or 7 times" a season, says Abram. Several farmers have traveled to China to buy chemicals and upon their return sold them to other farmers in the colony "for what they wanted. . . . Yes, from China."[22] Johann Fehr, of course, has already checked out China for his own needs and purchased a Chinese-made peanut-shelling machine. The Old Colony church does not allow computers, but Johann explains that "a Bolivian who is my friend, he found it on the internet, as you know everything is on the internet these days." And while places like neighboring Argentina produce excellent machinery, Johann describes the Chinese machines as "better, smaller and cheaper."[23]

China may be the most recent country the Old Colony Mennonites have visited, but the old homeland of their grandparents, Canada, remains important, even if it is a country that solicits ambivalent feelings. As children born to Canadian parents abroad, many of the Old Colonists have their stories of imperialist Canada forcing their children to study in English, the language of patriotism and war. But they still secretly value the Canadian passport, which allows them to access Canada's labor market. Nettie Blatz's is a common story: she has thirteen siblings, most living in Bolivia, but as she says casually, "one sister stayed there [in Mexico] and one brother is in Canada."[24] Many other Riva Palacio farmers have sojourned in Canada even though their Old Colony church prohibits such a move. Abram Hamm, of Campo 34, made this move in order to earn money to pay for farmland that was being repossessed in Bolivia. As a cattle dealer and auctioneer, as well as a farmer, he knew many Bolivians from the outside world, but he found in Canada his financial lifeline. As he puts it cryptically, "I couldn't continue to pay for the land as I had planned to. I had to return the land and then I went to Canada."[25]

Moving to Canada in fact was complex. It entailed an expensive flight from Santa Cruz to Toronto via Lima, from there a drive to the commercial gardens and greenhouses of southern Ontario, and then negotiating a new life in a strange language. And yet Ontario, especially, was an ideal place for large Old Colony families who were willing to work hard at piecework rates. There, many Bolivian Mennonites joined tens of thousands of other Low German–speaking Mennonites from Latin America, all hoping to succeed through hard work in the cucumber and tobacco fields of their former homeland.[26]

Marie Hamm, mother to several teenaged children, says that making the Canadian connection was not especially difficult for her. In fact, it was "easier for men in Canada than on a colony" in Bolivia; "all they have to do in Canada is please their bosses; here [in Bolivia] they have to worry about the harvest [and] it's also easier for women in Canada; they do the same work in both places, but [with modern appliances] it's easier in Canada."[27] Other sojourners to Canada emphasize how seasonal work in the Global North provided enough earnings to purchase land in Bolivia. Young, landless Gerhard Martens, for instance, who works in Peter Wiebe's peanut-processing plant, knows the formula for escaping his plight from watching his impoverished brother-in-law, so poor that he "barely had his buggy." He went north to the United States to find work, "the first time for about two years . . . came back and bought his land, put in his well, his tractor, and made a good house. Then he went again . . . and bought two more tractors, and a baler [to do custom work], and bought a little more land. And that's how he got started." Gerhard knows he could achieve similar results in Canada, where his father has many relatives. Still, he is dubious about Canada's quick-paced life: in Bolivia, by contrast, if you have a few dairy cows and a pasture "you can relax and do well."[28]

Other farmers are repulsed by stories of cold and frigid Canada. Peter Wall, who left Canada for the South in 1952, recalls life in Alberta's far northern Fort Vermillion region, a "cold country . . . cold, really cold." And about his job of bucking trees in deep snow he says, "Oh . . . that was rough for a person like me." He contrasts Canada to Bolivia, where he has assembled a fifty-four-hectare farm, raised kaffir, corn, and soybeans, built up a herd of about fifty cattle, and of course, come to own a US-made combine harvester.[29] Canada has provided a lifeline to some Bolivians, but its fast pace of life and intemperate climate have reaffirmed for the majority the miracle of exotic Bolivia.

Of all the places in the world, Bolivia has given thousands of Old Colony Mennonites a chance to become settler farmers. They have imagined themselves in a wider world defined by their peoples' historical migrations from Canada to Mexico to Bolivia, each with a distinctive climate, with forays into Paraguay, Argentina, Brazil, the United States, and even China. In these disparate places they have found the tools to farm Bolivia's tropical land. Ironically, they have secured a separate farm society by negotiating the wider world.

Iowa: Exports of Corn and Agricultural Knowledge

A significantly different connection to a global community is expressed by the acculturated Mennonites of southeastern Iowa. When farmers here reflect on their own histories as agriculturalists in a global context, it is with lenses afforded citizens of the mighty United States. Indeed, it is as US farmers that Iowa Mennonites see their vocation "to feed the world." And they do so in various ways: as export-oriented soybean and corn producers; as donors to international aid programs; and as agronomists and rural-development workers for NGOs in places like Bolivia.

As farmers, says Ed Hershberger, Iowa Mennonites have been globally linked. "Whether it was grain, or whether it was meat, milk and eggs . . . our products are going globally. So we're feeding people a long ways away." Upon reflection, he reasserts the importance of grain and meat, specifically corn and hogs, the commodities that made post–Second World War Iowa the center of the American breadbasket. Ed applauds government policy that has encouraged the export market. In fact, he is rather critical of any action by

Washington that has interrupted global trade and cites as an example the time "when Jimmy Carter was president, as good a man as he . . . was, he . . . put an embargo on grain [destined for Russia], and . . . that really hurt, because we lost our export markets; everything just hit the tank." To Ed's mind, more important than international trade policy in feeding the world is simply the quality and quantity of the nation's soil. He is unequivocal: for "this country, one of the primary and most valuable resources is productive land, productive farmland." He contrasts the United States to China and Japan, rich with labor but limited in terms of land.[30]

Other farmers point to US scientific knowledge as an indicator of the country's global reach. Gilbert Gingerich, a member of a Mennonite Church USA congregation in Iowa County, married to Sandy, a veteran schoolteacher, is pleased that his grandson is farming, having chosen a nationally recognized vocation focused on "feeding the world." He cites rough figures indicating that while a third of US corn is used for animal feed and a third is turned into ethanol, the final third is exported, as are fully half of all soybeans, and mainly to China. But what especially impresses him is the staying power of US exports. Competing soybean-producing countries like Brazil, says Gilbert, have tried to challenge this ratio, but the United States has prevailed because of its superior technical know-how. Indeed, he is confident that because of American-developed GMO technologies, other countries will always depend on US agriculture. When Sandy Gingerich, who is not involved in the farm's operational side, says she hopes "to see our farm at some point raise non-GMO corn and soybeans" and reminds Gilbert that "European countries don't buy our" GMO commodities, he has a quick rebuttal: "They don't have to at this point . . . but they're coming over." He explains that genetic modification promises larger crops using fewer chemicals, and he repeats that "without GMO, we are unable to feed the world." The only thing technology can't secure is fertile farmland, and here too the United States has the advantage. Gilbert says that history demonstrates that where humans have attempted to make more farmland, they have failed: in Africa water has been the issue, in South America clearcutting has exacerbated climate change.[31] Ultimately the United States has the advantage of both useful innovation and the world's richest soils.

Washington County Mennonites recount not only how as American farmers they have expanded agriculture to meet global needs but also how as Mennonites committed to service they have sought to transform global agriculture. Bob Miller, of Kalona, supports a variety of international charities, including evangelical organizations such as Wycliffe Bible Translators, but he especially endorses agricultural-outreach programs in foreign countries, in part because he has participated in them. He applauds Mennonite Central Committee (MCC) and Mennonite Economic Development Associates (MEDA), "our two Mennonite organizations that I feel have a tremendous track record." Bob's interest in global service dates from the time he participated in a Study-Service Term at Indiana's Goshen College. During this memorable time, he worked on an agricultural-research farm in Belize, and did so with a concern for local peasants. He thought, "They can't afford all of the inputs: fertilizers and chemicals. But if we can make seed more stable, they'll get higher yields." Interventions of this kind became his passion. Later, after he married Pam, a nurse, he took this vision to Bangladesh, where he worked for MCC as an agronomist in soybeans. Again he taught farmers to work without expensive inputs, emphasizing seed selection and soil management, not maximum profits. Bob also has indirect experience as a visitor to a MEDA site in eastern Bolivia, where his brother ran a similar program with Indigenous highland farmers who had migrated down to the lowlands: "These [highland] guys didn't know anything about farming down there [in the lowlands] and they found that if they incorporated beans into a rotation, the beans added nitrogen for their corn. . . . So they . . . form[ed] co-ops to get the right bean seed and fertilizer, etc., and also a marketing mechanism to market the grain."[32] Exported know-how from the United States was transformative in Bolivia.

A number of Washington County farmers who have served with MCC in Bolivia agree. For Bruce Hostetler, an organic farmer from near Wellman, motivation to serve overseas came from several channels. He had always been impressed that his father had served with MCC's Pax program, rebuilding war-torn Europe in the late 1940s. Then he was stimulated by his first introduction to global agriculture as a student at Eastern Mennonite University in Virginia. And finally, when he met his wife, Edie, a nurse, he found a soulmate who shared his passion for overseas service. MCC headquarters in Akron, Pennsylvania, then directed the young couple to Bolivia as rural-development workers, and there Bruce was able to make use of insights gained from his training in the United States. Like other MCC workers, Bruce worked with Andean migrants to the lowlands; he directed them in grafting trees, introducing cacao, improving seeds, and cattle breeding. He loved relating to the five villages that he was assigned to supervise, but he derived special satisfaction from living in one particular village, as "I had the most influence there."[33]

This confidence also showed itself in the way Bruce and Edie related to Old Colony Mennonites they happened upon in Bolivia. They found the Old

Colonists primitive and say that even the national Bolivians could tell that though "long ago we came from the same" stock, "we didn't behave like a Colonist." Edie adds that the Old Colonists were "something to see with their overalls and . . . the women had bright scarves and dresses." It was as if the Iowa and Bolivia Mennonites had nothing in common. Bruce, in particular, was rather critical of their approach to farming. At Santa Rosa Colony, founded in 1969, undue cultivation and seeking two crops per year on sandy soil "wore out" the soil; at least this is what locals told him.[34] Clearly, both the Indigenous Bolivian highlanders and the horse-and-buggy Old Colonists had things to learn from Northern know-how.

The satisfaction of transforming agriculture in the Global South was not the only motivation of MCC or MEDA workers. The young farmer Mark Gingerich, from just outside Iowa City, indicates that the scarcity of farmland has also led young Iowans with an agricultural aptitude to go abroad to serve. Often the Iowa farms were too small to accommodate more than one or two of the children who wished to farm. Overseas service is a pathway Mark has considered. His base of operations is a relatively small, 375-acre parcel still owned by his parents, who have yet to decide whether to keep the farm in the family. His parents have encouraged him to travel abroad, especially "within the Mennonite world." They themselves served with MCC in the 1970s, in part because their own parental farm was too small to accommodate them. Upon returning home from studying at Goshen College in Indiana, Mark's dad found that "the farm was occupied" by a younger brother and that the farm didn't have enough work for two brothers. Thus the idea of serving with MCC in Bolivia arose. Mark's dad committed for one term of three years; and then, when he married, he went for another two terms.[35]

For Mark this problem of the small family farm's viability is a not a uniquely US Mennonite problem. He imagines that Mennonite farms in Western Europe, Ukraine, or Canada are similar to farms in the United States, all seeking to succeed with "tractors, on a larger scale." But he has a hard time imagining agriculture among Mennonite farmers in the Global South: "I have to think about what it would look like for African Mennonite farmers to be farming. Nepali farmers in their fields . . . I wouldn't be able to identify [with them]."[36] And yet, identifying with struggling farmers in the Global South has become the vocation of many a redundant Mennonite farm daughter or son stateside.

This vocational pathway, however, has attracted but a minority of Washington County Mennonite farmers. Members of the more conservative but car-driving Beachy Amish, for example, have worked hard in aiding a suffering world, but it has more often been a world closer to home. David Miller, for one, readily shares stories of the time when he served as the local Mennonite Disaster Service (MDS) coordinator, addressing environmental damage within the United States. And when he thinks internationally, he speaks of his daughters leaving home for places like Ireland and Kenya, where the Beachy Amish have mission outposts, though without an agricultural component.[37] The pastor Delmar Bontrager recalls that Beachy Amish foreign missions once did involve agriculture: a dairy farm at a mission in Romania, perhaps one too in Ireland, and farms in Costa Rica and Belize. But these posts were meant to support missionaries and not expand modern agriculture among local populations.[38]

The idea of offering scientific expertise to reform the agriculture of the Global South is one propounded by farmers from the more liberal Mennonite Church USA congregations. With their university and college education, their vision is for a reformed agriculture in the Global South. They have traveled abroad, learned the appropriate languages, and experimented with new forms of agriculture. But alongside those farmers who remained in Iowa to actually farm, they have developed an identity as agriculturalists that "feed the world." The homebodies have traveled less, but as American farmers they have all imagined how the rich soils of the Midwest might serve the wider world. Iowa farmers have found more than one way to "feed the world."

Friesland: Rural Refugees and Assertions of Farm Technology

The dichotomy between US farmers exporting agricultural know-how to Bolivia to save the world and Bolivian farmers importing US technology to save theirs from that world is stark. Indeed, it points to a basic divide in the Mennonite world between the assimilated and the traditionalist farmers. This story becomes more complex when the five other communities—Friesland, Manitoba, Java, Matabeleland, and Siberia—are added to the mix. A close reading of this relationship reflects each place's particular corporate memory, its relative degree of isolation, and its level of integration in the wider world. The two communities most like that of Iowa are, of course, the two others in the "West," Friesland and Manitoba. Both were once the breadbaskets of imperial powers, integrated into highly developed capitalist markets, and tied via trade agreements to powerful neighboring nation-states. Where they have differed from each other, it has been in degree and not substance.

Friesland Mennonites may seem especially engaged with the wider world. Their ancestors served as whalers in the sixteenth century, exported butter to England in the seventeenth, settled in South Africa and North America in the nineteenth, and reached out to aid fellow Mennonites in the Soviet Union even before the founding of MCC in 1920. Farmers today have few oral stories to tell of these early linkages, but they do have stories of suffering severe stress from a ruthless occupying military force during the Second World War and then, shamefully, of their own involvement as invaders in the colonial war in Indonesia in the late 1940s.

Indeed, memories of the world wars are at the front of many farmers' thoughts when questions of international linkages are raised. The Second World War, for example, reinforced Dutch Mennonite ties to a wider world. At the farm level, many have personal stories of encountering the "shouting Krauts." In fact, as the dairy farmer Menno de Vries, of Witmarsum, Menno Simons's own birthplace, explains, the war reoriented Friesland's international ties, now focused on surviving a military occupation. While Menno says that locals infrequently encountered German soldiers, farmers were painfully aware of the war. For example, like many farm families, the de Vries family responded to the Dutch resistance's requests for farmers to hide Jews. He tells of the terrifying day the Nazis showed up to search the de Vries farm for hidden Jews, which involved poking long steel prods into their piled hay inside the hay mow, an eventuality for which the de Vries family had dug a cellar under the hay pile. Menno also tells how in the final days of the war a German contingent arrived to requisition a horse to pull an army cooking wagon. Disregarding their own safety, the de Vrieses provided the cook with their most spirited horse, which bolted upon being harnessed, damaging the wagon and sending the Nazi cook sprawling on the ground. Only when he threatened to begin shooting did the de Vries family cooperate, harnessing their slowest Belgian to their own wagon. Menno's family watched in amusement as the hapless cook ambled down the farm lane and out of sight. Both lighthearted and terror-laden events pointed to a new reality: the midcentury Frisian world was more than an economic hinterland of Amsterdam; the wider world had come to Friesland. It was a reality reinforced in the postwar period, during which the de Vries family tapped Marshall Plan funds to tear down an old house and use its bricks to construct a new cattle barn.[39] Now agriculture itself began shifting in significant ways.

Other farm families found in the very circumstance of the war opportunity to connect with the global Mennonite community, in particular its settler diaspora. The Dutch, of course, had hosted the Third Mennonite World Conference in 1936, but now in the aftermath of the war, local Dutch farmers had their own opportunity to build on a global ethnoreligious network. An espe-cially meaningful meeting occurred when the Dutch Doopsgezinden spon-sored Mennonite refugees escaping the oppressive farm collectives in the Soviet Union. The farmer Jan Idsardi, of Holwerd, for example, recalls that his grandparents housed refugees from Russia, preparing them for their eventual departure for farms in Paraguay,[40] while Ytsje Huisman-van der Meer, of Grou, recalls that her parents took in Russian Mennonite refugees bound for Brazil. Other possible international nexuses of the Doopsgezinden and non-Dutch Mennonites failed to materialize. Ytsje notes, for example, that in the after-math of the war many Frisian Doopsgezinden, including a nephew, migrated overseas, especially to places in the United States and Canada, but rarely con-nected with Mennonite churches in those regions, as they were deemed to be overly conservative.[41] Ironically, the postwar period was also a time when the Dutch Mennonites began to come to terms with their colonial past. Griet Buiteveld, of Nes, recalls that her father refused to join Hitler's forces but nev-ertheless "volunteered in Indonesia because that was Dutch territory . . . supposedly to go free Indonesia." Family lore has Griet's grandmother con-fronting him, saying, "You volunteer, because your crazy queen says that you have to free Indonesia. . . . What do you think that the Indonesians want? Su-karno also wants his country free!"[42] Griet does not visit the irony that her father, a Dutch Mennonite farmer, joined Dutch forces in their fight against the Indonesian independence movement, a conflict that itself was supported by militant Mennonite youth from Margorejo. in rural Java.

Significantly, when Dutch Mennonite farmers think of their coreligionists elsewhere, they see a distant and more conservative folk living in close-knit, agrarian communities, unknown in Friesland. Margriet Faber, of Tzum, com-pares the Mennonites from the Netherlands with those in the diaspora and says that the Dutch sense of a religiously defined Mennonite community is minimal at best. Ironically, few Dutch Mennonites have even visited Menno Simons's birthplace in Witmarsum. According to Margriet, "The rest of the world gets all emotional when the topic is brought up . . . 'ooh, Menno's soil,' yes. Well, about us, I don't know if we are too down-to-earth or . . . too indif-ferent, or too busy."[43] The Frisian farmers have been largely oblivious to the historical meaning that the global Mennonite community has given their soil.

When Dutch Mennonite farmers have developed strong links to the out-side world, it has been in their role as highly integrated citizens of the Neth-erlands engaging in political and environmental issues, and not as Doopsgez-inden. They see a world that has pitted the Netherlands against the wider world, especially the regulatory world of the Brussels-based Common Market. The retiree Hendrik Bosscha, of Akkrum, protests that "measures laid down by Brussels, we have to keep to them," and that it is unfair that rules for Dutch farmers differ from those applied to Belgian, German, French, and Italian farmers: "One has these requirements, and the other has . . . another directive. It's enough to make you sick."[44] Margriet Faber, who is similarly skeptical about Brussels, pines for a time in the past when "there was still freedom" and hales the French for resisting Common Market regulations: they "stand in the square, with their manure spreaders and their tomatoes," something unthinkable in the Netherlands.[45] The independent-minded Frisian farmer disdains directives from the outside world, even with Frisians' civic history of joint action on land management.

Other farmers look even further afield and object not to regulation but to a historically unmitigated neoliberal marketplace without a moral compass. Yme Jan Buiteveld, of Poppenwier, mentions thirteen different countries in his critique of global agriculture, which centers on Dutch know-how and good sense. He begins by asserting that Dutch farmers have always been at "the bottom layer of the population" because they "can't say, 'ok people, I've had so much expenses, so you need to pay, that much money for a liter of milk.'" In his view, a high-quality agriculture can be assured only if there is also a just income. To make this point, he argues that tractors from Finland are of better quality and built with more expensive labor than those from Hungary. To illustrate his argument, he tells how primitive ships built in East Germany and Poland during the Soviet era with cheap labor had to be rebuilt using highly paid Dutch technicians. Marquee labor produced good product, but good product also fetched high prices. Yme is critical of the Dutch cheese factory cooperative in Hoogeveen, which once faltered by making cheap "bulk cheese for the Russian market" and regained its footing by turning to high-quality cheese packaged for the Albert Heijn company and shipped to destinations able to pay high prices.

And because Dutch society has valued research, it has led the world in agricultural technology. Yme highlights dairy innovations, including the free-stall system of 1978, the mobile field milking wagons of 1985, and then all "those robots," developed by the Dutch alongside the Swedes, another society that pays its researchers well. He also points to Dutch leadership in commodity processing, adding value to soy hulls, maize gluten, and honey. He praises inventiveness generally: "increases of scale with bigger machines and bigger plows" from "all around the world." He tells of the time he joined a

Dutch delegation to Ireland to see a new plastic soil-heating foil that increased microbial life, a technology, he offers, that would be well suited for both Canada and France. He ends with the biblical injunction to use "your talents . . . because we [as farmers] need to feed humanity." And then he adds a footnote about underpaid miners in Uzbekistan and Argentina, a social inequity that led to violence in Vietnam and Korea, where in the past they "cut all the rich guys' heads off."[46] Yme's multination rendition puts a high-quality, justly priced Dutch agriculture at the center of global concern.

And key to this transformation is the Dutch farmer, indelibly linked to the international community for centuries but since the Second World War even more economically or morally enmeshed in the wider world. That interaction has spawned new technologies making agriculture more efficient. It has linked the Doopsgezinden to other Mennonites, especially to Soviet Mennonites bound for postwar settlement in Canada, Brazil, and Paraguay but also to roving pilgrims from afar visiting "Menno's soil." Most often, however, their global ties as farmers have not been as Mennonites but as Dutch citizens.

Manitoba: Global Labor Markets and Shipping Overseas

Manitoba farmers have shared in this global story. They too tell few stories of their farms as breadbaskets in a colonial and imperial power, but they do tell of immigration and difficulty on the new frontier and even question how their coming would have dislocated Indigenous peoples.

The retired farmer Benno Loewen tells an upbeat version of this settler story: "When you look at the history of Mennonites . . . how they have moved from one country to another . . . they are the most aggressive" of the various groups of ethnoreligious farmers. They are among those who "start very poor and they work hard . . . building it up. . . . I think it's our tradition and our culture that does it."[47] This consciousness of hard work as pioneers shapes the way the Rhineland farmers think of their own history, but other farmers have a more critical view of it. They see Mennonite farmers in global terms as settlers, party to the global displacement of Indigenous people. The farmers Grant and Leona Nickel are blunt about Mennonite complicity in this dispossession. Grant says that his Mennonite ancestors "wanted to make [the Indigenous people] go over here [to Roseau River First Nation], so they cleared land," dispossessing the local Anishanaabe people. It's also what later Rhineland farmers did who emigrated from Manitoba to Paraguay in the 1920s and 1940s. "You can't do it here no more because it's all done, so they go to Paraguay and do it there," displacing Indigenous people just as large-scale Brazilian farmers have more recently displaced the Amazon Indigenous. Leona Nickel, is more specific, recalling how in 1948 her relatives migrated from Manitoba to eastern Paraguay, where they toiled, accomplishing "in 25 years what had taken 150 years for the Mennonites here," but only by "exploiting the people that were living there. . . . For 5 cents a day, to build . . . [and to] clear their land." Leona says it is a continuation of a pattern begun by her great-grandparents in Manitoba in the 1870s.[48]

Other farmers, however, speak of even more recent immigrations, those in the 1980s and 1990s, that enabled a more intensely cultivated agriculture in Rhineland. They point to the recent return to Canada of the descendants of the Mennonite migrants to Mexico and Paraguay and even more recent immigrants from Kazakhstan and Siberia via Germany, who began arriving under the Manitoba Provincial Nominee Program in the 1990s. The market gardener Joe Braun has employed both groups of immigrants on his labor-intensive vegetable operation. He has appreciated the newcomers' work ethic. He recalls both a Mrs. Giesbrecht and a Mrs. Heinrichs from Mexico, who worked for him in the 1980s with their children, supplementing the salaries their husbands earned in local industry. Then in the 1990s he hired a few Russian and Kazakh-German immigrants, the so-called *Aussiedler* migrants, and especially recalls a Mrs. Pauli and a Ms. Gerstein, who helped to process cucumbers. He remembers thinking of the rich ethnic mix of workers on his farm. It was a reality brought home one day when he was having his tractor fixed at Willy Funk's shop, where he met Willy's granddaughter's boyfriend, from London, England, and it suddenly dawned on him: "Well what is this here? . . . the United Nations?"[49] Rhineland was no longer the homogeneous Low German Mennonite society of the past; its robust rural economy and diverse agriculture were increasingly made possible by a global labor market.

The wider world has also presented itself to the Mennonites as a global marketplace, especially enabled by neoliberal policies of recent governments. For Melvin Penner, of H&M Farms, near Altona, the international aspect of agriculture is about larger global markets. They made the dissolution of the Canadian Wheat Board in 2012 a "non event," he says. He explains that "if I want to sell stuff directly into South Africa or New Mexico, I have the ability to do that. . . . I follow our beans, where they go, they are into Ethiopia and I've been to Kenya to see where our wheat ends up, and I mean it's just . . . a small world today."[50] He applauds the fact that local farmers can now speculate on the Chicago Board of Trade. For other farmers, global engagement entails importing knowledge. Grant Nickel, who has experimented widely with yellow mustard, lentil, and soybeans, has been on several farm tours—in Brazil, Argentina, Chile, and Peru—to see how other places farm. He was especially impressed by Brazil and its megafarms: "Like . . . you're thinking that H&M's 30,000 acres [is big]: no. That's just a . . . very small operation in Brazil." Leona Nickel shares that farmer Ernie Kehler went to Kazakhstan and learned about farming "really well" in the former Soviet Union.[51] In Grant and Leona's view, the world has become an interconnected village of farms in different continents.

Like the Iowa farmers, the Manitobans also have worked to reform agriculture in the Global South. Ray Siemens recalls vacations to California with his father, the agricultural reformer J. J. Siemens, in 1947 and 1948 and how impressed they were that wheat could be planted in the fall, contrasting it to Manitoba's spring wheat. But he especially remembers his father traveling to help Old Colony Mennonites with agricultural reform in Mexico. Even though these plain people "didn't like outsiders," Ray's father would "buy some land just outside of the colony and then he would hire some people and . . . start growing these crops and do irrigation, and he figured in no time at all they would" learn from his experimentation.[52] It's a vision that elderly Peter Hildebrandt also has for his Mennonite relatives in Mexico. At first he mentions the seasonal differences between Chihuahua and Manitoba, the former's brown and dry winters and the latter's green, well-watered summers. But in his travels south to visit these relatives, he also has put on his reformist hat. "I couldn't understand the [Mexico Mennonite] way of doing things and their thinking. Because their philosophy was 'we've gotta keep things the way they are . . . don't change anything.'" As he sees it, this way of thinking is simply unsustainable.[53] Steel-wheeled tractors and horse-and-buggy transportation systems have not only kept farms simple in Mexico, they have kept them poor.

Manitoba farmers have connected to the wider world as employees, grain marketers, tourists, and agrarian reformers. They don't speak with quite the same national self-importance of either their Iowa counterparts or their Frisian coreligionists. But they do speak from a position of privilege, harnessing the world's resources to advance a modern agriculture.

Matabeleland: Struggling Farmers and the Rich Americans

The final three communities—one each in Asia, Africa, and Eurasia—have faced much greater economic, political, and social hurdles than have the communities in the Americas and Western Europe. Those obstacles have defined their wider world and shaped the ways farmers have cultivated their lands.

Of these three communities, the one affected most decidedly by recent war and revolution, Matopo Mission in Zimbabwe, has also had the closest ties to a global community. Its foreign missionaries left relatively recently, and many of its youth have dispersed throughout other countries. After independence in 1980, as Zimbabwe redefined its ties to England, South Africa, and the United States, so too did the farmers of Matopo. In addition, the financial crisis that plagued Zimbabwe in the following generation and Africa's AIDS epidemic have taken their toll, with the result that most farmers in Matopo know someone who resides in South Africa or England and rely on remittances from the diaspora. These ties are indicative that agriculture at Matopo has become weakened, increasingly a vocation for the elderly to survive in an unstable economy.

Even the oldest stories are narratives of sorrow. Ndebele farmers occasionally recall the proud history of the Zulu-descendant king Mzilikazi invading Matabeleland in the nineteenth century and establishing an agriculture-based society there.[54] Zandile Nyandeni recounts the unusual story of her paternal grandparents, the Ndlovus, arriving from South Africa and becoming enslaved by the Moyo kingdom, part of the Lozwi Empire, which eventually succumbed to Mzilikazi.[55] As Zandile tells the story, her people arrived at Matopo perhaps two centuries ago and began producing crops of rapoko, sorghum, wheat, and beans; eventually her particular clan came to be known as the Sibhula, meaning "the threshers" and referencing the most difficult of farm tasks. Despite their subordinate role, she seems proud of her clan's earthy tenacious history in southern African soil. By happenstance, old lore also suggests that the best rains came from South Africa. As Zandile puts it, "If it comes from the opposite side . . . it will just blow wild fruits away." And yet, any allusion to South Africa is bittersweet: its economic strength now reminds her of Zimbabwe's failed economy, too weak to yield any South African rands, the common currency of the rural Matabeleland black market. Furthermore, South Africa has become the place where Ndebele children disappear. Her own son, looking for work, has ended up there, and she can only hope that "maybe one day he will come home."[56]

Other Ndebele farmers share this narrative of children leaving for far-flung cities. John Masuku, the former Bulawayan, is comfortable with city life and tells of his fortunate sons in England, one even having "an office of his own." The third son, in South Africa, however, has been less fortunate, and John is glad that at least "I am still talking to him."[57] Their migrations abroad seem permanent; farming with children is a thing of the past. John's neighbor Alfred Gumpo has his story of children leaving for South Africa, sequentially, one after the other. At first Alfred encouraged them to leave Matopo but to remain tied to the land at home in some way. The farm's vegetable exports were able to support one son, whom Alfred put in charge of a market stall in Bulawayo. And yet, uncertain how this venture could be sustained, Alfred encouraged him to attend driving school. Even when "he wanted to give up and come with us to the rural areas . . . I told him that you are going nowhere until you get the license." As Alfred constructs this story, a child's return to the rural districts is not a triumph for the farm but a sign of failure in the city. He tells of the time his impoverished ex-daughter-in-law came to Matopo to demand the *lobola*, the "bride price," which his son had not yet paid to her family. On that day, Alfred was working peacefully in the field when he "heard her shouting 'my money, my money'"; when he hurried to the kraal he found the young woman "crying, asking me to beg my son on her behalf." Alfred was deeply saddened at the breakdown of the couple's marriage and felt obliged to welcome the young woman to join his clan's farm household; he could offer her no other help. Nor could he do much more than counsel his seven children, all scattered in South Africa, to maintain their dignity and independence. He told his firstborn, a taxi driver, "not to always work for someone, but to also buy his own" car, and he is pleased that "he now has his."[58] The farm has been unable to offer his children a good life; it has become but a stopgap for the most desperate in a transnational matrix of poverty.

In turn, the rural district of Matopo has sometimes found an economic backstop in foreign aid, including from Mennonites in North America. True, there are local NGOs, such as ORAP, the Organization of Rural Associations for Progress, that distribute aid to local residents as determined by the village head. Denny Ndlovu, a village head himself, says that as such he "would know best the poor as they . . . come to you seeking help."[59] But other farmers invoke the names of NGOs based in North America. Eldah Siziba says that as an old person she relies on her children, who help her farm; but she also acknowledges the US-based World Vision workers stationed at Maphisa, who taught her how to manage the soil, to understand soil types, and to determine the most suitable crops for her land.[60] Effie Moyo, born in 1948, also has fond memories of a foreign NGO and says in passing that her sweet potatoes paid for the education of her ten children, with the exception of "one of my children who was very intelligent [and] got a scholarship from the Mennonite Central

Committee and later from a well-wisher in Holland." She also recalls a program in which Canadians sponsored goats for poor and orphaned Zimbabwean children, and another in which "sacks of relief food aid from Canada" arrived with the condition "that we were not meant to brew beer with it."[61] International aid has come free, but not without a cultural price.

Indeed, the idea that rich North America has a history of offering aid to southern Africa has not pleased some farmers. As John Masuku sees it, African confidence has been challenged by North American wealth. He insists that the North Americans "also need our help at times," and he adds, "I think, as blacks, we have an inferiority complex; we look down upon ourselves and we think we are poorer than they are. We have turned ourselves into beggars, but we mustn't. Giving is not limited to money because we also can contribute something with our brains . . . on to how to improve our lives as a whole."[62] A number of John's neighbors echo the sentiment that Africans have also contributed to the well-being of North Americans. Zandile Nyandeni recalls the Mennonite World Conference in Bulawayo in 2003, when thousands of Mennonites from around the world flocked to Matabeleland. She says with pride that the visitors were "amazed at how we cook[ed] our food," and they queued up to eat isitshwala.[63] They learned how to cook with simple, nourishing ingredients. Other farmers, such as Denny Ndlovu, have heard that rich America also has its own poor, who might benefit from African help. He speaks as a farmer, proud and independent, often praying to God "to take care of them," the Americans, "as both our thoughts and theirs are known by him." Denny imagines that poor Americans might learn from his own measure of well-being: "I might say I'm rich [as] I have never asked for salt from anyone."[64]

The farmers in Matabeleland have their myths about rain coming from South Africa, but lately their stories are mostly about the difficulty of keeping youth on the land at Matopo. Unfortunately, the wider world, with its relative largess and history of aid to Africa, has underscored the relative poverty of Zimbabwean farmers. To secure their own sense of worth, they have focused on local Ndebele ways; as survivors on the land, they have procured their own source of "salt."

Siberia: Being "Othered" on the Isolated *Sovkhoz*

The Mennonite farmers in Apollonovka, Siberia, hold yet another distinctive global perspective. Indeed, the farmers here had almost no interaction with the wider world during the fifty-year period of the *kolkhoz* and *sovkhoz*. They even had few ties to other parts of the Soviet Union. As Andrei Pauls, a herder brigadier of the Sovkhoz Novorozhdestvenskii, puts it in cultural terms, it wasn't until 1989, with the collapse of Communism, "when the borders opened," that people "realized that there were educated people, even Pushkin, who were believers, that educated people, even writers and great scientists like Einstein . . . worshiped God."[65] In social terms, the end of the Soviet Union also meant that the worlds of Apollonovka residents expanded exponentially, especially as many Mennonites chose to move to Germany and farm partners arrived from Canada. These experiences with the wider world introduced new ways of relating to the environment.

These are but recent developments. In the years before 1989 the very idea of a foreigner in Apollonovka was an internal construction in which Mennonite descendants were the "foreigners." Indeed, it was an identity, embedded in internationally rooted ethnicity. Margarita Drei, the *sovkhoz* milk-machine operator at Novorozhdestvenskii, for example, lived in two worlds, a secondary one that was Communist and Russian and her own, primary world, which was evangelical Protestant and German. It was an identity bolstered by the history of the Second World War, even though by the 1970s "we were not enemies, it was just that we were Germans, we were here," speaking Low German, meeting as "believers" in worship.[66]

With this identity came a particular appreciation for the environment and nature. As noted in a previous chapter, a common escape from this ethnic and religious stigma on the *sovkhoz* was simply to keep quiet and work hard. But as German "believers" they also found refuge in divinely appointed nature: outdoors, in different seasons, in the birch forests and in their household gardens. Elena Petrovna Dik recalls that her father began attending church when she was about six and they would walk together outdoors to church; "he would take my hand and we went together in the bad weather . . . and back together, and the walk [was good] for the soul."[67] Numerous other Apollonovka residents express this refuge from their stigma as foreigners in the outdoors. The Novorozhdestvenskii cattle herder Ivan Peters, for example, recalls the interminable tension arising from his identity as a German believer in his youth, when he served in the Soviet Army, and also later on the state farm. His answer lay in the outdoors, a zone out of the reach of the Communist state. "During our youth," says Ivan, "we would often go together into nature," into the birch forest that stretched endlessly beyond the village, a place "where we could sing about God," for "we lived in a country that prohibited praising God." After marrying, he and his wife also found refuge in another outdoor venue, their own garden, allowed by the *sovkhoz* authorities.

Ivan highlights the potatoes they planted in fertile soil after official work hours: "It was fresh, great land that had rested for a long time. In the autumn we saw [our] potatoes, large, they were yellow, that was the kind. They weren't small, but large."[68] As a "foreigner" he found respite in the nation's soil.

With the collapse of the Soviet Union and Novorozhdestvenskii's transformation into a joint stock company in 1989, the Mennonites in Apollonovka quite suddenly entered a much bigger world. For one, the Soviet Union threw open its borders to foreign investors, and the Canadian farmer Walter Willms, of Fort St. John, British Columbia, founded the Villok Farm corporation, based in Apollonovka. Willms invited several local farmers to join him in a partnership dedicated to profit-oriented farming.[69] David Epp, born in 1962, married to Anna Tevs and father to five, was one of Willms's two local partners from the start. And David has high praise for what he dubs a Canadian approach to farming, synonymous with an "advanced" agriculture that respected farmers' views rather than the dictates of so-called experts. He recalls that this system of state-sanctioned "experts" was so imbedded in the Soviet mentality that even after 1989 "the methods used in the *sovkhoz* continued" in the post-Soviet joint stock company that succeeded the state farm.[70]

David recalls first meeting Willms and disagreeing with him on many points. But Willms was ultimately right, insists David, patiently using logic and persuasion to introduce new agricultural approaches to the Siberian steppe: "For instance, how deep to seed. We seeded deeper than he did. Then in terms of the time to seed, he wanted to seed earlier. We would seed on the 15th of May, not earlier." Shallow and early seeding was more profitable, producing stronger plants that were more competitive against weeds and ripened before the cold fall set in. But the new system also required a greater reliance on chemicals to eradicate weeds. Indeed, Willms emphasized synthetic fertilizers and disparaged the practice of black summerfallowing. In short, Willms "told us we should work with chemicals and over time we did."[71] Soviet agriculture may have developed a reputation for an overreliance on science and industry, but in his pursuit of increased yields and profits it was Willms, the Canadian Mennonite, who taught the Russians to value the chemical.

If the Canadian connection was welcomed at Apollonovka in the post-1989 era, residents were ambivalent about other transnational elements such as ties to Germany. In fact, many local residents who moved to Germany, including David Epp, returned in the 1990s. Similarly, those who remained, while thinking of themselves in positive terms as German, were deeply skeptical of Germany itself. Valya Pauls, a former *sovkhoz* milk technician, found comfort in the fact that the joint stock company that succeeded the *sovkhoz* in 1989 had "Germans in leadership positions." But as for Germany itself, no matter that it was a highly developed and welcoming country after the collapse of the Soviet Union, it was still an artificial place, killing body and soul. Valya says that in Russia they always "ate natural products," while the emigrants to Germany have "used more chemicals than we ever had" and as a result suffer "more illness." She recalls that in Soviet times her father was rarely sick, but just a few years after emigrating to Germany he died. She blames modern Germany, which lured the emigrants away from a balanced life rooted in nature. In Germany, she says, "they study to become an accountant, but they don't know anything else. Here, you work as an accountant, but you go home to something else and you have . . . a range of interests," including the garden and the birch forests.[72]

The Apollonovka residents, however, have a surprisingly positive perspective about the closest neighbor, Kazakhstan, whose national border lies only thirty kilometers to the west. In fact, Valya Pauls and her husband, Andrei Pauls, have encountered the Kazakhs in life-giving ways even though a recent wave of impoverished, strange Kazakh newcomers to the village has worried many residents. Valya recalls that before the fall of the Soviet Union it was a black market–based Kazakhstan lifeline that fed her and Andrei's legal allotment of two pigs and two cows and thus sustained their tiny farm. "Someone would come at night, a knock at the window, a car from nearby Kazakhstan, would arrive, unload sacks and if you needed, for instance, ten sacks, you would go there at night. That was the only way to buy it." In contrast, the recent Kazakh newcomers to the village are in no position to "bring things to us"; fate has turned matters around, and now, following the collapse of Soviet era, if "they need something, we try to help."[73]

Valya's husband, Andrei, the former *sovkhoz* sheepherder, has an equally sympathetic perspective on the Kazakhs. He shares Valya's dismissal of Russian leadership on the state farms and dismisses Germany as a place where folks don't even gather for thanksgiving. He says that even unbelievers at Apollonovka joined the Mennonites at the annual fall thanksgiving feast, the village's "biggest celebration," for the simple reason that "everyone understood it was necessary to thank God for the harvest." Like Valya, Andrei puts Kazakhstan into a distinct category, certainly a foreign place culturally and socially. But he has good rapport with the Kazakh cattle herder at the joint-stock-company farm, even though he treats the worker with sweets and sunflower seeds rather than the requested cigarettes. But Andrei also recalls his father, an outgoing man who married a Ukrainian woman in 1946 and enjoyed hawking the family's farm commodities at the marketplace, making a good profit selling mutton to Muslim "Kazakhs who lived around us . . . for the celebration of Eid al-Adha." Andrei himself has a certain respect for these Muslim neighbors and recalls an especially difficult drought in 1982, when the Christians, knowing that God "can do global things," prayed fervently for rain and even fasted. And then the rain did come, on June 26: "It was the first time in our family life that I can remember that God answered the prayers." In telling this story Andrei adds nonchalantly, "I know that Muslims and Kazakhs went to the cemetery and sacrificed a lamb and prayed that there would be rain. And we had fasted and prayed, and the Lord answered so well that there was a good harvest afterwards."[74] Ironically, the border between the Kazakhstan Muslims and the Siberia Mennonites is more porous in some ways than the cultural barrier between the Mennonites and secular Germany.

When Valya Pauls thinks of post-Soviet Russia, she still sees it as a country in which state farms have "fallen apart," in which "lots of land remains stagnant," a country "of green tomatoes [where] . . . nothing is successful." And yet it is a place where every family has a garden and every "believer" has access to the Siberian forest. Despite being critical of Russia, she has no intention of leaving it and praises the Russian government's social program, which encourages large families and "helps people build homes."[75] In comparison with other countries, Russia is an earthy, gritty country, close to nature. Although advanced Canadian agricultural practices have been welcomed, Germany's sophisticated economy and society are pilloried by many as artificial. Siberia has met the wider world and received it with ambivalence.

Java: Rice Paddies within a World unto Itself

Of the Mennonite farmers in the seven places, those of Margorejo have the fewest memories of a global community. The reason at first seems basic enough: the island of Java is densely populated and in many ways is a world unto itself. With 140 million people, the island has more than twenty times the population of either Zimbabwe, Bolivia, or the Netherlands, and it is smaller geographically than Iowa, which has only 3 million people. Furthermore, historically the Javanese Mennonites were not taught a European language, and thus unlike the Ndebele and the Dutch, most of whom are able to speak English, and the Siberians and Bolivians, who all speak a form of German, the Javanese have fewer linguistic ties to the outside world.

Even when the Javanese farmers recall their history, they speak in only vague terms of the monumental effect the global community once had on Java, that time of Dutch colonial rule. Several farmers retell stories of a Dutch past when the missionary Pieter Anton Jansz came to the Margorejo area to spread Christianity, but each of the stories is told with the haze of distant history. The farmer Sugiman, father of two children, says he has never heard the term *Mennonite* and that what matters is a "belief in God, and if we believe in God, then why concern ourselves with streams of [various] churches." And yet when asked if has heard of the Dutch missionary Jansz, Sugiman obliges, "Yes . . . our pastor told that to us, but I forgot: sorry."[76] It is the same response from the 60-year-old twins Sutarmi and Sumarmi: they have heard of "Mr. Jansz," but the former says, "I don't know; the older people maybe know," while the latter recalls that although she was told about Jansz by her parents-in-law, she doesn't "know much about him."[77]

Even farmers who do retell stories about the Dutch within the physical space of Margorejo in specific moments place the encounters in a decidedly distant past. Yoga, born in 1972, has heard only that before the Second World War "we obtained aid from Mr. Jansz and [other] Dutch missionaries, but it was stopped" after the war.[78] Other farmers vaguely recall stories that could be interpreted as casting the Dutch missionary as either a martyr or an interloper. Sagi, born in 1939, apologizes that he has heard only one story about "Mr. Jansz" and doesn't even know if it is true. In it Jansz was teaching at Brojol, later renamed Bumiharjo, and one "time when he was walking from Brojol to Margorejo . . . along the way, someone stopped him and spit on his face."[79] There is no indication whether the hostility arose from Javanese nationalism, revulsion over Jansz's farm-village management, or something else. Only one of the interviewed farmers, the elderly Sunardi, born in 1930, recalls an incident of Dutch agricultural practice, and it too is a passing comment. He remembers a simple agriculture at Margorejo in the past, a time of the handheld *ani-ani* threshing tool and the mutual assistance that came with it. And then he links this mutuality also to the "the time of the Dutch," when the rainy season would bring farmers together to clean the canals and raise the dikes in anticipation of flooding. Although he adds that at the time he was just a student in elementary school, too young to fully understand the communitarian nature of Margorejo, he associates its culture with Dutch rule.[80] The elderly Kliwon also recalls the Dutch era, but only in that "when I was a child, we called the leader 'queen,' not 'president,'" presumably a reference to Queen Wilhelmina, who abdicated in 1948. But this is the extent of his memory of the Dutch era.[81]

Similar to their vague corporate memory of Dutch colonial times, farmers also speak in only general terms of other countries or global markets. The farmer Purwanto, for example, makes mention of no country other than Indonesia in describing the world rice market. He speaks of low rice prices and attributes them to "following the global price." He explains that he has been offered only eight thousand rupiah, when he knows his rice is worth ten thousand. He insists that "the farmer must follow the market."[82] Other farmers speak in even more ambiguous terms of a market price set someplace beyond their own world. Adi Retno thanks his mother for teaching him about markets but distrusts markets he can't see. He outlines a historical process, one of an expanding marketplace that generally disadvantages the farmer: "In the past my mother was the rice seller, besides selling our rice, she also bought [rice] from the farmer. Now we sell our rice to wholesalers directly . . . in the form of husked rice. . . . I think the price is typical. But the price doesn't favor the farmer. . . . [This inequity is the] same as everywhere else."[83] It's an imprecise reference to a wider market. Dwi Sabdono, a watermelon grower born in 1970, has a more specific sense of the wider market based in Bandung, Jakarta, and Yogyakarta, places where "I can meet many wholesalers."[84] These Javanese cities mark the outside social boundary of Margorejo rice farmers' imagined wider world.

GITJ (Mennonite) farmers in Margorejo take a break from guarding their rice paddies from birds to host Royden Loewen in June 2015. The farmers are, *from left to right*, Parlan, Suwarno, Sulikin, Hartono, Pitono, and Pariman. Photo by Danang Kristiawan, Jepara.

Indeed, for Margorejo farmers far-off places are usually sites within Indonesia, which is a massive island country, thirty-five hundred kilometers from one end to the other. Often sojourns in these distant places have spelled failure. A number of farmers tell stories of the large island of Sumatra, a thousand kilometers west of Margorejo. It's not a foreign country, but farmers have gone there to work and seek their El Dorado, sometimes as laborers but also as transmigrants. Sugiman, for example, whose parents were so poor "we didn't have a field . . . only a square lot for our house," went to Sumatra and found work at a shrimp levee at the early age of 15. But he was ambitious and knew that if he wanted to purchase good, profitable wetland he would need to raise the stakes. Thus, first he rented a shrimp levee at Maringgi; failing there, he rented a large, nine-hectare one at Kalimantan, but the land proved too acidic and he failed a second time. After returning home to Java to rejuvenate, he tried a third time at Jambi, in central Sumatra; there he succeeded but lost everything again with the 2004 Boxing Day Tsunami. "Then I went home and tried to work in farming until now," but only on the dryland of his parents-in-law, raising cassava and watermelon.[85] He had been unable to reach his goal of becoming an independent rice farmer in the wider world; ultimately work within his old village allowed him to earn enough for the bricks, sand, and mortar to build a house.

Unfortunately, places beyond Margorejo often signify virtual foreign places where those unable to succeed as farmers have gone. Sumatra, for example, has only fleeting significance for other farmers: for Suratman it is a nexus of business, a source of new strains of cassava,[86] but for Kasdi it is a place of ambivalence, clawing at the social fabric of Margorejo.[87] Yoga, who has extended family in Jakarta and Semarang on Java and in Bengkulu on Sumatra, says that often when someone leaves Margorejo for a faraway place, they sell their inherited parcel of land to a sibling. These ex-Margorejo residents have given up their dream of farming; they "don't become farmers anymore."[88]

The one social boundary that all Margorejo farmers are keenly aware of is the religious line between Java's Muslim majority and its Christian minority. Kliwon says he was born in nearby Juwana to a Muslim family but that he converted to Christianity when he moved to Margorejo, for one reason: "When I lived here I couldn't be a Muslim anymore because all of my neighbors here were Christian. No one can live alone; we need other people, so we need to harmonize with the people so that we can live peacefully; that is the reason." Perhaps other villages are pluralistic—Muslim, Christian, Buddhist—with people cooperating as neighbors, explains Kliwon, but in Margorejo the farmers are all Christian, so he converted to avoid being alone and feeling alienated. His world is also philosophically divided between this earth and the afterlife. Eternity is real for Kliwon; he quotes a Muslim *kyai* who preached that "people love to enter heaven and dislike hell." The *kyai* explained further that "when people do the right thing they can hope to go to heaven," and avoid hell, which "many people hate, as according to stories, hell is very hot, there is a big fire." To avoid hell and enter heaven, says Kliwon, people must be ethical, live in honesty, and "if you are Christian, you have to go to church, and if you are a Muslim you go to the mosque."[89] The only foreign land Kliwon cares about is heaven.

Margorejo is a Christian farm village, separated now from its Dutch past, focused on growing rice for domestic consumption. Religious piety, not economic acumen, provides a farmer with ideas of a world beyond the local village.

Conclusion

The farmers in all the seven places have been affected in some way by the global community. In each place they have encountered, directly or indirectly, world markets, geopolitically determined food policies, global climate change, and the hegemony of agricultural science, among various other broad developments in farming. Yet, inherent to this matrix are local cultural factors—old lore, kin loyalties, religious teachings, and ethnic identities—that have conditioned farmers' engagement with the world community. Sometimes these experiences facilitated linkages to the global, but other times they served as obstacles to it.

Global agricultural histories have elaborated on this dialectic between the farmer in the local village and the global community, each affecting the other. But this chapter has sought to contribute to an understanding of this relationship simply by reporting on interviews with local farmers in the seven communities. Certainly, calls by leading environmental historians to demonstrate the common elements of transnational relationship, to highlight globally interconnected phenomena or those happening simultaneously in myriad places all over the world, are foundational to this inquiry. But the ways in which farmers have spoken about their experiences with the global community, from within their locales, is required as well. How have they engaged it, imagined it, bothered with it? In particular, what stories have they told of how global markets affected what they grew, how they cultivated their land, how they processed agricultural knowledge?

Because these processes reflect a particular local dialectic with the land, no common experience is apparent. By happenstance, where the Bolivian Old Colony Mennonites imported technology from the American Midwest, thus bolstering their communitarian agricultural culture, the Iowa Mennonites exported knowledge to places like Bolivia in acts of religious duty to help modernize them, thus "feeding the world." Ultimately, both the Bolivian importers and the Iowa exporters engaged the globe to reinforce their particular approaches to agriculture within their respective natural environments. Friesland and Manitoba, the other two places in highly developed capitalist societies, have pursued relationships with the global community that parallel Iowa's. Confidently, they too have imagined themselves as both participants in and reformers of a global agriculture, even with an awareness of an imbalance of power, of extranational policies affecting transnational engagement, and of technologically advanced agriculture skewing questions of development.

The three agrarian communities least connected to the outside world, those in Southern and Northern Asia, have also been affected by the global expansion of agriculture. Matopo in Zimbabwe, with its long history of racial segregation followed by debilitating civil war and then a crises-ridden economy that sent its young abroad, has been compelled to keep its farms small and its eye on survival. Apollonovka, after three generations of Communist rule that left it isolated from the wider world, quite suddenly discovered Germany and Canada, places of hope but also places that could create concern about how modern science shaped the environment. Margorejo, in Java, the world's most populous island, has faced nationalist postcolonial economics geared to feeding one of the world's fastest-growing countries, Indonesia itself; this rural community, by virtue of its location, has been most firmly shaped by forces within its nation's borders.

If there is a divide among Mennonite farmers, it is more than merely between those of the Global North and those of the Global South. It is a division specifically between farmers financially secure enough to "play" the global markets and simultaneously "reform" global agriculture, on the one hand, and those who have struggled to survive on the land within dramatic, globally ordered, often postcolonial social change, on the other. The first group of farmers have seen their outreach to the wider world increase through their access to advanced technology and globally assertive governments. The second group have seen their farms circumscribed by war against colonial power, by political upheaval with international ramifications, and by structural obstacles to economic development. When they reflect on the global community, it is thus often not as citizens from within particular countries but as farmers seeking survival: as parents worried about their children's place on family farms; as commodity producers concerned about debt loads; as tillers of the land seeking guarantees that agricultural change dictated by an amorphous outside world is sustainable. Global relations have not allowed for the rise of a world-scale homogeneous Mennonite farm village, merely seven Mennonite "villages" each adapting in distinctively local ways to global forces.

Conclusion

In this book I have argued that an understanding of agriculture is predicated on a knowledge of its local dimensions. It is at the local level that the daily challenge of coaxing the soil to produce the plants required for human sustenance becomes most clear. Here "personal knowledge" and "tacit assent," to reference Michael Polanyi's 1958 classical work, result from "a life-long chain of experiences encountered and shaped" by ordinary persons in daily life "seeking to make sense of . . . it as best" they can.[1] Here we find farmers quietly bringing inherited ideas of nature into conversation with lived realities, and further, into critical dialogue with ideas introduced by forces outside the household and village. A comparison of these dimensions in seven far-flung Mennonite communities makes for a complex story of agricultural history. But it also adds an understanding of "the interplay of local and global" forces, the promise of comparative history generally.[2]

By focusing on seven *Mennonite* communities I have sought to highlight the crucial role of culture, and lived religion in particular, in the everyday environmental relations of one particular group of farmers. But by choosing Mennonite communities I have also sought to make a seven-way comparative study of place and sustainability around the world doable and manageable. My own connections to a global Mennonite social network and associations with seven young, skilled researchers enabled a meaningful entrance into the heart of these disparate places. Moreover, this concentration on seven communities also served to define and delimit a common subject of inquiry, in this case the *global* farmer in identifiable places. The result may well be a global Mennonite history, literally from the ground up, illuminating how farmers from this religious tradition have filtered historical teachings on place, peace, and love in the everyday. As importantly, it is my hope that this comparative study will result in a fuller understanding of the farmers' challenges at the local level in the history of food growing everywhere.

The very chapter subjects—cultural transfers, modernization, religion, gender, biopower, climate change, and transnationalism—are meant to suggest that these disparate farmers shared common concerns. But just below the surface of these broad themes are stories of immense complexity, containing the minutiae of history of specific places. At their foundation, they are affected by the particularities of local experience with nature. Consider, for example, the link between specific land quality and agricultural practice. The rich subterranean soils of Friesland, protected by dikes and laced with a network of canals leading to major cities, propelled a profit-oriented, civic-minded agriculture bent on experimentation and innovation. The abundance of fertile loess soils in southeastern Iowa, on the so-called American frontier, within a vibrant metropolis's hinterland, allowed for an agriculture that was simultaneously self-sufficient and increasingly specialized. The vast, open prairie in Manitoba, within a difficult environment of a short growing season and challenging lacustrine, black clay soils, facilitated the birth of a close-knit settler community of transplanted farm villages. The low-lying yellow and red silt soils, as well as higher-ground volcanic grumosols, carved out from the Javanese jungle supported an even more closely knit, religiously oriented agricultural community, but one also informed by an ancient rice-growing culture. The scattered spring-fed gardens that lay among massive granite rocks and semiarid, treed pastures in Matabeleland encouraged a kin-based agriculture of vegetables and cattle, surviving a stridently racialized colonialism. The chernozem soils of Siberia, protected by birch forests buffeting both icy northerly winds and searing desert-made southerlies, allowed for a self-sufficient, diasporic community, but one easily overrun by a repressive, centrally planned Soviet-era agriculture. Finally, the soils under the vanquished forests of eastern Bolivia, in the ecotone of humid Amazonia and dry bushland Chaco, interspersed with sandy and clay alluvia, permitted a state to seek food security and, simultaneously, an agrarian community to resist modernity.

The microhistories within the various chapters of this book certainly distinguish each of the seven communities from the others. But they also suggest distinctive dialectics of environment and culture. If culture is defined as those symbols, performances, and "constructable signs" that serve to make sense of life in nature, then agriculture is never simply some product of environmental determinism.[3] Rather, local cultures, deeply enmeshed in the history of place, shaped agriculture in particular environments. In most of these communities, old communitarian rhythms of agriculture were unsettled by postwar technologies and land consolidations, but this transformation was different for the profoundly integrated Friesland, the oldest of the communities, than it was for the more recently established places. In the settler societies, agriculture reflected a particular transplantation, a plow culture from Europe but challenged in the postwar era by geopolitical realities, but again those patterns were distinctive, even for the farmers of neighboring Canada and the United States. In Matabeleland and Java, Western teaching demystified the natural environment, yet ancient mystical ecological ideas persisted and combined with postcolonial food-security policies to generate a particular hybrid culture. Still, the farmers in rice paddies and those in cattle kraals were tested by white colonial rule in different ways and then reasserted distinctively after respective wars of liberation. In a number of these communities national revolutions disparaged old ways and expanded agriculture: in Siberia, farmers were turned into technicians, answerable to *sovkhoz* experts; in Bolivia, communitarian farmers were granted freedom to clear-cut forests, resulting in environmental catastrophe.

As Mennonites, these farmers came to the land with a particular religious outlook. True, the historic Anabaptist ideas of humility and nonviolence, rooted in sixteenth-century teachings on *Nachfolge Christi*, a literal following of Christ, did not automatically extend to a relationship with the land in a serendipitous extension of nonviolence. While hardly akin to peasant uprisings outlined in James C. Scott's works, certain "hidden transcripts" of agrarian resistance to capitalist encroachment were nonetheless embedded in the seven places, in corners and passing moments, as quiescent resistances of the everyday.[4] In Friesland the plain Janjacobsgezinden contested modernity until the 1850s, while in Iowa the Amish distinguished themselves from acculturating Mennonites especially after the 1920s. In Manitoba, creative energies birthed the 1930s cooperative-based Rhineland Agricultural Society's campaign for soil stewardship and alternate crops. In Siberia, direct resistance to collectivization gave way a quiet disdain for state farm officials amid subversive religious devotion. Even more broadly based resistance to a "high modern" agriculture can be seen in the Global South, where local, animistic spiritual leaders, as well as later children of converts to Christianity, rediscovered old nature-based ways and critiqued various acquiescences to colonialism. At their very core, the immigrant settlements of Riva Palacio, Bolivia, marked resistance to modernity, to both an integrative nationalism and an incipient modernization. Myriad other such moments of resistance, some fleeting, others enduring, infuse this history of seven Mennonite communities.

These stories are also simply accounts of a people seeking a lasting rural life. Informed by Anabaptist teachings that variously emphasized community, humility, charity, nonviolence, and simplicity, these farmers venerated the idea of sustainability. *Sustainability* is, of course, a complex term. With reference to agriculture it can mean an environment's ability to sustain life, a system of production that seeks equitable relationships, and a religiously ordered community's link to a common life-giving spirituality.[5] But more generally the term is a possible "linguistic shell that legitimizes the exploitation of nature," in Radkau's words, as well as the dispossession of Indigenous lands and even an embrace of profit as the sole determining factor in agricultural systems. The meaning of the term, though, argues Radkau, can be illuminated by comparative historical inquiry.[6]

For Mennonite farmers a commitment to community was an especially important aspect of sustainability. This particular aim was seen in the value given to circumscribed farm size and a celebration of restraint and contentment. But the ways these attitudes were expressed differed widely within the diverse global Mennonite community. They are reflected in Gerald Yoder's statement that he could have "loaded up on land" in Washington County when it was only three thousand dollars an acre and in Sukarman's assertion in Margorejo that he didn't need the village's historical maximum of three and a half hectares to find contentment. Both Yoder and Sukarman express in their everyday life the spiritual virtues sometimes associated in the depiction of the Dutch peasant gracing the pages of the seventeenth-century *Martyrs Mirror*, accompanied by only the words "work and hope."[7] A commitment to the sustainable community also expressed itself in the idea that the *local* place served the central resource for community wholeness. From Washington County to Riva Palacio Colony, farm families benefited from church-supported programs that helped young families acquire their own land; and, of course, Rhineland in Manitoba and Waldheim in Siberia represented such places for earlier generation of farmers from southern Russia.

A commitment to the idea of cultural sustainability also revealed itself when members spoke of a spiritual link with the local wonders of nature. Testimonials in this book associate specific physical features to the divine in each of the places: in Friesland, flowery meadows; in Apollonovka, the birch forests; in Matabeleland, the massive granite baboliths; in Manitoba, the secluded vegetable gardens; in Iowa, a lower creek's watershed; in Java, a mountain shadow by a shallow sea; in Bolivia, January's warm rains. Each of the communities—but especially the less technologically sophisticated, like Matopo, Margorejo, and Riva Palacio—emphasized a transcendent God located beyond creation but sustaining it nevertheless, answering prayers for rainfall or good crops and responding to the values of submission and humility. But in more recent times environmental stewardship has also been linked to Mennonite identity and faith commitment. It could be seen, for example, when in February 1980 Iowa Mennonite School hosted a seminar on "profitable farming without pesticides," and it was apparent when Mennonites in Friesland embraced the Menno Simons Groen exhibition in Witmarsum in June 2018, celebrating a historical *Mennonite* environmental concern.[8]

The historical struggles of farmers around the world to grow food in sustainable ways, however, was not so much a conference or exhibition topic as the tacit imperative of everyday life. In this endeavor all farmers relied in some fashion on their neighbors, finding in tight social networks the resources to meet the challenge of the cycle of seasons and climatic variations. They found in these social nexuses also the resources to pursue an agricultural ethic—to keep the soils healthy for the next generation—albeit a commitment variously understood. This rural ethic always included some form of social leveling, such as supporting young farmers, sharing the bounty of harvest with the poor, or resisting limitless growth for the health of community.[9] In these pursuits farmers drew deeply from inherited and locally ordered traditions of intervening in nature, in particular of working the precious natural resource, soil, to sustain human existence. The stories in this book speak to a common challenge faced by Mennonite farmers around the world, but they also relate to all farmers compelled by the question of sustainable agriculture.

Methodology

Accessing Seven Points
on Earth

At the heart of this book's methodology is the idea of a research team coming to understand the local agricultural community within a global context. This idea inspired the research-grant proposal to the Social Sciences and Humanities Research Council of Canada in 2012 titled "Seven Points on Earth: Mennonites and Farm Culture in the Twentieth Century World." I chose to focus on Mennonites because in history they have been a disproportionately rural people who have given a great deal of thought to sustainable agriculture. I chose seven places because, as noted in the introduction, I imagined a comparative history of a number of places around the world, enough to account for significant diversity but not overwhelming in complexity, and of course because the number related to "wholeness" in a number of global cultures. The criteria for the seven communities chosen included that they be places from around the world and would welcome our research team, but the happenstance of encounter also played a part. Gaining access to these seven communities was definitely facilitated by my position as Chair in Mennonite Studies and the global network that developed as a result of this position; indeed that social network produced the distinctive pathways into each of the seven places presented itself. But key to traveling those pathways was a team of committed, gifted, and empathetic young scholars. Their linguistic ability, empathy with local farmers, and burning curiosity took them into the heart of each community.

Gaining access to the Mennonite places in Europe and North America was the most straightforward of tasks. Indeed, Friesland came into purview, as it has for Mennonites from around the world, when my family and I visited Menno Simons's birthplace in Witmarsum in 1993 and 2000. In successive years I met the "Dutchies," professors at Dutch universities, who gave me a rudimentary orientation on the history of Friesland. Through such networks and other personal ties, the geographer Hans Peter Fast, a citizen of both the Netherlands and Canada, became available as the research associate in Friesland. In July 2013 Hans Peter and I visited three pastors to whom Piet Visser, of Amsterdam, had directed us, and they in turn offered names of their parishioners to Hans Peter. Later, Anne Kok, a doctoral student in environmental studies at Amsterdam's Vrije Universiteit (VU), whose MA thesis was based on the Mennonites, undertook supplementary research on Friesland, again assisted by Piet Visser.

The Seven Points on Earth research team at the concluding project workshop in Neubergthal, Manitoba, 2017. *From left to right:* Hans Peter Fast, Aileen Friesen, Belinda Ncube, Danang Kristiawan, Royden Loewen, John Eicher, Susie Fisher, Ben Nobbs-Thiessen. Photo by Ruth Dyck, Winnipeg.

A study of the two North American sites was, of course, the easiest for me to undertake. Iowa's Washington County and environs became possible through John Eicher, a Mennonite from Indiana, who was completing his PhD on Mennonites and transnationalism at the University of Iowa when I met him at a Mennonite studies conference in Winnipeg. Initially I asked John to help me study central Kansas, using Bethel College as a base and thus building on my own doctoral work, but he persuaded me to make Washington County, Iowa, the second point of this project instead. A visit there in 2014 with my wife, Mary Ann, made it clear that this home to Swiss–South German Mennonites and Amish would well serve as a counterpoint to the Dutch–North German Mennonite story elsewhere. Manitoba, of course, is my homeland, and although our own farm is located on the Mennonite East Reserve, it is only a hundred kilometers from the West Reserve, on the other side of the Red River, the third site. When Susie Fisher, a doctoral student at the University of Manitoba, chose to focus her highly innovative dissertation on the history of emotion and horticulture in the eastern half of the West Reserve (or Rhineland Municipality), and on the village of Neubergthal in particular, it was clear that her expertise on this community would fit the contours of this project well.

Gaining access to places farther afield was more complex. Java became a possible site of study because of my friendship with the Javanese church historian Adhi Dharma, of Semarang, a tie that developed when both he and I were authors in the Mennonite World Conference's five-volume Global Mennonite History Series. When my wife and I visited him in 2011, he introduced me to his executive assistant, Utami, who took us to visit the historic mission farm village of Margorejo. Later, Utami nominated Danang Kristiawan, a young Mennonite pastor in Jepara who had spent some time at the VU in Amsterdam, to become the Javanese team member. A study of Matabeleland developed because Eliakim Sibanda, a scholar of Matabeleland, refugee from Zimbabwe, and fellow historian at the University of Winnipeg, connected me to his sister-in-law Connie Nkomo in 2003 on the occasion of the fourteenth Mennonite World Conference in Bulawayo. Connie graciously facilitated a visit to her brother's farm at Matopo, and later Eliakim introduced me to the rural-development worker Belinda Ncube, of Swiss Church Aid of Bulawayo, our Zimbabwe team member. Siberia became a possible research site after Paul Toews, of Fresno Pacific University in California, and I, aided by Paul's Ukrainian wife, Olga Shmakina, and by Tatiana Smirnova, of the Dostoevsky State University, co-convened the 2010 "Germans in Siberia" history conference in Omsk. After the conference we toured the Low German–speaking farm villages west of Omsk, including Apollonovka (formerly Waldheim). Along the way I met the local historian Reverend Peter (Piotr) Epp and became more acquainted with Aileen Friesen, a Russian-speaking Canadian PhD student who was completing her research on Siberian settler societies in Omsk; Aileen naturally became our Siberian team member.

Finally, I was introduced to Riva Palacio when as Chair in Mennonite Studies I paid a fraternal visit to Bolivia in 2004 and later resided there for a few months in 2012 when drafting *Horse and Buggy Genius*, an oral history that included the Old Colony Mennonites in the *South*. On both occasions I benefited from the friendship of Mennonite Central Committee personnel, who introduced me to a number of Riva Palacio farmers and leaders. Ben Nobbs-Thiessen, whom I had come to know for his MA on Mennonites in South America and later as a doctoral student at Emory University writing on Bolivian lowland settlers, was a natural fit as the Bolivian team member. Later, Kerry Fast, a Low German speaker from Toronto with a doctorate in gender and religion, undertook supplementary research among Riva Palacio women, who generally were not as versant in Spanish as the men.

Team members not only provided the pathways into seven sets of agricultural knowledge but also, as noted in the introduction, met in Amsterdam in 2013 to create the interview instrument. The goal was to seek accounts from farmers regarding the way religion and land intersected and the ways in which this relationship changed over time. The team agreed on twenty-four sets of questions in the general areas of personal and farm history, social interaction, cultural meaning, and environmental particulars (see appendix B). Team members agreed to be flexible in the interviews—not to force questions, to skip questions that seemed inappropriate in the time and space provided, and to dig deeper on questions farmers seemed to want to answer. The interviews varied in length, from half an hour to three, depending on how much time a particular farmer had or the farmer's interest in the process.

Halfway through the project, in the summer of 2015, team members presented their preliminary findings at conferences: Belinda, Danang, Hans Peter, and John presented at the Mennonite World Conference in Harrisonburg, Pennsylvania, in July 2016, while Aileen, Ben, and Susie presented at the "Artefacts of Agraria" conference in Guelph, Ontario, in September 2016. Final papers from each of the seven researchers were presented at the "Mennonites, Land and the Environment" history conference at the University of Winnipeg in October 2017; the papers were published in the 2018 issue of the *Journal of Mennonite Studies*.

All interviews used for this book were transcribed, and those in Javanese, Dutch, Ndebele, Low German, Spanish, and Russian were translated by the interviewers themselves. Some interviews are not cited because of time constraints of individual researchers; they include several interviews in Dutch, English, and Russian. All interview transcripts will be deposited, with the permission of the interviewees, in the Mennonite Heritage Archives in Winnipeg, Canada.

Seven Points on Earth Interview Questions

I. Life History

1. Biography: Where were you born, when, to whom? When did you marry? Where did you first live and work, and when did you begin farming? How many children do you have?

2. Religion: When did you join the Mennonite [variously, Amish, Baptist, Brethren in Christ, Doopsgezind, GITJ, etc.] Church? Were you born Mennonite [again, variously as above], or did you become one? What has been your involvement in the church?

3. Farm history: Please tell me about the history of your farm: what do you produce today and how has it changed over time, and how did it come into your family?

II. Farming

4. Seasons: What are the seasons of your farming/gardening? How does your farm change over the course of the year: when do you plant, harvest, place poultry, oversee calving, etc.? How do you know just *when* to do these tasks?

5. Farm methods: How have farm methods changed in your lifetime? What kinds of new machines, chemicals, or seeds have been introduced and when? Did new machines make a difference in how you do your work? What was good and what was bad about all of this? Was there any opposition to these new ways?

6. Traditional (dualistic) ways: What has been your experience with subsistence and commercialized or monocrop specialized farming? Is your farm's history a matter of transitioning from one to the other, or is it a matter of combining the two, and if so, how has this "dualistic" farming culture changed over time?

7. Education: How and from whom did you learn farming? Have you ever heard folklore that may have been passed down from your grandparents, such as the time to plant, when storms might be approaching, how much seed to use, the relationship of one type of weather with another? Did you learn agriculture or food production in school, the mission school, the university? How has your learning about farming differed from that of your parents? Has it made you a better farmer, gardener, steward of the land?

III. Social Relations

8. Gender: What was the role of men and women on your farm? In what ways did they work together? And how has this changed over time? Has farming or gardening made men feel manly or women feel womanly in any way?

9. Family: What has been the role of family networks on your farm or in your garden? Do you work with your children, and if so, in what ways? Do you work together within a wider family network, among siblings, or intergenerationally with your parents? What did you learn about farming from your parents or grandparents? How have your parents helped you establish the farm? Did sibling rivalry or kin-based competition play a role?

10. Community: In what ways have members of the local community cooperated, collaborated, supported one another, helped out the poor, or helped out young farmers? Can you remember a really bad year or tragedy in which the community came together to offer assistance? Has your community worked with other farm or rural communities; if so, in what way, and how successfully? How has this dynamic changed over time?

11. Prestige (status): Who has been or was the most respected person in your farm community during your lifetime, and why? How has farming success or failure affected the arrangement of power in your community? How have prestige and status in the community changed over time?

12. Class: Have you ever had workers on your farm, or have you ever been a worker? Can you tell me about your experience with class relations? Was it a good experience—why or why not? Do you think Mennonites are good employers? Do you think poor people have as much chance to get ahead today as in the past?

IV. Religion and Identity

13. Religion: In what ways (if any) would you say that religion has affected the way you see nature, farming, or gardening? Are there past incidences in which you saw God in nature? Over the years, have you prayed about farming? Have there been religious rituals for seeding and harvest? In your life, have you felt growing food was a form or worship? Have you felt that land itself is divine in any way? Has this viewpoint changed over time?

14. Mennonite identity: As a Mennonite (Baptist in Russia, Brethren in Christ in Zimbabwe, Doopsgezind in the Netherlands, etc.) you come from a faith tradition that has emphasized nonviolence, pacifism, simplicity, and community cohesiveness at some point in your history. In what ways has this set of practices affected your farming? How has the link changed over time?

V. Social Boundaries

15. Government: How have the government policies and programs and laws changed the way farming is done? Did it make a difference who owned the land or how commodities were marketed? Has the government had a good or bad effect on agriculture in your community? What kinds of government representatives have you encountered—

politicians, agents, bureaucrats, inspectors? In what ways has the state passed on agricultural knowledge about farming to you?

16. Marketing: How has the way farmers market their crops changed? Which crops and products were marketed, and which ones were consumed by your family? Which ones were sold locally, and which ones were prepared for a global market? How has the way you have been paid for crops changed over time?

17. Transnationalism: When you think of Mennonites from other countries, how do you think of them? Do you see people in need or people who are rich? Do you identify with them in some way? How have you connected to them? Have you ever donated money to overseas projects? Have you ever considered moving to another country? why and to where?

18. Urban relations: Do you think city people understand farmers? Has the city come closer to you over the years? Are city people more interested in farming than in the past?

VI. Nature

19. Weather: Can you remember storms of years ago? which were the worst ones? what happened? Would you say that the climate today is any different than climates of decades past? In what way is it different? Have you seen evidence of climate change or global warming? Have you encountered a changing environment, such as soil fertility, forest cover, or insects and wildlife?

20. Animals: Has the way farmers relate to animals changed over time? Did you have a favorite animal, or one you disliked especially? Do you think people respect animals as much today as in the past or more? in what ways? Have you ever seen wild animals on your farm; when? what happened?

21. Foodways: How do you get your food? What kinds of foods do you eat, and how are they different from the past? How is food prepared differently today than in the past? How do you eat to be healthy, and how has this changed over time?

22. Soil: How has soil, water quality, air quality, etc., changed over time? What steps do you take to keep it from eroding or deteriorating? Do you feel that the environment has become endangered somehow? What steps have you taken to safeguard the environment?

23. Wilderness: Do you enjoy nature? How do you interact with nature? How has this changed over time?

24. Environmentalism: Do you think Mennonites [variously, Amish, Baptists, Brethren in Christ, Doopsgezinden, GITJ members] have had a close relationship to the land? Have they been good stewards? Have you been a good steward?

Introduction

1. *Mennonite* in this book is used broadly and includes the Amish in the United States, Brethren in Christ Church in Zimbabwe, and certain congregations of Mennonite descent among the Evangelical Baptists in Siberia.

2. Berry, "People, Land and Community," 194. For studies that define "local culture" with reference to agriculture, see other essays in *The Art of the Commonplace*, as well as Berry, "The Work of Local Culture." In my reading of Berry I have been guided especially by Joseph Wiebe, *Place of Imagination*; and Hettinger, "Ecospirituality."

3. Radkau, *Nature and Power*, xi.

4. From Emmanuel Le Roy Ladurie's work on fourteenth-century Montaillou to John Shover's mid-twentieth-century study *First Majority, Last Majority*, historians have argued that grand change cannot be understood without listening to local voices, "the direct testimony of the peasants themselves," or considering how broad transformations might have effected "family farming and the village life that existed symbiotically with it." Le Roy Ladurie, *Montaillou*, vii; Shover, *First Majority, Last Minority*, xii.

5. For a classic study outlining changes to the very definition of community, see Nisbet, "Community as Typology."

6. Stewart, "Number Symbolism."

7. McNeill, *Something New under the Sun*, 193, 227.

8. Levine, "Is Comparative History Possible?," 340, 347.

9. Hughes, "Global Dimensions of Environmental History," 91, 97.

10. Simmons, "World Scale," 532, 534.

11. Ong, *Orality and Literacy*, 39, 42, 104.

12. The term *case study* can, of course, refer not only to specific cultural groups but to any phenomenon applied globally; for an excellent study of global agricultural history through case studies of a pathogen, see McCook, *Coffee Is Not Forever*.

13. These are common themes in Menno Simons's "Foundation Book."

14. One example of this teaching is Heinrich Balzer's "Verstand und Vernunft" (Understanding and Reason). This sermon by a Dutch-Russian Mennonite minister in 1833 was published by the Swiss-American Mennonite publisher J. G. Stauffer, of Quakertown, Pennsylvania, in 1886. See Balzer, "Faith and Reason," a translation by Robert Friedman.

15. Bender, "Anabaptist Vision," 87. Three teachings in particular arise from this general outlook: first, "a transformation of life through discipleship," that is, a literal enactment of Christ's ethical teachings; second, the centrality in these ethics of "nonresistance, or biblical pacifism"; third, an emphasis on "a brotherhood of love," or a congregational community that emphasizes love above all virtues. Bender has been criticized by theologians for essentializing Anabaptism, but more recent works, including W. Klaassen's *Anabaptism*, Snyder's *Anabaptist History and Theology*, J. Denny Weaver's *Nonviolent Atonement*, and others, more or less agree with Bender's outline, even as they historicize, nuance, and politicize these distinctive Anabaptist teachings.

16. Fountain, "Mennonites and Anthropology," 17. For an opposing view, see A. Neufeld, *What We Believe Together.*

17. Redclift, "In Our Own Image," 113, 118.

18. See Genesis 1:28 and 2:15. For expressions of this teaching, see also various essays in Harker and Bertsche Johnson, *Rooted and Grounded*, as well as Mennonite Creation Care Network annual reports from 2009 to 2019, https://mennocreationcare .org/wp-content/uploads/2017/02/MCCN-Annual-Report-2009-FNL.pdf.

19. Worster, "History as Natural History," 5; Schama, *Landscape and Memory*, 13, 14; Downes, *Nature and Art Are Physical*, x, 146; Mosley, *Environment in World History*, 58.

20. Worster, "World Without Borders," 11.

21. Richards, *Unending Frontier*, 4.

22. Shiva, *Making Peace with the Earth.*

23. Sutter, "World with Us," 94–110.

24. Lee and Newfont, *The Land Speaks.*

25. See, e.g., Portelli, *Death of Luigi Trastulli.*

26. The Amsterdam workshop heard presentations from Piet Visser and Anna Voolstra on Dutch Mennonite history, Alexander von Plato on oral-history methodology, and Martine Vonk on environmental history.

Chapter 1. • *Sect and Settler in the North*

1. Dop, "'Terwyl in de Kalkwyk Veele dier Doopsgezinden Woonen.'"

2. H. de Jong, "Bread for the Poor."

3. Bieleman, *Five Centuries of Farming*. De Jong's "Bread for the Poor" suggests that Anabaptist refugees fleeing from southern Germany in the 1720s introduced the potato to Friesland itself.

4. de Hann, "Geert Veenhuizen."

5. Crosby, *Ecological Imperialism*, 306.

6. John C. Weaver, *Great Land Rush.*

7. Richards, *Unending Frontier*, 2, 4.

8. For one view of the history of Dutch Mennonite integration into the modern world, see Urry, *Mennonites, Politics, and Peoplehood.*

9. P. Visser, email message to author, 2 April 2020. Visser refers especially to Jan Buisman's *Duizend Jaar Weer, Wind en Water in de Lage Landen V: 1650–1750* (Baarn, Netherlands: Van Wijnen, 2005).

10. A historical study of changing weather patterns for the Low Countries, focused on the Netherlands, suggests that winter temperatures dropped from a mean of 2.3°C

in about 1050 to 1.5° during the "Little Ice Age" of the early seventeenth century but then rose to 2.1° in 1900 and 2.7° in 2000. The average summer temperatures were much more steady, dropping from 16.4°C in 1400 to 16.1° in 1600, then rising to 16.3° in 2000. See van Engelen, Buisman, and Ijnsen, *Millennium of Weather.*

11. Bieleman, *Five Centuries of Farming,* 41.

12. Bieleman, *Five Centuries of Farming,* 44, 41, quotation on 44.

13. See Paping, "General Dutch Population Development."

14. Bieleman, *Five Centuries of Farming,* 36, 38, 45, 80, 82, 103, 173.

15. Bender and Yoder, "Rural Life"; P. Visser, email message to author, 5 April 2020.

16. See Krahn, *Dutch Anabaptism,* 9, 13, 19; and Neff, "Brethren of the Common Life."

17. Stayer, Packull, and Deppermann, "From Monogenesis to Polygenesis," 115. See also Deppermann, "Melchior Hoffman," 188.

18. Bender, "Brief Biography of Menno Simons," 4; Krahn and Dyck, "Menno Simons."

19. Van Braght, "Tjard Reynders," 455.

20. Menno Simons, "Foundation of Christian Doctrine," 223.

21. Menno Simons, "Foundation of Christian Doctrine," 173; Menno Simons, "True Christian Faith," 367.

22. Menno Simons, "True Christian Faith," 379.

23. Van der Zijpp, "Jacobsz, Jan (1542–1612)."

24. Schroeder, "'Parks magnificent as paradise,'" 14. For an interpretation that the Dutch Golden Age undermined the very idea of humility and simplicity, see Sprunger, "Mennonite Capitalistic Ethic."

25. Trompetter, "Mennonite Capital and Rural Transformation," 41, 43. See also Trompetter, *Agriculture, Proto-Industry and Mennonite Entrepreneurship.*

26. The specific number in this community was 1,848 Doopsgezind persons.

27. Groenveld, "Doopsgezinden in Tal en Last," 100; I thank Cor Trompetter for directing me to this reading. Of the seven municipalities, Smallingerland (i.e., Oudega and Drachten area), Opsterland (i.e., Duurwoude), Schoterland (Heerenven and surrounding area), and Haskerland (i.e., Joure, Nes, and surrounding area) were in the southeastern quadrant of the province; Harlingen lay in the southwest; Daatumadeel (i.e., Dokkum and surrounding area) was in the north-central part; and Leewarden was in the central part.

28. P. Visser, email message to author, 9 February 2016.

29. Piet Visser makes this conclusion based on an 1839 map published in Blaupot ten Cate, *Geschiedenis der Doopsgezinden in Friesland.* P. Visser, email message to author, 9 February 2016.

30. Kasdorf, "Work and Hope"; Weaver-Zercher, *Martyrs Mirror,* 63.

31. Mutel, *Emerald Horizon.* See also Sherow, *Grasslands of the United States.*

32. Mutel, *Emerald Horizon,* 4, 8, 10, 15, 77.

33. Hall, *Uncommon Defense,* 3, 6.

34. Foster, *Indians of Iowa,* 14.

35. Turner, "Significance of the Frontier," 119, 127.

36. Neils Conzen, "Peasant Pioneers"; Luebke, "Ethnic Group Settlement"; Saloutos, "Immigration Contribution to American Agriculture."

37. Gjerde, *Minds of the West*, 138.

38. Cronan, *Nature's Metropolis*, 180, 212, 221, 230, 283.

39. Packull, "Origins of Swiss Anabaptism."

40. Snyder, "Margret Hottinger of Zollikon," 46.

41. Geiser, "Switzerland."

42. See Nafziger, "Mennonites and the Conestoga Massacre of 1763."

43. MacMaster, *Land, Piety, Peoplehood*, 81, 90, 95.

44. Reschly, *Amish on the Iowa Prairie*, 34.

45. For a study that links such innovation to a class of Mennonite "peasant merchants," see Konersmann, "Middle-Class Formation in Rural Society," 116ff.

46. See Page Moch, *Moving Europeans*, 11, 105–7.

47. Blosser Yoder, *Same Spirit*, 5.

48. Guengerich, "First Mennonites in Iowa," 2. For a survey of Washington County soils, see United States Department of Agriculture, Soil Conservation Service, *Soil Survey of Washington County, Iowa.*

49. Guengerich, "First Mennonites in Iowa," 5.

50. Guengerich, diary, 1866–1889.

51. Neils Conzen, "Making Their Own America."

52. Guengerich, diary, 1, 4 July, 4 January, 11 April, 11 January, 29 April, 20 August 1866.

53. Yoder Lind, *From Hazelbrush to Cornfields*, 490.

54. Yoder Lind, *From Hazelbrush to Cornfields*, 489.

55. "Administrators Record and Account Book of the Estate of John Reber Deceased."

56. Yoder Lind, *From Hazelbrush to Cornfields*, 492, 545.

57. Yoder Lind, *From Hazelbrush to Cornfields*, 546.

58. For insightful histories of the Mennonites in Manitoba, see Francis, *In Search of Utopia*; and J. Friesen, *Building Communities.*

59. Daschuk, *Clearing the Plains.*

60. Klippenstein, "Manitoba Metis and Mennonite Immigrants," 477, 479, 481, 482.

61. Warkentin, *Mennonite Settlements of Southern Manitoba*, 32.

62. For Canadian paradigms, see Careless, "Frontierism, Metropolitanism, and Canadian History"; and McKay, "Liberal Order Framework." For American critiques of the frontier thesis, see essays by Donald Worster, Gerald Thompson, Michael P. Malone, and Walter Nugent in Limerick, Milner, and Rankin, *Trails.*

63. Stunden Bower, *Wet Prairie*; Eyford, *White Settler Reserve.*

64. G. Friesen, *Canadian Prairies*, 301. See also Carter, *Imperial Plots.*

65. See A. Ens, *Subjects or Citizens?* For samples of village histories in this area, see M. Neufeld, *Schoenthal Revisited*; and Bergen, *Sommerfeld Village.*

66. Warkentin, *Mennonite Settlements of Southern Manitoba.*

67. Alexander Henry, quoted in Warkentin, *Mennonite Settlements of Southern Manitoba*, 15.

68. For a history of Mennonite agricultural adaptation in northern Poland, see Rybak, "Agricultural Achievements of the Mennonites."

69. For a history of the Bergthaler migration to Manitoba, see G. Wiebe, *Causes and History.*

70. Podolsky, "Canada-Manitoba Soil Survey," 7.

71. For a geographic survey of the East Reserve, see Braun and Klassen, *Historical Atlas of the East Reserve*. For a typical explanation for the migration west across the Red River, see J. Dyck, "Edinburg, 1879–1947," 271.

72. G. Ens, *Volost and Municipality*, 60.

73. MacFayden, *Flax Americana*.

74. G. Ens, *Volost and Municipality*, 63.

75. See Epp-Tiessen, *Altona*; and F. G. Enns, *Gretna*.

76. Fisher, "Seeds from the Steppe," 57.

77. Elias, *Voice in the Wilderness*, 39.

78. Elias, *Voice in the Wilderness*, 147, 150.

79. Down, "Report on Colonization in Manitoba," 23, 25.

80. Van Dyke, "Among the Mennonites in Manitoba, 1880," 41.

81. Bitsche, "Mennonites in Manitoba," 72.

82. Rae, "Mennonites and Icelanders in Manitoba," 62.

83. Warkentin, *Mennonite Settlements of Southern Manitoba*, 91.

84. Warkentin, *Mennonite Settlements of Southern Manitoba*, 1.

85. A. Friesen, *Colonizing Russia's Promised Land*; Sokolsky, "Taming Tiger Country." For examples of the more common association of Siberia with exile and not agriculture, see Pries, *Exiled to Siberia*; and A. Toews, *Siberian Diary*.

86. For an overview of the Siberian settlements, see Krahn, "Siberia (Russia)." For an overview of the Mennonite Brethren Church, including its foray into Siberia, see J. A. Toews, *History of the Mennonite Brethren Church*.

87. Crosby, *Ecological Imperialism*, 36.

88. A. Friesen, *Colonizing Russia's Promised Land*, 12.

89. Crosby, *Ecological Imperialism*, 38.

90. Werner, "Siberia in the Mennonite Imagination," 165.

91. Moon, *Plough that Broke the Steppes*, 7. See also Moon, "Cultivating the Steppe."

92. Moon, *Plough that Broke the Steppes*, 7, 139, 280, 1, 4, 285.

93. Staples, *Cross-Cultural Encounters*, 107, 114, 121. More generally, see the chapters "Adaptation on the Land-Rich Steppe," 45–86, and "The Great Land Drought," 87–106. See also the more recent Staples, "Afforestation as Performance Art."

94. John B. Toews, "Mennonites and the Siberian Frontier," 84, 86, 93.

95. Werner, "Siberia in the Mennonite Imagination," 160. The papers include the *Odessa Zeitung* and *Der Botschafter*, both published in Russia, and the *Mennonitische Rundschau* and *Zionsbote*, published in the United States

96. Werner, "Siberia in the Mennonite Imagination," 165.

97. Werner, "Siberia in the Mennonite Imagination," 164, 165, 169.

98. Werner, "Siberia in the Mennonite Imagination," 165. The reported yield would be equivalent to almost forty-five bushels per acre of wheat, a very high yield for the time by western Canadian standards.

99. Werner, "Siberia in the Mennonite Imagination," 169.

100. A. Dyck, "Waldheim [Wissenfeld, Beresowka, Beckerschotor]."

101. A. Dyck, "Waldheim [Wissenfeld, Beresowka, Beckerschotor]," 155.

102. A. Dyck, "Waldheim [Wissenfeld, Beresowka, Beckerschotor]," 157.

103. Rahn, "Lebens Erinnerungen," 13.

104. Rahn, "Lebens Erinnerungen," 13.

105. Rahn, "Lebens Erinnerungen," 15.
106. A. Enns, "Das Menschen Leben," quotations on 2.
107. J. Epp, *Von Gottes Gnade Getragen,* 9.
108. J. Epp, *Von Gottes Gnade Getragen,* 10.

Chapter 2. • *Peasant and Piety in the South*

1. For biographies of the two men, see L. Yoder, "Tunggul Wulung (Tunggulwul-ung), Ibrahim (d. 1885)"; and Neff, "Jansz, Pieter (1820–1904)."
2. Sukoco and Yoder, "Way of the Gospel in the World of Java," 67. All references to this impressive work are to an unpublished version generously provided by Lawrence Yoder; the book was published in 2020 by Goshen College.
3. Sukoco and Yoder, "Way of the Gospel in the World of Java," 48.
4. Sukoco and Yoder, "Way of the Gospel in the World of Java," 42, 63, 67.
5. Crosby, *Ecological Imperialism,* 6.
6. For a concise history of Margorejo (also Margoredjo), see Amstutz and Matthijssen, "Margoredjo Mennonite Mission." For a history of the more urbanized, predominantly Chinese-Javanese church, see L. Yoder, *Muria Story.*
7. For basic information on Mount Muria and the Muria volcano, see "Muria Volcano, Island of Java, Indonesia."
8. See Government of Indonesia, Planning Department of the Department of Forestry, *Vegetation Map of Indonesia.*
9. See Soil Research Institute, Bogor, *Soil Associations of Java*; and ISRIC, World Soil Information, *Exploratory Soil Map of Java and Madura.*
10. World Bank, Climate Portal, "Average Monthly Temperatures for Indonesia"; World Bank, Climate Portal, "Average Monthly Rainfall for Indonesia at location."
11. See, e.g., MacKinnon, "Javans Fired Up."
12. Geertz, *Agricultural Involution,* 6, 28, 84, 99.
13. Sukoco and Yoder, "Way of the Gospel in the World of Java," 107, 111, 112.
14. Sukoco and Yoder, "Way of the Gospel in the World of Java," 140.
15. Sukoco and Yoder, "Way of the Gospel in the World of Java," 142, 143, 144.
16. Nijdam, Golterman, and Yoder, "Java (Indonesia)," identifies the date as August 1881, but a more recent piece, L. Yoder, "Villages of Tunggel Wulung," 163, gives the date as August 1882.
17. Jansz, "Regulations." Note that his copy differs somewhat from what is found in Sukoco and Yoder, "Way of the Gospel in the World of Java," 150, 151.
18. Strangely, an earlier version elaborated on "idol worship" and forbade "contacting satans and spirits of the dead, using divination, calculating auspicious days," that is, practices deemed to be "in conflict with the Christian faith."
19. I thank Danang Kristiawan for explaining these terms to me in an email message on 1 April 2020.
20. Sukoco and Yoder, "Way of the Gospel in the World of Java," 155, 159.
21. Sukoco and Yoder, "Way of the Gospel in the World of Java," 166.
22. E. Fast, "Fast, Johann (1861–1941)."
23. See: Van der Zijpp, "Johann Klaassen." On this link between Margorejo and the Mennonite colonies in Russia, see also Hoekema, *Dutch Mennonite Missions in Indonesia,* 92.

24. Sukoco and Yoder, "Way of the Gospel in the World of Java," 165.

25. Thiessen, "Im Weinberg des Herrn," 15 April 1906.

26. Thiessen, "Im Weinberg des Herrn," 5 August 1906.

27. J. Klaassen, "Im Weinberg des Herrn," 10 May, 31 May 1908.

28. J. Klaassen, "Im Weinberg des Herrn," 15 January 1914.

29. Sukoco and Yoder, "Way of the Gospel in the World of Java," 170, 171, 187.

30. Sukoco and Yoder, "Way of the Gospel in the World of Java," 180.

31. For early histories of the Ndebele in what later became Rhodesia, see Rasmussen, *Mzilikazi*; and C. Ndlovu, "Missionaries and Traders in the Ndebele Kingdom."

32. For a history of Matopo Hills, see Ranger, *Voices from the Rocks*. For a history of the BICC at Matopo, see S. Ndlovu, *Brethren in Christ Church*.

33. S. Ndlovu, *Brethren in Christ Church*, 16.

34. Sider, "Davidson, Hannah Frances (1860–1935)."

35. H. Sibanda, "Sustainable Indigenous Knowledge Systems."

36. UNESCO, "Nomination Dossier," 7.

37. Government of Zimbabwe, Department of Research and Special Services, *Provisional Soil Map of Zimbabwe Rhodesia*.

38. World Bank Group, *Climate Change Knowledge Portal, Zimbabwe*.

39. Phiri, "El Nino and drought."

40. "Landslide, impassable bridges."

41. South Africa National Biodiversity Institute, PlantzAfrica.

42. Natural History Museum of Zimbabwe, "Matobo Hills World Heritage Landscape."

43. Natural History Museum of Zimbabwe, "Matobo Hills World Heritage Landscape," 8, 12.

44. Ranger, *Voices from the Rocks*, 15.

45. Ranger, *Voices from the Rocks*, 19, 22.

46. E. Sibanda, "Voices from the Hills," 205, 206.

47. E. Sibanda, "Voices from the Hills," 219.

48. Davidson, *Record of Fifteen Years' Missionary Labors*, 7, 35, 22.

49. Davidson, *Record of Fifteen Years' Missionary Labors*, 33, 43, 55.

50. Davidson, *Record of Fifteen Years' Missionary Labors*, 47, 57.

51. Davidson, *Record of Fifteen Years' Missionary Labors*, 60, 110.

52. Davidson, *Record of Fifteen Years' Missionary Labors*, 205, 230.

53. Davidson, *Record of Fifteen Years' Missionary Labors*, 64, 94, 127, 138.

54. Davidson, *Record of Fifteen Years' Missionary Labors*, 180, 181, 182.

55. For histories of this migration, see Sawatzky, *They Sought a Country*; and Loewen, *Village Among Nations*.

56. See Nobbs-Thiessen, *Landscape of Migration*.

57. See Loewen and Nobbs-Thiessen, "Steel Wheel."

58. Vera, *Country Pasture/Forage Resource Profiles: Bolivia*.

59. Wright et al., *Report on the Soils of Bolivia*.

60. Nobbs-Thiessen, *Landscape of Migration*. The book is based on Nobbs-Thiessen, "Cultivating the State." For an insightful study of Mennonite-Indigenous environmental history in neighboring Paraguay, see Canova, "Negotiating Environmental Subjectivities."

61. Nobbs-Thiessen, "Cultivating the State," 418, 435, 11.

62. J. Peters, "Tagebuch, 1968."

63. Johann Wiebe, "Meine Erinnerungen und Miterleht Aufschreiben."

64. "Send auch auf meinen Wegen / Mir deinen Engel zu / und spricht du selbst den Segen / zu allem was ich thu / Herr, sende du mir Kraefte / von deiners Himmelschoeh."

65. Lanning, "Old Colony Mennonites of Bolivia," 48, 54, 66, 70.

66. Lanning, "Old Colony Mennonites of Bolivia," iii, v.

67. Lanning, "Old Colony Mennonites of Bolivia," 42, 16, 114, 80, 97, 108.

Chapter 3. • *Something New under the Mennonite Sun*

1. *Kalona (IA) News*, 14 September 1922, 11 January 1923.

2. *Kalona (IA) News*, 14 September 1922.

3. *Kalona (IA) News*, 20 January 1921.

4. *Kalona (IA) News*, 17 November 1921.

5. McNeill, *Something New under the Sun*, 212, 216.

6. For an overview of the history of Mennonites and animals, see Loewen, "Come Watch this Spider."

7. Dan Miller, account book, 1919–39.

8. Kinsinger, diary, 1915–85.

9. Peterson, diary, 1956–2000.

10. *Kalona (IA) News*, 22 May 1913.

11. *Kalona (IA) News*, 28 October 1937, 2 June 1938.

12. *Kalona (IA) News*, 1 October 1942, 18 March 1940, 24 July 1941, 5 June 1947, 2 June 1949.

13. *Kalona (IA) News*, 29 January 1959, 13 April 1950, 1 September 1960.

14. *Kalona (IA) News*, 14 January 1960, 15 June 1961.

15. *Kalona (IA) News*, 17 February 1994. See also Bill Jr., "Flint Ridge, Bill jr."

16. *Kalona (IA) News*, 14 April 1983, 14 February 1985.

17. For a history of organic agriculture in the United States, see Obach, *Organic Struggle.*

18. *Kalona (IA) News*, 21 February, 1 May, 31 July 1980, 18 March 1993.

19. For an overview of agriculture among Canadian Mennonites in western Canada, see Regehr, *Mennonites in Canada*, 101–47.

20. See Warkentin, *Mennonite Settlements of Southern Manitoba*, 207, 214, 215; and G. Ens, *Volost and Municipality*, 204, 233. Warkentin notes that the average farm size on the West Reserve increased slowly, from 166 acres in 1921 to 204 acres in 1956, while in the Rural Municipality of Rhineland only 2 percent of farmers had more than 560 acres in 1951.

21. For several sample diaries from this period, see Braun and Braun, Altona, MB, diary, 1958–73; Anna Friesen, Winkler, MB, diary, 1929–53; and Jacob W. Friesen, Rosenheim, MB, diary, 1939–45.

22. Klassen, unpublished memoir.

23. For an insightful commentary on class and land in the West Reserve, see A. Wiebe, *Salvation of Yasch Siemens.*

24. Epp-Tiessen, *Altona*, 226.

25. *Altona (MB) Echo*, 15 January 1947.

26. *Altona (MB) Echo*, 27 February 1952.

27. *Red River (MB) Echo*, 3 July 1957. The *Altona Echo* was renamed *Red River Echo* in 1955, when it amalgamated with the *Morris Herald*. Epp-Tiessen, *Altona*, 253. It ceased publication in 2020.

28. *Red River (MB) Echo*, 24 July 1957.

29. *Red River (MB) Echo*, 18 August 1971, 7 September 1966, 6 September 1967.

30. *Red River (MB) Echo*, 4 February, 31 March 1992.

31. For a history of Altona's largest corporation and a sense of the entrepreneurial ethos of the town, see Thiessen, *Manufacturing Mennonites*.

32. *Red River (MB) Echo*, 5 September 1995.

33. Van der Zijpp, "Annaparochie, Sint (Friesland, Netherlands)."

34. Sjoerd de Jong to Johannes Hoogland, 5 January, 15 May 1915.

35. Sijke Hoogland to Johannes Hoogland, 8 May, 12 December 1919.

36. Sjouke Hoogland to Johannes Hoogland, 9 September 1923.

37. de Stoppelaar, *Door Zon en Wind*.

38. Visser and Pol-Visse, *Meniste Jistertinzen*, 19. I thank Piet Visser, of Amsterdam, the son of Renske Visser-Oosterhof, for translating portions of these poems from the Dutch into English for my benefit.

39. Bieleman, *Five Centuries of Farming*, 243.

40. Vos and van der Zijpp, "Friesland (Netherlands)."

41. Visser-van Dam, *Ut it Libben fan Siebe Peenstra*.

42. For a general history of Mennonites in the Soviet Union, see John B. Toews, *Czars, Soviets and Mennonites*.

43. J. Epp, *Von Gottes Gnade Getragen*, 3, 11.

44. J. Epp, *Von Gottes Gnade Getragen*, 12.

45. J. Epp, *Von Gottes Gnade Getragen*, 14.

46. Aileen Friesen, email message to author, 13 April 2020. The information on the villages that made up Kirov is based on Peter (Piotr) Epp's work on the German Baptists in Siberia, *100 let pod krovom Vsevyshnego*.

47. J. Epp, *Von Gottes Gnade Getragen*, 18, 19, 20.

48. *Der Praktischer Landwirt* (Moscow), 15 May, 1 August, 1 July, 1 December 1925.

49. *Der Praktischer Landwirt* (Moscow), 1 October 1925.

50. Sovkhoz Medvezhdenskii Commission, "Order from 9 March 1957."

51. See Scott, *Seeing Like a State*, 196ff.

52. Executive Committee of the Regional Council of People's Deputies for the District of Isil'kul, "Land Balance Report," notes that of the 239,416 hectares under its directive as "state farm" land, 144,000 hectares (60%) were devoted to cereal production, 12,800 hectares (5%) to hayland, 27,000 hectares (11%) to grazing land, 29,000 hectares to "bush" and 200 hectares to "protective belts" (i.e., 13% to trees), 9,000 hectares to marshes, 3,000 hectares to water, 1,500 hectares to ravines or gullies, and 500 to forest clearing (6% undisturbed landscape). In short, the "state farm" land was 75 percent cultivated meadow land and 20 percent natural lands. Then 2,000 hectares (or 1.4% of the total land) were devoted to roadways (in contrast to perhaps 6.3% in arable portions of the North American West). "Gardens and personal land use of the individual collective farmers, workers and employees, and other non-members of the

collective farms" accounted for 4,900 hectares (3%), and 3,000 hectares in the "long-term" were deemed to have become "unusual for agriculture," raising the specter of soil salinization, erosion, or some other economic equation, but the cause is not noted. Finally, 33,000 hectares were for collective farming; 4,400 hectares, "subsidiary farming"; 4,300 hectares, "state land fund"; 12,000 hectares, "state forest fund;" 2,000 hectares, "urban land;" 1,900 hectares, "industrial land, rail and road transport, special purpose, and other organizations." A further note that there were another 33,000 hectares within collective farms, 4,400 hectares in "subsidiary farming" of some sort, 4,300 hectares in the "state land fund," 12,000 hectares in the "state forest fund," 2,000 hectares in "urban land," and another 1,900 for "industrial land, rail and road transport, special purpose, and other organizations," not detailed, suggests nevertheless the centrally planned approach to land use in Soviet Siberia.

53. Sovkhoz Medvezhinskii, "(75) Report."

54. Executive Committee of the Regional Council of People's Deputies for the District of Isil'kul, "Document of Agriculture, 1967."

55. Executive Committee of the Regional Council of People's Deputies for the District of Isil'kul, "Decision: 17.07.1968 #131."

56. Sovkhoz Novorozhdestvenskii, "Fond 2633: Historical Information." A third source documenting the rise and history of the Medvezhinskii Collective is a set of 100 photos, mostly officially produced, of the technological advancement of the collective in the possession of Mennonite descendants. The photos, from the 1940s to the 1980s, invariably are posed, with women or men standing, expressionless, in front of machinery or barns, looking at the camera.

57. Sovhkoz Novorozhdestvenskii, "Requirement of Pesticides, 1979."

58. Sovhkoz Novorozhdestvenskii, "Measures to Combat Smut, 1979."

59. See Gill, *Peasants, Entrepreneurs, and Social Change*, 32–38. For the failure of the plan to incorporate Indigenous farmers, see Fabricant, "Ocupar, Resister, Producir." For a recent history of this process in neighboring Brazil, see Klein and Vidal Luna, *Feeding the World*.

60. *Steinbach (MB) Post*, 9 May, 1 June 1923, 17 December 1924.

61. See Loewen, "Competing Cosmologies."

62. *Mennonitische Post* (Steinbach, MB), 21 April, 4 August 1977.

63. *Mennonitische Post* (Steinbach, MB) 8, 20 October 1977.

64. J. Enns, "Tagebuch."

65. For a discussion of the term *Kagal*, see Warkentin, *Mennonite Settlements of Southern Manitoba*, 37–38.

66. *Pendataan Jemaat, Kelompok Wakil: Kusnadi*, . . . ; *Pendataan Jemaat, Kelompok Wakil: Sudarno*, . . . ; *Pendataan Jemaat, Kelompok Wakil: Suparjono*, . . . ; *Pendataan Jemaat, Kelompok Wakil: Suratman*, . . .

67. Another 15 of the 369 households list the husband, and 10 of these the wife too, as simply a *buruh* (laborer). It is possible that these are also farm laborers, in which case about 10 percent of the farming populace would appear to be listed as *working* on a farm as opposed to farming.

68. Only 33 households listed a housewife; only 18, one or more persons as self-employed;18, an entrepreneur; 13, a laborer; 11, a pensioner; 7, a teacher; 7, either a *sopir* or a *pengemudi*, two categories of drivers; 4, a trader; and 1, a tailor.

69. Significantly in households with only one member baptized only a slight majority of the non-members were men, reflective of the relative gender equality inscribed in Javanese culture.

70. Sukoco and Yoder, "Way of the Gospel in the World of Java," 244.

71. Sukoco and Yoder, "Way of the Gospel in the World of Java," 235. The 168-hectare figure is at variance with the 179 hectares the state demanded back upon the maturation of the seventy-five-year lease in 1957, as noted in Sukoco and Yoder, 289.

72. Sukoco and Yoder, "Way of the Gospel in the World of Java," 261, 251, 252.

73. Sukoco and Yoder, "Way of the Gospel in the World of Java," 269, 282, 277, 278, 284, 290.

74. Sukoco and Yoder, "Way of the Gospel in the World of Java," 347, 358.

75. Sukoco and Yoder, "Way of the Gospel in the World of Java," 357, 360, 361, 362, 365.

76. David Hall to Mr. and Mrs. John Hall, 26 July 1932.

77. David Hall to Mr. and Mrs. John Hall, 1 July 1935.

78. Engle, Climenhaga, and Buckwalter, *There Is No Difference*, 23, 35, 30, 91.

79. Engle, Climenhaga, and Buckwalter, *There Is No Difference*, 39, 40, 61, 62.

80. Matopo Staff Minutes, 1933–1956, February 1934, item 9.

81. Matopo Staff Minutes, 1933–1956, October 1937. item 13, report of the strike.

82. Matopo Staff Minutes, 1933–1956, August 1940, item 6.

83. Matopo Staff Minutes, 1933–1956, "Matopo Training Institute Library Books for Grant Approval, 1936."

84. Matopo Staff Minutes, 1933–1956, item 6.

85. Matopo Staff Minutes, 1933–1956, March 1942.

86. Mann et al., "Go My Spirit," 17, 35.

87. Bundy, Journal.

88. S. Ndlovu, *Brethren in Christ Church*, 9, 248, 250, 257.

Chapter 4. • *Making Peace on Earth*

1. Broer, "Neveldag," 22. I thank Piet Visser, of Amsterdam, for translating portions of Broer's articles from the Dutch into English for my benefit.

2. Broer, "De Natuur en ons Godsdienstig Leven."

3. The Bhagavad Gita 9:4 says, "All this visible universe comes from my invisible Being"; the Koran 2:22 says, "Who . . . sends down rain . . . with it subsistence . . . therefore do not set up rivals to Allah"; the Old Testament, Psalm 24, says that "the earth is the Lord's and fullness thereof"; and Matthew 6:28 says, "Consider the lilies of the field, how they grow; they toil not, neither do they spin." See also Brueggermann, *The Land*.

4. White, "Christian Myth and Christian History"; Berry, *Art of the Commonplace*.

5. C. Redekop, *Creation and the Environment*, xvii.

6. W. Klaassen, "Pacifism, Nonviolence," 140.

7. Hiebert, "Creation, the Fall, and Humanity's Role," 121; D. Weaver, "New Testament and the Environment," 137. For an earlier version of such a call, see W. Janzen, "In Quest of Place."

8. C. Redekop, *Creation and the Environment*, 68, 76, 94.

9. Harker and Bertsche Johnson, *Rooted and Grounded*, xxiv, xxv. This collection is based on a conference at the Anabaptist Mennonite Biblical Seminary in Elkhart,

Indiana, in November 2015; see essays by D. Ezra Miller, Douglas D. H. Kaufmann, and Rebecca Horner Shenton. See also Guthrie, "Fidelity and Fecundity," based on a talk originally given at the Faith, Food and Agriculture Conference in in Washington County, Iowa, March 2009.

10. Harker and Bertsche Johnson, *Rooted and Grounded*, xiii, xv.

11. Harker and Bertsche Johnson, *Rooted and Grounded*, 219, 215, 237.

12. Vonk, *Sustainability and Quality of Life*, 23, 24.

13. McConnell and Loveless, *Nature and the Environment*, 66, 179, 186, 221, 236. For a similar interpretation of the indirect results of "long-standing values" among the Amish, see Kraybill, Johnson-Weiner, and Nolt, *The Amish*, 288; for an interpretation that implicitly argues the opposite, see Kline, *Great Possession*, which links the Amish to a "nonviolent way of farming" (xix).

14. Berry, "People, Land and Community," 182, 187, 189, 194.

15. For a recent institutional history of Mennonites in Friesland, see Trompetter, *Doopsgezinden in Friesland*. For the standard, early history of Mennonites in the Netherlands, see Krahn, *Dutch Anabaptism*.

16. Van der Zijpp and Brüsewitz, "Netherlands."

17. Verbeek and Hoekema, "Mennonites in the Netherlands," 75, 60.

18. Verbeek and Hoekema, "Mennonites in the Netherlands," 77. Contrary to the authors' statement, Friedeshiem opened its doors on 28 July 1929, not 1931; this, according to Piet Vellings, *75 jaar Fredeshiem*, 9.

19. On the role of the Vredesgroep in criticizing the Dutch government for its colonial policies in Indonesia, see Hoekema, *Dutch Mennonite Missions in Indonesia*, 6.

20. Verbeek and Hoekema, "Mennonites in the Netherlands," 77, 85, 86, 95. The title given here is the translated title of the Dutch version of the book; the book was subsequently published as *Jorwerd: The Death of the Village in Late Twentieth-Century Europe*.

21. Jelke Hanje and Roelie Hanje, interview by Hans Peter Fast, 20 November 2014.

22. For works that for reasons of space or non-Iowan authorship were not considered, see "Faith and Agricultural Forum Proceedings"; Dilly, "Religious Resistance to Erosion"; Reid, *Henry Ellenberger*; and Reschly, *Amish on the Iowa Prairie*.

23. Gingerich, *Mennonites in Iowa: Marking the One Hundredth Anniversary*. For yet another work, see Sanford Calvin Yoder's 1949 *Days of my Years*.

24. Gingerich, *Mennonites in Iowa*, 208, 216, 329, 338.

25. Gingerich, *Mennonites in Iowa*, 205–12, 223, 284.

26. F. Yoder, *Opening a Window to the World*, 19, 28, 33, 251, 254.

27. Gerald Yoder and Brent Yoder, interview by John Eicher, 23 June 2014.

28. For other histories of the Bergthaler Mennonites, see K. Peters, *Bergthaler Mennonites*; J. Dyck, *Bergthal Gemeinde Buch*; and Epp-Tiessen, *Altona*. For a history of the Old Colonists, see Plett, *Old Colony Mennonites*.

29. Ens, Peters, and Hamm, *Church, Family and Village*, 185.

30. Gerbrandt, *Adventure in Faith*, 95.

31. Ens, Peters, and Hamm, *Church, Family and Village*, 129.

32. Heppner, *Search for Renewal*, 47.

33. M. Neufeld, *Prairie Pilgrim*, 132, 133.

34. Norma Giesbrecht, interview by Tracey Ruta, 21 October 2004. I thank Susie Fisher for bringing this interview to my attention.

35. Sukoco and Yoder, "Way of the Gospel in the World of Java," 198, 199, 201, 202.

36. Danang Kristiawan, email message to author, 25 January 2021.

37. Sukoco and Yoder, "Way of the Gospel in the World of Java," 241, 264. I thank Danang Kristiawan (email message to author, 1 May 2020) for clarifying the context of this name change.

38. Sukoco and Yoder, "Way of the Gospel in the World of Java," 256, 266, 277.

39. Sukoco and Yoder, "Way of the Gospel in the World of Java," 288, 301, 311, 352, 354.

40. Sukarman, interview by Danang Kristiawan, 1 December 2015.

41. Engle, Climenhaga, and Buckwalter, *There Is No Difference*, 184, 196, 195.

42. Dube, Dube, and Nkala, "Brethren in Christ Churches in Southern Africa," 103.

43. Dube, Dube, and Nkala, "Brethren in Christ Churches in Southern Africa," 118.

44. Josephine Siziba, interview by Belinda Ncube, 18 May 2015.

45. Peter Epp, " Short History of the Omsk Brotherhood," 118, 119, 120, 122, 123.

46. Peter Epp, "Short History of the Omsk Brotherhood," 123, 128.

47. Margarita Janzen Drei, interview by Aileen Friesen, 12 May 2015.

48. J. Hamm et al., "Eine Erinnerung und Antwort," 1.

49. J. Hamm et al., "Eine Erinnerung und Antwort," 3.

50. J. Hamm et al., "Eine Erinnerung und Antwort," 4, 6.

51. J. Hamm et al., "Eine Erinnerung und Antwort," 7, 9.

52. Johann Fehr, interview by Ben Nobbs-Thiessen, 7 May 2014.

Chapter 5. • Women on the Land

1. Krismiati, interview by Danang Kristiawan with Royden Loewen, 12 June 2014.

2. Sturgeon, *Ecofeminist Natures*, 23.

3. Gaard, "Ecofeminism Revisited," 44.

4. Dixon, *Rural Women at Work*, 2; Hershatter, *Gender of Memory*.

5. Berninghausen and Kerstan, *Forging New Paths*.

6. Tickamyer and Kusujiarti, *Power, Change, and Gender Relations*, 8, 11, 24.

7. Sumarmi and Sutarmi, interview by Danang Kristiawan with Royden Loewen, 14 June 2014.

8. Suyatno, interview by Danang Kristiawan, 15 June 2014.

9. Sutari, interview by Danang Kristiawan, 23 March 2015.

10. Shumba, "Women and Land."

11. Mazingi and Kamidza, "Inequality in Zimbabwe," 362.

12. Ncube, "Mothers, Soil and Substance." For a more thorough study of gender relations, which Ncube echoes here, see Urban-Mead, *Gender of Piety*.

13. Davidson, *Record of Fifteen Years' Missionary Labors*.

14. For an argument that Davidson was de facto the lead missionary despite the presence of male superintendents, see S. Ndlovu, *Brethren in Christ Church*, 183, which quotes the BICC church historian E. Morris Sider as describing Davidson as "capable, energetic and in many ways the effectual leaders of the mission from the beginning" and saying that she left the mission when Henry P. Steigerwald became mission superintendent, as "he had difficulty in conceiving that a woman, even including Frances Davidson, should be anything else except submissive to the leadership of a man."

15. Capturing this idea of cultural power, S. Ndlovu, *Brethren in Christ Church*, 194, quotes the missionary Anne Engle as saying that "in pagan Africa, girls were consid-

ered to be worth little more than property by which man's wealth could be increased . . . his beast of burden, the bearers of his children, or merely instruments of lust. . . . Such conditions loudly called to the early missionaries. . . . The British government, however, was just beginning to make its power felt in the country and as in all countries into which it entered, sought the uplift of the people."

16. Heta Mlilo, interview by Belinda Ncube, 1 May 2015.

17. Eldah Gumpo, interview by Danang Kristiawan, 14 November 2015.

18. Josephine Siziba interview, 18 May 2015.

19. Priscilla Sibindi, interview by Belinda Ncube, 8 May 2015.

20. Alfred Gumpo, interview by Belinda Ncube, 24 April 2015; Denny Ndlovu interview, 15 May 2015; John Masuku interview, 29 April 2015; Timothy Sibanda interview, 24 April 2015; Elliot Gumpo, interview by Belinda Ncube, 14 November 2015.

21. Eldah Gumpo interview, 14 November 2015.

22. Eldah Gumpo interview, 14 November 2015.

23. Eldah Siziba, interview by Belinda Ncube, 24 April 2015.

24. Effie Moyo, interview by Belinda Ncube, 10 January 2016.

25. Zandile Nyandeni, interview by Belinda Ncube, 24 April 2015. See also Ncube, "Mothers, Soil and Substance," 2017, 232.

26. See Conquest, *Harvests of Sorrow*; and Viola, "Bab'i bunti and Peasant Women's Protest."

27. M. Epp, *Women without Men*. See also M. Epp, "Semiotics of Zwieback."

28. M. Janzen, "Meine Lebensgeschichte," 5, 6, 7.

29. M. Janzen, "Meine Lebensgeschichte," 11, 12.

30. M. Janzen, "Meine Lebensgeschichte," 15, 18, 22.

31. M. Janzen, "Meine Lebensgeschichte," 38, 41.

32. Bekker, "Meine Lebensgeschichte," 5, 12, 13.

33. Bekker, "Meine Lebensgeschichte," 89.

34. Bekker, "Meine Lebensgeschichte," 143, 148, 152, 166, 198, 200.

35. Bekker, "Meine Lebensgeschichte," 212, 215.

36. Bekker, "Meine Lebensgeschichte," 244, 249, 269, 273.

37. Bekker, "Meine Lebensgeschichte," 319, 330, 352, 340.

38. Salamon, *Prairie Patrimony*.

39. Fink, *Open Country, Iowa*.

40. Barker Devine, *On Behalf of the Family Farm*.

41. Nettie Wittrig to "Dear Sister and Family," 3 August 1911, 4 January 1913.

42. Preheim, "Memories of the Joseph Wittrig Family." Nettie, born on 23 March 1896, would have been 15 in 1911 and 17 in 1913; she later married Adolph Preheim. Her siblings include Mary (b. 1876), Sarah (b. 1877), Fannie (b. 1879), Katherina (b. 1881), John (b. 1883), Emma (b.1885), Anna (b. 1887), Margaret (b. 1890), Myrtle (b. 1892), Susie (b. 1893, married at 15, died in the flu epidemic at age 25, while husband Roy died on the day of her funeral), and Nettie, the youngest (b. 1896).

43. Wittrig to her sister, 3 August 1911, 1, 2, 12, 13.

44. Wittrig to her sister, 4 January 1913, 1, 3, 6, 7.

45. Brenneman, diary, 1942.

46. "Ancestors of Audrey Brenneman Miller's Line."

47. Brenneman, diary, 1942.

48. Brenneman, diary, 1942.

49. Green County Farm Bureau Women's Club, minutes, 2 October 1959.

50. Green County Farm Bureau Women's Club, minutes, 27 September 1962.

51. Green County Farm Bureau Women's Club, minutes, 22 November 1963.

52. Green County Farm Bureau Women's Club, minutes, 1 June 1961, 2 March 1962.

53. Green County Farm Bureau Women's Club, minutes, 19 April 1977.

54. Green County Farm Bureau Women's Club, minutes, 28 November 1978.

55. Green County Farm Bureau Women's Club, minutes, 24 May 1994.

56. Ericksen and Klein, "Women's Roles and Family Production," 285.

57. Reschly and Jellison, "Production Patterns, Consumption Strategies," 162. See also Jellison, "Amish Women and the Household Economy," 99, 100.

58. K. Friesen, "Tagebuch," 1979.

59. K. Friesen, "Tagebuch," 2002.

60. Nettie Blatz, interview by Kerry Fast, 30 July 2014.

61. Anne Peters, interview by Kerry Fast, 31 July 2014.

Chapter 6. • *Farm Subjects and State Biopower*

1. Elias, "Emigration from Russia to Manitoba," 43.

2. Elias, "Emigration from Russia to Manitoba," 46.

3. Fehr, "From Russia to Canada in 1875," 39.

4. Belich, *Replenishing the Earth*, 56.

5. Belich, *Replenishing the Earth*, 81, 82, 147.

6. Scott, *Seeing Like a State*, 6.

7. Scott, *Seeing Like a State*, 186.

8. Loo, "High Modernism," 38, 43.

9. Radkau, *Nature and Power*, 9, 153, 52, 232.

10. Radkau, "Nature and Power: An Intimate and Ambiguous Connection," 327.

11. For example, compare works on Manitoba, including S. Klassen, "Heroes of a Flat Country," and Stunden Bower, *Wet Prairie*, with works on Siberia.

12. See Taylor, *Fashioning Farmers*; Carter, *Imperial Plots*; and Sandwell, *Powering Up Canada*.

13. Massie, *Forest Prairie Edge*, 102; Stunden Bower, *Wet Prairie*, 2, 4.

14. Driedger, "Native Rebellion and Mennonite Invasion," 290.

15. Fisher, "(Trans)planting Manitoba's West Reserve"; Joseph Wiebe, "On the Mennonite-Metis Borderland."

16. William Janzen, *Limits on Liberty*; A. Ens, *Subjects or Citizens?*

17. Russell, *How Agriculture Made Canada*, 233, 235.

18. Morton, *Manitoba*, 189.

19. Peter Wiens, quoted in Zacharias, *Reinland*, 59.

20. G. Ens, *Volost and Municipality*, 46, 48, 84.

21. H. Hamm, *Sixty Years of Progress*, 10, 24.

22. H. Hamm, *Sixty Years of Progress*, 18, 21.

23. Epp-Tiessen, *Altona*, 223, 253.

24. *Altona (MB) Echo*, 22 June 1957.

25. *Altona (MB) Echo*, 1 September 1948, 2 April, 30 June 1952; *Red River (MB) Echo*, 15 November 1957, 29 April 1959.

26. *Red River (MB) Echo*, 29 April 1959.

27. *Red River (MB) Echo*, 25 March 1964, 21 September 1966.

28. *Red River (MB) Echo*, 11 November 1978.

29. Ray Siemens, interview by Susie Fisher, 20 February 2017.

30. Joe Braun, interview by Susie Fisher, 17 June 2017.

31. Grant Nickel and Leona Nickel, interview by Susie Fisher, 17 August 2017.

32. Melvin Penner, interview by Susie Fisher, 22 September 2017.

33. Benno Loewen and Mary Loewen, interview by Susie Fisher, 12 October 2017.

34. Melvin Penner interview, 22 September 2017.

35. Sukoco and Yoder, "Way of the Gospel in the World of Java," 263.

36. Kristiawan, "Javanese Wisdom, Mennonite Faith," 177.

37. Suko Sukarman, interview by Danang Kristiawan, 16 June 2014.

38. Kristiawan, "Javanese Wisdom, Mennonite Faith," 82.

39. Adi Retno, interview by Danang Kristiawan, 14 June 2014.

40. Sugiman, interview by Danang Kristiawan, 25 January 2016.

41. Kasmito Kliwon, interview by Danang Kristiawan, 14 June 2016.

42. Sugiman interview, 25 January 2016.

43. Adi Retno interview, 14 June 2014.

44. Adi Retno interview, 14 June 2014.

45. J. Hamm et al., "Eine Erinnerung und Antwort."

46. Johann Boldt, interview by Ben Nobbs-Thiessen, 5 April 2014.

47. Cornelius Froese, interview by Ben Nobbs-Thiessen, 6 May 2014.

48. Peter Klassen, interview by Ben Nobbs-Thiessen, 29 July 2014.

49. Johann Fehr interview, 7 May 2014.

50. Isaak Peters, interview by Ben Nobbs-Thiessen, 30 April 2014.

51. Cornelius Peters, interview by Ben Nobbs-Thiessen, 11 May 2014.

52. Jakob Buhler, interview by Ben Nobbs-Thiessen, 21 July 2014.

53. Gerhard Martens, interview by Ben Nobbs-Thiessen, 21 July 2014.

54. Abram Reimer, interview by Ben Nobbs-Thiessen, 3 August 2016.

55. Peter Friesen, interview by Ben Nobbs-Thiessen, 21 April 2015.

56. Johann Boldt interview, 5 April 2014.

57. Abram Thiessen, interview by Ben Nobbs-Thiessen, 17 April 2014.

58. Bieleman, *Five Centuries of Farming*, 158, 239.

59. Bieleman, *Five Centuries of Farming*, 243, 244.

60. Bieleman, *Five Centuries of Farming*, 270.

61. Yme Jan Buiteveld, interview by Hans Peter Fast, 20 December 2014. He seems to be referring to the Fries Rundvee Stamboek, founded in 1879.

62. Griet Buiteveld and Jan Buiteveld, interview by Hans Peter Fast, 26 December 2014.

63. Klaas Hanje and Tiny Hanje, interview by Hans Peter Fast, 28 November 2014.

64. Margriet Faber, interview by Hans Peter Fast, 12 June 2015.

65. Douke Odinga, interview by Hans Peter Fast, 3 June 2015.

66. Jelke Hanje and Roelie Hanje interview, 20 November 2014.

67. Yme Jan Buiteveld interview, 20 December 2014.

68. Baukje Bosscha and Hendrik Bosscha, interview by Hans Peter Fast, 5 November 2014.

69. Helms, *Conserving the Plains*, 59.

70. Smith-Howard, "Ecology, Economy and Labor," 45.

71. Bob Miller, interview by John Eicher, 7 June 2014.

72. Calvin Yoder, interview by John Eicher, 10 June 2014.

73. Carolyn Bontrager and Delmar Bontrager, interview by John Eicher, 22 July 2014.

74. Carolyn Miller and Perry Miller, interview by John Eicher, 17 June 2014.

75. S. Ndlovu, *Brethren in Christ Church*, 217, 236, 255.

76. Sikhanyisiwe Masuku, interview by Belinda Ncube, 1 May 2015.

77. John Masuku, interview by Belinda Ncube, 29 April 2015.

78. Eldah Siziba interview, 24 April 2015.

79. Alfred Gumpo interview, 24 April 2015.

80. For a recent study of Mennonite suffering during the Russian Revolution and the related civil war, see Patterson, *Makhno and Memory*.

81. Neufeldt, "Reforging Mennonite Spetspereselentsky"; Letkemann, "Mennonite Victims."

82. Peter Epp, quoted in the film *Seven Points on Earth*.

83. P. Epp, "My Experience Working on the Soviet Collective."

84. Peter Epp, quoted in the film *Seven Points on Earth*.

85. Ivan Ivanovich Peters, interview by Aileen Friesen, 14 May 2015.

86. Abram Dirksen, interview by Aileen Friesen, 13 May 2015.

87. Vladimir Friesen, interview by Aileen Friesen, 13 May 2015.

88. Abram Dirksen interview, 13 May 2015.

89. Vladimir Friesen interview, 13 May 2015.

90. Abram Dirksen interview, 13 May 2015.

91. Abram Dirksen interview, 13 May 2015.

92. Abram Dirksen interview, 13 May 2015.

93. Radkau, *Nature and Power*, 153.

Chapter 7. • *Vernaculars of Climate Change*

1. Mary Dube, interview by Belinda Ncube, 19 April 2015.

2. Mary Dube interview, 19 April 2015.

3. Coen, "Big is a Thing of the Past," 306.

4. Bashford, *Global Population*, 355.

5. Gupta, *History of Global Climate Governance*.

6. For a wider debate on the political aspects of climate change, see Boyd, "Working Together on Climate Change"; and Palm, Lewis, and Feng, "What Causes People to Change their Opinion about Climate Change?"

7. John Masuku interview, 29 April 2015.

8. John Masuku interview, 29 April 2015.

9. John Masuku interview, 29 April 2015.

10. Zandile Nyandeni interview, 24 April 2015.

11. Aaron Tshuma, interview by Belinda Ncube, 8 May 2015.

12. Moses Tshuma, interview by Belinda Ncube, 8 May 2015.

13. Elliot Gumpo interview, 14 November 2015.

14. Timothy Sibanda, interview by Belinda Ncube, 24 April 2015.

15. Sibonginkosi Dube, interview by Belinda Ncube, 10 January 2016.

16. Mary Dube interview, 19 April 2015.

17. See Hedberg, *Outside the World*.

18. Jakob Buhler interview, 21 July 2014.

19. Johann Fehr interview, 7 May 2014.

20. Abram Enns, interview by Ben Nobbs-Thiessen, 7 May 2014.

21. Peter Friesen interview, 21 April 2015.

22. Franz Rempel, interview by Ben Nobbs-Thiess, 26 June 2014; Abram Rempel, interview by Ben Nobbs-Thiess, 26 June 2014.

23. Johann Boldt interview, 5 April 2014.

24. Jakob Giesbrecht, interview by Ben Nobbs-Thiessen, 17 April 2015.

25. Jelke Hanje and Roelie Hanje interview, 20 November 2014.

26. Klaas Hanje and Tiny Hanje interview, 28 November 2014.

27. Jelke Hanje and Roelie Hanje interview, 20 November 2014.

28. Margriet Faber interview, 12 June 2015.

29. Douke Odinga, interview by Hans Peter Fast, 3 June 2015.

30. Jan Idsardi, interview by Hans Peter Fast, 13 November 2015.

31. Baukje Bosscha and Hendrik Bosscha interview, 5 November 2014.

32. Jelke Hanje and Roelie Hanje interview, 20 November 2014.

33. Klaas Hanje and Tiny Hanje interview, 28 November 2014.

34. Brent Yoder and Gerald Yoder, interview by John Eicher, 23 June 2014.

35. Bob Miller interview, 7 June 2014.

36. Gilbert Gingerich and Sandy Gingerich interview by John Eicher, 4 June 2014.

37. Gilbert Gingerich and Sandy Gingerich interview, 4 June 2014.

38. Gilbert Gingerich and Sandy Gingerich interview, 4 June 2014.

39. Marlin Miller and Mary Miller, interview by John Eicher, 23 July 2014.

40. Carolyn Bontrager and Delmar Bontrager interview, 22 July 2014.

41. Inez Yoder and Vernon Yoder, interview by John Eicher, 18 July 2014.

42. Joe Braun interview, 17 June 2017.

43. Melvin Penner interview, 22 September 2017.

44. Grant Nickel and Leona Nickel interview, 17 August 2017.

45. Peter Hildebrandt, interview by Susie Fisher, 26 October 2017.

46. Leonard Wahl, interview by Susie Fisher, 26 October 2017.

47. Kevin Nickel, interview by Susie Fisher, 19 June 2017.

48. Benno Loewen and Mary Loewen interview, 12 October 2017.

49. Barnett, "Theology of Climate Change."

Chapter 8. • *Mennonite Farmers in "World Scale" History*

1. Johann Fehr interview, 7 May 2014.

2. Johann Fehr interview, 7 May 2014.

3. Simmons, "World Scale," 532, 533.

4. Worster, "World Without Borders," 10, 12.

5. See Vertovec, *Transnationalism*; Bayly et al., "AHR Conversation: On Transnational History"; and Darian-Smith and McCarty, "Global Studies as a New Field."

6. For one such recent thorough and rigorous study, see Klein and Vidal Luna, *Feeding the World*.

7. For a classical study of phenomenology and rurality, see Swierenga, "New Rural History."

8. Simmons, "World Scale," 534.

9. Loewen, *Horse-and-Buggy Genius.*

10. Johann Fehr interview, 7 May 2014.

11. Hensen, *Agrarian Revolt*, 7; see also R. Janzen, *Liminal Sovereignty.*

12. Abram Reimer interview, 3 August 2016.

13. Elizabeth Thiessen, interview by Kerry Fast, 6 August 2014.

14. Isaak Peters interview, 30 April 2014.

15. Abram Redekop, interview by Ben Nobbs-Thiessen, 27 July 2014.

16. Johann Fehr interview, 7 May 2014.

17. Abram Thiessen interview, 17 April 2014.

18. Abram Klassen, interview by Ben Nobbs-Thiessen, 23 April 2015. According to Nobbs-Thiessen, *Landscape of Migration*, Marinkovic was linked to the fascist Ustashe organization, led by a family that is still one of the largest soybean processors in Bolivia.

19. Johann Boldt interview, 5 April 2014.

20. Enrique Wiebe, interview by Ben Nobbs-Thiessen, 14 May 2015.

21. Cornelius Peters interview, 11 May 2014.

22. Abram Klassen interview, 23 April 2015.

23. Johann Fehr interview, 7 May 2014.

24. Nettie Blatz interview, 30 July 2014.

25. Abram Hamm, interview by Ben Nobbs-Thiessen, 3 April 2015.

26. For a full history of this phenomenon, see Loewen, *Village Among Nations.*

27. Marie Hamm, interview by Kerry Fast, 5 August 2014.

28. Gerhard Martens interview, 21 July 2014.

29. Peter Wall, interview by Ben Nobbs-Thiessen, 21 July 2014.

30. Ed Hershberger and Marge Hershberger, interview by John Eicher, 9 June 2014.

31. Gilbert Gingerich and Sandy Gingerich interview, 4 July 2014.

32. Bob Miller interview, 7 June 2014.

33. Bruce Hochstetler and Edie Hochstetler, interview by John Eicher, 16 June 2014.

34. Bruce Hochstetler and Edie Hochstetler interview, 16 June 2014.

35. Mark Gingerich, interview by John Eichler, 28 July 2016.

36. Mark Gingerich interview, 28 July 2016.

37. David Miller and Martha Miller, interview by John Eicher, 7 July 2014.

38. Caroline Bontrager and Delmar Bontrager interview, 22 July 2014.

39. Menno de Vries, interview by Royden Loewen, 7 June 2018.

40. Jan Idsardi interview, 13 November 2015.

41. Ytsje Huisman-van der Meer, interview by Hans Peter Fast, 22 June 2015.

42. Griet Buiteveld and Jan Buiteveld interview, 26 December 2014.

43. Margriet Faber interview, 12 June 2015.

44. Baukje Bosscha and Hendrik Bosscha interview, 5 November 2014.

45. Margriet Faber interview, 12 June 2015.

46. Yme Jan Buiteveld interview, 20 December 2014.

47. Benno Loewen and Mary Loewen interview, 12 October 2017.

48. Grant Nickel and Leona Nickel interview, 17 August 2017.

49. Joe Braun interview, 17 June 2017.

50. Melvin Penner interview, 22 September 2017.

51. Grant Nickel and Leona Nickel interview, 17 August 2017.

52. Ray Siemens interview, 20 February 2017.

53. Peter Hildebrandt interview, 26 October 2017.

54. Not one of the twenty farmers interviewed recalled this history, although S. Ndlovu, *Brethren in Christ Church*, makes Mzilikazi's rural kingdom central to the BICC story.

55. The context of this story was provided by Eliakim Sibanda in an email message to the author on 8 June 2018.

56. Zandile Nyandeni interview, 24 April 2015.

57. John Masuku interview, 29 April 2015.

58. Alfred Gumpo interview, 24 April 2015.

59. Denny Ndlovu, interview by Belinda Ncube, 15 May 2015.

60. Eldah Siziba interview, 24 April 2015.

61. Effie Moyo interview, 10 January 2016.

62. John Masuku interview, 29 April 2015.

63. Zandile Nyandeni interview, 24 April 2015.

64. Denny Ndlovu interview, 15 May 2015.

65. Andrei Pauls, interview by Aileen Friesen, 17 May 2015.

66. Margarita Janzen Drei, interview, 12 May 2015.

67. Elena Petrovna Dik, interview by Aileen Friesen, May 2015.

68. Ivan Ivanovich Peters interview, 14 May 2015.

69. For a fuller story, see B. Yoder, "Siberian Success."

70. David Epp, interview by Aileen Friesen, 18 May 2015.

71. David Epp, interview, 18 May 2015.

72. Valya Pauls, interview by Aileen Friesen, 5 November 2015.

73. Valya Pauls, interview, 5 November 2015.

74. Andrei Pauls Jr., interview by Aileen Friesen, 22 May 2015.

75. Valya Pauls interview, 5 November 2015.

76. Sugiman interview, 25 January 2016.

77. Sumarmi and Sutarmi interview, 14 June 2014.

78. Yoga Saptoyo, interview by Danang Kristiawan with Royden Loewen, 12 June 2014.

79. Sagi, interview by Danang Kristiawan, 25 February 2015.

80. Sunardi, interview by Danang Kristiawan, 23 March 2015.

81. Kasmito Kliwon interview, 14 June 2016.

82. Purwanto, interview by Danang Kristiawan, 25 February 2015.

83. Adi Retno interview, 14 June 2014.

84. Dwi Sabdono, interview by Danang Kristiawan, 25 February 2015.

85. Sugiman interview, 25 January 2016.

86. Suratman, interview by Danang Kristiawan, 25 February 2015.

87. Kasdi, interview by Danang Kristiawan, 11 March 2016.

88. Yoga Saptoyo interview, 12 June 2014.

89. Kasmito Kliwon interview, 14 June 2016.

Conclusion

1. Polanyi, "Articulation," 97. For similar ideas on so-called practical knowledge, see also Scott, *Seeing Like a State*, 307ff.

2. Levine, "Is Comparative History Possible?"

3. Geertz, *Interpretation of Cultures*, 14.

4. Scott, *Domination and the Arts of Resistance*, x.

5. See, e.g., Kuo, "Identifying Sustainability"; Wright and Annes, "Farm Women and the Empowerment Potential"; Cunfer and Krausmann, "Adaptation on an Agricultural Frontier"; Anderson and Kenda, "What Kinds of Places Attract and Sustain Amish Populations?"; and McDonald, "Settler Life Writing."

6. Radkau, *Nature and Power*, xii; Flachs, *Cultivating Knowledge*, 172.

7. Weaver-Zercher, *Martyrs Mirror*; Kasdorf, "Work and Hope."

8. *Kalona (IA) News*, 21 February 1980; *Menno Simons Groen*.

9. Eicher, "'Every family on their own.'"

Primary Sources

Transcribed Interviews

INTERVIEWS BY JOHN EICHER IN WASHINGTON COUNTY
AND NEIGHBORING COUNTIES IN IOWA

Transcripts by Eicher, in possession of author.
Bontrager, Carolyn, and Delmar Bontrager. Kalona. 22 July 2014.
Erb, Terry. Wellman. 10 June 2014.
Gingerich, Gilbert, and Sandy Gingerich. Parnell. 4 June 2014.
Gingerich, Mark. Iowa City. 28 July 2016.
Hershberger, Ed, and Marge Hershberger. Kalona. 9 June 2014.
Hochstetler, Bruce, and Edie Hochstetler. Parnell. 16 June 2014.
Miller, Bob. Kalona. 7 June 2014.
Miller, Carolyn, and Perry Miller. Kalona. 17 June 2014.
Miller, David, and Martha Miller. Kalona. 7 July 2014.
Miller, Marlin, and Mary Miller. Kalona. 23 July 2014.
Preheim, Marilyn. Iowa City. 30 May 2014. With Royden Loewen.
Yoder, Brent, and Gerald Yoder. Wellman. 23 June 2014.
Yoder, Calvin. Wellman. 10 June 2014.
Yoder, Inez, and Vernon Yoder. Kalona. 18 July 2014.
Yoder-Short, David, and Jane Yoder-Short. Iowa City. 10 June 2014.

INTERVIEWS BY HANS PETER FAST IN FRIESLAND, NETHERLANDS

Transcripts translated from Dutch and Frisian by Hans Peter Fast, in possession of
author.
Bosscha, Baukje, and Hendrik Bosscha. Akkrum. 5 November 2014.
Buiteveld, Griet, and Jan Buiteveld. Nes. 26 December 2014.
Buiteveld, Yme Jan. Poppenwier. 20 December 2014.
de Jong, Marie Louise. Ijlst. 21 November 2014.
Faber, Margriet. Tzum. 12 June 2015.
Hanje, Jelke, and Roelie Hanje. Haskerhorne. 20 November 2014.
Hanje, Klaas, and Tiny Hanje. Offenwier. 28 November 2014.
Huisman-van der Meer, Ytsje. Grou. 22 June 2015.

Idsardi, Jan. Holwerd. 13 November 2015.
Odinga, Douke. Pingjum. 3 June 2015.

INTERVIEWS BY KERRY FAST IN RIVA PALACIO, SANTA CRUZ DEPARTMENT, BOLIVIA

Transcripts translated from Low German by Kerry Fast, in possession of author.
Blatz, Nettie. Schoenwiese. 30 July 2014.
Blatz, Tien. Schanzenfeld. 25 July 2014.
Buhler, Anne. Felsenthal. 25 July 2014.
Buhler, Susanna. Lichtfeld. 2014.
Enns, Helen Peters. Steinreich. 7 August 2014.
Enns, Marie. Rosengart. 29 July 2014.
Enns, Marie Goertzen. Steinreich. 6 August 2014.
Enns, Tina Hamm. Waldheim. 7 August 2014.
Hamm, Anne Peters. Rosenthal. 6 August 2014.
Hamm, Marie. Waldrich. 5 August 2014.
Peters, Agatha. Schanzenfeld. 29 August 2014.
Peters, Anita. Felsenthal. 25 July 2014.
Peters, Anne. Rosenthal. 31 July 2014.
Peters, Gertrude Neufeld. 7 August 2014.
Peters, Justina Wiens. Schanzenfeld. 27 July 2014.
Thiessen, Elizabeth. Schanzenfeld. 6 August 2014.
Wiebe, Anne Friesen. Rosengard. 29 July 2014.

INTERVIEWS BY SUSIE FISHER IN RURAL MUNICIPALITY OF RHINELAND
AND ENVIRONS, MANITOBA, CANADA

Transcripts by Fisher, in possession of author.
Braun, Joe. Altona. 17 June 2017.
Ens, Abe. Reinland. 20 July 2018.
Hildebrandt, Jeremy, and Megan Hildebrandt. Gretna. 17 August 2018.
Hildebrandt, Peter. Gretna. 26 October 2017.
Langalotz, Jonah. Rosenfeld. 26 July 2018.
Loewen, Benno, and Mary Loewen. Altona. 12 October 2017.
Mierau, Terry. Neubergthal. 12 March 2019.
Nickel, Grant, and Leona Nickel. Altona. 17 August 2017.
Nickel, Kevin. Rosenfeld. 19 June 2017.
Penner, Melvin. Altona. 22 September 2017.
Plett, Marlene. Altona. 31 July 2018.
Rempel, Wayne. Winkler. 27 July 2018.
Siemens, Ray. Altona. 20 February 2017.
Spain, Kalynn. Roland. 4 November 2018.
Wahl, Leonard. Halbstadt. 26 October 2017.

INTERVIEWS BY AILEEN FRIESEN IN APOLLONOVKA, SIBERIA, RUSSIA

Transcripts translated from Russian by Friesen, in possession of author.
Dik, Elena Petrovna. May 2015.
Dirksen, Abram. 13 May 2015.

Dirksen, Ivan Yakovich. 27 May 2015.
Dirksen, Petr Petrovich. 12, 28 May 2015.
Dirksen, Yakov. 18 May 2015.
Drei, Margarita Janzen. 12 May 2015.
Epp, David. 18 May 2015.
Epp, Ivan Petrovich. 15 May 2015.
Friesen, Vladimir. 13 May 2015.
Pauls, Andrei. 17 May 2015.
Pauls, Andrei, Jr. 22 May 2015.
Pauls, Valya. 5 November 2015.
Peters, Abram. 20 May 2015.
Peters, Ivan Ivanovich. 14 May 2015.
Peters, Olga. 20 May 2015.
Wilms, Agatha. 14 May 2015.

INTERVIEWS BY DANANG KRISTIAWAN IN MARGOREJO, JAVA, INDONESIA

Transcripts translated from Javanese by Kristiawan, in possession of author.
Dwi Sabdono. 25 February 2015.
Kasdi. 11 March 2016.
Kasmito Kliwon. 14 June 2016. With Royden Loewen.
Krismiati. 12 June 2014. With Royden Loewen.
Ngadiman. 11 March 2015.
Purwanto. 25 February 2015.
Retno, Adi. 14 June 2014. With Royden Loewen.
Sagi. 25 February 2015.
Solikon. 11 March 2015.
Sudarti. 16 June 2014.
Sugiman. 25 January 2016.
Sugioni. 16 June 2015.
Sugito. 11 March 2015.
Sukarman. 1 December 2015.
Suko Sukarman. 16 June 2014.
Sumarmi and Sutarmi. 14 June 2014. With Royden Loewen.
Sunardi. 23 March 2015.
Suratman. 25 February 2015.
Sutari. 23 March 2015.
Suyatno. 15 June 2014.
Yoga Saptoyo. 12 June 2014. With Royden Loewen.

INTERVIEWS BY BELINDA NCUBE IN MATOPO, MATABELELAND, ZIMBABWE

Transcripts translated from Ndebele by Ncube, in possession of author.
Dube, Mary. 19 April 2015.
Dube, Sibonginkosi. 10 January 2016.
Gumpo, Alfred. 24 April 2015.
Gumpo, Eldah. 14 November 2015.
Gumpo, Elliot. 14 November 2015.

Masuku, John. 29 April 2015.
Masuku, Sikhanyisiwe. 1 May 2015.
Mlilo, Heta. 1 May 2015.
Moyo, Effie. 10 January 2016.
Ndlovu, Denny. 15 May 2015.
Nyandeni, Zandile. 24 April 2015.
Sibanda, Timothy. 24 April 2015.
Sibindi, Priscilla. 8 May 2015.
Siziba, Eldah. 24 April 2015.
Siziba, Josephine. 18 May 2015.
Siziba, Thokozile. 10 January 2016.
Tshuma, Aaron. 8 May 2015.
Tshuma, Moses. 8 May 2015.

INTERVIEWS BY BEN NOBBS-THIESSEN IN RIVA PALACIO AND CHIHUAHUA COLONIES,
SANTA CRUZ DEPARTMENT, BOLIVIA

Transcripts translated from Spanish by Nobbs-Thiessen, in possession of author.
Boldt, Johan. Campo 11, Riva Palacio. 5 April 2014.
Buhler, Jakob. Campo 17, Riva Palacio. 21 July 2014.
Enns, Abram. Campo 21, Riva Palacio. 7 May 2014.
Fehr, Johan. Campo 27, Riva Palacio. 7 May 2014.
Friesen, Peter. Campo 1, Riva Palacio. 21 April 2015.
Froese, Cornelius. Campo 21, Riva Palacio. 6 May 2014.
Giesbrecht, Jakob. Campo 28, Riva Palacio. 17 April 2015.
Hamm, Abram. Campo 34, Riva Palacio. 3 April 2015.
Klassen, Abram. Campo 1, Riva Palacio. 23 April 2015.
Klassen, Peter. Campo 14, Riva Palacio. 29 July 2014.
Martens, Gerhard. Campo 35, Riva Palacio. 21 July 2014.
Peters, Cornelius. Campo 31, Riva Palacio. 11 May 2014.
Peters, Isaak. Campo 27, Riva Palacio. 30 April 2014.
Redekop, Abram. Campo 9, Riva Palacio. 27 July 2014.
Reimer, Abram. Campo 32, Riva Palacio. 3 August 2016.
Rempel, Abram. Central, Chihuahua Colony. 26 June 2014.
Rempel, Franz. Central, Chihuahua Colony. 26 June 2014.
Siemens, Enrique. Campo 11, Riva Palacio. 8 May 2014.
Thiessen, Abram. Campo 19, Riva Palacio. 17 April 2014.
Wall, Peter. Campo 35, Riva Palacio. 21 July 2014.
Wiebe, Enrique. Campo 9, Riva Palacio. 14 May 2015.

INTERVIEW IN ENGLISH BY BEN NOBBS-THIESSEN IN SANTA CRUZ DEPARTMENT, BOLIVIA

Transcript in possession of author.
Enns, Abram. Campo 21, Riva Palacio. 2014.

INTERVIEWS BY ROYDEN LOEWEN IN FRIESLAND, NETHERLANDS

Transcripts in possession of author.
De Vries, Menno. Witmarsum. 7 June 2018.

Trompetter, Cor. Wolvega. 7 June 2018.
Visser, Flora. Makkum. 31 July 2013.
Visser, Piet. Amsterdam. 5 June 2018.

INTERVIEW BY TRACEY RUTA IN STEINBACH, MANITOBA, CANADA, FOR THE MENNONITE
HERITAGE VILLAGE GARDEN PROJECT
Transcript in possession of the Mennonite Heritage Village.
Giesbrecht, Norma. Neubergthal, MB. 21 October 2004.

Other Primary Sources
"Administrators Record and Account Book of the Estate of John Reber Deceased, David
Reber and Samuel Guengrich Administrators." Mennonite Historical Society,
Kalona, IA.
"Ancestors of Audrey Brenneman Miller's Line: Oral and Gertrude Brenneman Yoder;
Audrey's Parents." Unpublished typescript in possession of Anita Miller Beachy,
Welman, IA.
Balzer, Heinrich. "Faith and Reason: The Principles of Mennonitism Reconsidered in a
Treatise of 1833." Trans. Robert Friedman. *Mennonite Quarterly Review* 22 (1948):
75–93.
Bekker, Elizaveta Dik. "Meine Lebensgeschichte." Unpublished, handwritten manu-
script, 1978. Original in possession of Peter P. Epp, Isul'kul, Russia.
Bill Jr. [Don W. Yoder]. "Flint Ridge, Bill Jr." *Kalona News*, 1964–87. In possession of
Lisa Lammer, Kalona, IA.
Bitsche, Fr. Jean-Théobold. "The Mennonites in Manitoba, 1883." In *The Outsiders' Gaze:
Life and Labour on the Mennonite West Reserve*, ed. Jacob E. Peters, Adolf Ens, and
Eleanor Chornoboy, 67–76. Winnipeg: Manitoba Mennonite Historical Society, 2015.
Blaupot ten Cate, Steven. *Geschiedenis der Doopsgezinden in Friesland.* Leeuwarden,
Netherlands: W. Eekhoff, 1839.
Blosser Yoder, Holly. *The Same Spirit: History of Iowa-Nebraska Mennonites.* Freeman,
SD: Central Plains Mennonite Conference, 2003.
Braun, Frank, and Tina Braun. Diary, 1958–73. Vol. 5707. Mennonite Heritage Archives,
Winnipeg.
Brenneman, Gertrude Yoder. Diary, 1942. In possession of Anita Miller Beachy,
Welman, IA.
Broer, Andries Lucas. "De Natuur en ons Godsdienstig Leven." Trans. Piet Visser.
Brieven: Uitgegevendoor de Vereeniging voor Gemeentedagen van Doopsgezinden 22:5
(1939): 101–4.
———. "Neveldag." In *Open Vensters*, trans. Piet Visser. Assen, Netherlands, 1929.
Bundy, G. E. Journal, August 1991. Brethren in Christ Archives, Messiah College,
Mechanicsburg, PA.
Davidson, Hannah Frances. *South and South Central Africa: A Record of Fifteen Years'
Missionary Labors among Primitive Peoples.* Elgin, lL: Brethren Publishing House,
1915.
de Jong, Sjoerd. Sjoerd (Sam) de Jong, Crookston, MN, to Johannes Hoogland,
St. Annaparochie, Netherlands, 5 January and 15 May 1915. Emigranten Brieven,
http://digicollectie.tresoar.nl/e_brieven/index.php?keuze=res&map=91&bnr=36.

Down, J. W. "Report on Colonization in Manitoba, 1876." In *The Outsiders' Gaze: Life and Labour on the Mennonite West Reserve*, ed. Jacob E. Peters, Adolf Ens, and Eleanor Chornoboy, 21–28. Winnipeg: Manitoba Mennonite Historical Society, 2015.

Dyck, Abram J. "Waldheim [Wissenfeld, Beresowka, Beckerschotor]." In *Mennoniten in der Umgebung von Omsk*, ed. Peter Rahn, 155–60. N.p., ca. 1975.

Dyck, John, ed. *Bergthal Gemeinde Buch*. Steinbach, MB: Hanover Steinbach Historical Society, 1993.

Dyck, John, and William Harms, eds. *1880 Village Census of the Mennonite West Reserve*. Winnipeg: Manitoba Mennonite Historical Society, 1998.

Elias, Peter A. "The Emigration from Russia to Manitoba." In *Church, Family and Village: Essays on Mennonite Life on the West Reserve*, ed. Adolf Ens, Jacob E. Peters, and Otto Hamm, 41–52. Winnipeg: Manitoba Mennonite Historical Society, 2001.

———. *Voice in the Wilderness: Memoirs of Peter A. Elias, 1843–1925*. Winnipeg: Manitoba Mennonite Historical Society, 2013.

Engle, Anna R., John A. Climenhaga, and Leona A. Buckwalter. *There Is No Difference: God Works in Africa and India*. Nappanee, IN: E. V. Publishing House, 1950.

Enns, Abram. "Das Menschen Leben." Unpublished handwritten memoir in possession of Aileen Friesen, Winnipeg.

———. "Tagebuch." 1994. In possession of Abram Enns, Campo 21, Riva Palacio, Bolivia.

Enns, Jacob. "Tagebuch." 1963. In possession of Abram Enns, Campo 21, Riva Palacio, Bolivia.

Epp, Johann. *Von Gottes Gnade Getragen*. Gummersbach, Germany: Verlag Friedens-stimme, 1984.

Executive Committee of the Regional Council of People's Deputies for the District of Isil'kul. "Decision: 17.07.1968 #131: Rules to Protect Public Health, Serving Brucellosis of Animals and State Farms." IAOO, F.r-2421, op.1, d.22 l. State Archives, Omsk, Russia.

———. "Document of Agriculture, 1967." IAOO F.r-1011, op.1, d.441, l.32. State Archives, Omsk, Russia.

———. "Land Balance Report." 1 November 1956. IAOO F.r-1937, op.1, d.524, l.75. State Archives, Omsk, Russia.

"Faith and Agricultural Forum Proceedings: March 12 and 13, 1982." Mennonite Historical Museum, Kalona, IA.

Fehr, Jacob. "From Russia to Canada in 1875." In *Reinland: An Experience in Community*, by Peter D. Zacharias, 33–41. Altona, MB: Reinland Centennial Committee, 1976.

Friesen, Anna. Diary, 1929–53. Vol. 5707:16–18. Mennonite Heritage Archives, Winnipeg.

Friesen, Jacob W. Diary, 1939–45. Vol. 4955. Mennonite Heritage Archives, Winnipeg.

Friesen, Katherina Wiebe. "Tagebuch." 1979. In possession of Abram Wiebe, Riva Palacio, Bolivia.

———. "Tagebuch." 2002. In possession of Abram Wiebe, Riva Palacio, Bolivia.

Government of Indonesia. Planning Department of the Department of Forestry. *Vegetation Map of Indonesia*. N.p., 1950. https://www.google.com/search?rls=en&sxsrf=ALeKk02nEzzgQiYOfd5p8dxJD3deGMAyug:1612617335608&source=univ&tbm=isch&q=Planning+Department+of+the+Department+of+Forestry++(Go

vernment+of+Indonesia).+Vegetation+Map+of+Indonesia.&client=safari&sa
=X&ved=2ahUKEwiP6cnbq9XuAhWIG80KHS71CEgQjJkEegQIBBAB&biw
=1440&bih=634.

Government of Zimbabwe. Department of Research and Special Services. *Provisional Soil Map of Zimbabwe Rhodesia*. N.p., 1979. http://esdac.jrc.ec.europa.eu/images/Eudasm/Africa/images/maps/download/afr_zw2006_so.jpg.

Green County Farm Bureau Women's Club. Minutes, 1959–96. Box 1. University of Iowa Women's Archive, Iowa City, IA.

Guengerich, Samuel D. Diary, 1866–1889. Copied by Mary A. Gingrich. State Historical Society of Iowa, Iowa City, IA.

———. "The First Mennonites in Iowa." 1924. Unpublished typescript comp. Mary A. Gingerich, 1956. State Historical Society of Iowa, Iowa City.

Hall, David. David Hall, Matopo Mission, Bulawayo, South Rhodesia, to Mr. and Mrs. John M. Hall, Upland, CA, 26 July 1932 and 1 July 1935. Brethren in Christ Archives, Messiah College, Mechanicsburg, PA.

Hamm, Johann F., Bernard Penner, Bernard F. Peters, Abram D. Wall, and Peter Wiebe. "Eine Erinnerung und Antwort von die Reinlaender Mennoniten Gemeinde auf die Aufforderung von Boliviansche Landsvolks." Typescript, 1997. In possession of Justina Peters, Schanzenfeld, Riva Palacio, Bolivia.

Hoogland, Sijke. Sijke Hoogland, Eldred, MN, to Johannes Hoogland, St. Annaparochie, Netherlands, 8 May and 12 December 1919. http://digicollectie.tresoar.nl/e_brieven/index.php?keuze=res&map=91&bnr=48.

Hoogland, Sjouke. Sjouke Hoogland, Adrian, MN, to Johannes Hoogland, St. Annaparochie, Netherlands, 9 September 1923. http://digicollectie.tresoar.nl/e_brieven/index.php?keuze=res&map=91&bnr=48.

ISRIC, World Soil Information. *Exploratory Soil Map of Java and Madura*. Wageningen, Netherlands, 1960. https://climateknowledgeportal.worldbank.org/watershed/161/climate-data-historical.

Jansz, P. A. "Regulations." Presented to the Dutch Mission Association Board, 18 March 1885. Doopsgezinde Archives, Amsterdam.

Janzen, Maria Derksen. "Meine Lebensgeschichte." 1991. Original in possession of Peter Epp, Isul'kul, Russia.

Kinsinger, Erlis. Diary, 1915–85. In possession of Nancy Halder, Parnell, IA.

Klaassen, Johann. "Im Weinberg des Herrn." *Die Friedenstimme*, 10, 31 May 1908, 15 January 1914.

Klassen, Menno. Unpublished memoir. Mennonite Heritage Archives, Winnipeg.

"Landslide, impassible bridges as heavy rains pound Matabeleland South." *Chronicle*, 3 February 2017. http://www.chronicle.co.zw/landslide-impassable-bridges-as-heavy-rains-pound-matabeleland-south/.

Matopo Staff Minutes, 1933–1956. VII-I-2.1. Missions, Zimbabwe. Mission Station Papers, Matopo. Brethren-in Christ Archives, Messiah College, Mechanicsburg, PA.

Menno Simons. "Foundation of Christian Doctrine." In *The Complete Writings of Menno Simons*, trans. Leonard Verduin. Scottdale, PA: Herald Press, 1956.

———. "The True Christian Faith." In *The Complete Writings of Menno Simons*, trans. Leonard Verduin. Scottdale, PA: Herald Press, 1956.

Miller, Dan. Account book, 1919–39. In possession of Ed and Henry Miller, Kalona, IA.

NASA Images. Accessed 16 October 2017. http://nasaimages.lunaimaging.com/luna /servlet/detail/nasaNAS-7-7-41736-145580:Muria-Volcano,-Island-of-Java,-Indo.

Natural History Museum of Zimbabwe. "Matobo Hills World Heritage Landscape: Management Plan, 2015–2019." http://naturalhistorymuseumzimbabwe.com/wp -content/uploads/2016/10/Matobo-Hills-WHS-Management-Plan-2015-2019.pdf.

Pendataan Jemaat, Kelompok Wakil: Kusnadi, Suntari, Sabari, Wargo Suwito. Margorejo, Indonesia: GITJ Margorejo, 2010.

Pendataan Jemaat, Kelompok Wakil: Sudarno, Sagi, Sukarto, Kristianto, Sarjono. Margorejo, Indonesia: GITJ Margorejo, 2010.

Pendataan Jemaat, Kelompok Wakil: Suparjono, Retnadi, Sutrisno. Margorejo, Indonesia: GITJ Margorejo, 2010.

Pendataan Jemaat, Kelompok Wakil: Suratman, Martono, Yoga Saptoyo. Jamin Daniel. Margorejo, Indonesia: GITJ Margorejo, 2010.

Peters, Johann K. "Tagebuch, 1968." In possession of Cornelio Peters, Campo 31, Riva Palacio, Bolivia.

Peterson, Duane. Diary, 1956–2000. In possession of Janet Peterson, Wayland, IA.

Preheim, Nettie Wittrig. "Memories of the Joseph Wittrig Family." Marilyn Preheim Rose Papers, Box 1, University of Iowa Women's Archive, Iowa City, IA.

Pries, Anita. *Exiled to Siberia.* Steinbach, MB: Derksen Printers, 1979.

Rae, W. Frazer. "Mennonites and Icelanders in Manitoba, 1881." In *The Outsiders' Gaze: Life and Labour on the Mennonite West Reserve,* ed. Jacob E. Peters, Adolf Ens, and Eleanor Chornoboy, 61–66. Winnipeg: Manitoba Mennonite Historical Society, 2015.

Rahn, Catarina Dueck. "Lebens Erinnerungen." Unpublished typed memoir in possession of Peter Epp, Isul'kul, Russia.

Soil Research Institute, Bogor. *Soil Associations of Java, Madura, and Bali.* Rome: Food and Agricultural Organization of the United Nations, 1959. https://esdac.jrc.ec .europa.eu/images/Eudasm/Asia/images/maps/download/PDF/ID2009_2SO.pdf.

South Africa National Bioversity Institute. PlantzAfrica. Accessed 25 November 2017. pza.sanbi.org/burkea-africana.

Sovkhoz Medvezhinskii. "(75) Report on the Number and Structure by Education of the Lead Workers and Specialists of the Sovkhoz 9.04.1963." IAOO, f.r-2421, op.1, d.20, l.75. State Archives, Omsk, Siberia.

Sovkhoz Medvezhdenskii Commission for the Acceptance of the Property of Kolkhoz Kirov. "Order from 9 March 1957." IAOO, f.r-1011, op.1, d.338, l. State Archives, Omsk, Russia.

Sovkhoz Novorozhdestvenskii. "Fond 2633: Historical Information." http://iaoo.ru/af /index.php?act=fund&fund=2000177309. State Archives, Omsk, Russia.

———. "Measures to Combat Smut, 1979." IAOO, F. r-2633, op.1, d.85, l.3. State Archives, Omsk, Russia.

———. "Requirement of Pesticides, 1979." IAOO, F. r-2633, op.1, d.85, l. and IAOO, F. r-2633, op.1, d.85, l.2. State Archives, Omsk, Russia.

Thiessen, Nicholai. "Im Weinberg des Herrn." *Die Friedenstimme,* 15 April, 5 August 2006.

Toews, Aron. *Siberian Diary of Aron P. Toews.* Winnipeg: CMBC, 1979.

UNESCO. "Nomination Dossier for the Proposed Matobo Hills World Heritage Area." Accessed 15 October 2017. http://whc.unesco.org/uploads/nominations/306rev.pdf.

United States Department of Agriculture, Soil Conservation Service. *Soil Survey of Washington County, Iowa*. Washington, DC, 1983. https://www.nrcs.usda.gov/Internet/FSE_MANUSCRIPTS/iowa/IA183/0/washington.pdf.

van Braght, Thielman. "Tjard Reynders." In *Martyrs Mirror of the Defenceless Christians*, trans. F. Sohm. Scottdale, PA: Herald Press, 1987.

Van Dyke, Henry J. "Among the Mennonites in Manitoba, 1880." In *The Outsiders' Gaze: Life and Labour on the Mennonite West Reserve*, ed. Jacob E. Peters, Adolf Ens, and Eleanor Chornoboy, 39–42. Winnipeg: Manitoba Mennonite Historical Society, 2015.

Vera, Raul R. *Country Pasture/Forage Resource Profiles: Bolivia*. Food and Agriculture Organization of the United Nations. Accessed 19 October 2017. http://www.fao.org/ag/agp/agpc/doc/counprof/Bolivia/Bolivia.htm#10.%20AUTHOR.

Visser-van Dam, Meintsje. *Ut it Libben fan Siebe Peenstra: Uit de Dageboeken van een Friese Veehouder, 1947–2010*. N.p., 2010.

Wiebe, Gerhard. *Causes and History of the Emigration of the Mennonites from Russia to America*. 1900. Trans. Helen Janzen. Winnipeg: Manitoba Mennonite Historical Society, 1981.

Wiebe, Johann. "Meine Erinnerungen und Miterleht Aufschreiben von die Auswanderung von Mexico nach Bolivian." Unpublished typescript in possession of Jakob Giesbrecht, Campo 28, Riva Palacio, Bolivia.

Wittrig, Nettie. Nettie Wittrig, Noble, IA, to "Dear Sister and Family," 3 August 1911 and 4 January 1913. Marilyn Preheim Rose Papers, Box 1, University of Iowa Women's Archive, Iowa City, IA.

World Bank, Climate Portal. "Average Monthly Rainfall for Indonesia at location (-6.44,110.99) from 1901–2015." Accessed 23 September 2017. http://sdwebx.worldbank.org/climateportal/index.cfm?page=country_historical_climate&ThisCCode=IDN.

———. "Average Monthly Temperatures for Indonesia at location (-6.44,110.99) from 1901–2015." Accessed 13 February 2021. https://climateknowledgeportal.worldbank.org/watershed/161/climate-data-historical.

World Bank Group. *Climate Change Knowledge Portal, Zimbabwe*. Accessed 5 September 2017. http://sdwebx.worldbank.org/climateportal/index.cfm?page=country_historical_climate&ThisRegion=Africa&ThisCCode=ZWE.

Wright, A. C. S., with Lucio Arce, Luis de Leon, and Rafael Pachoco. *Report on the Soils of Bolivia*. Rome: World Soils Resources Office, Food and Agriculture Organization of the United Nations, 1964. http://www.fao.org/3/a-15622e.pdf.

Yoder, Sanford Calvin. *Days of my Years*. 1949. Scottdale, PA: Herald Press, 1959.

Newspapers

Altona (MB) Echo
Die Friedenstimme (Russia)
Kalona (IA) News
Mennonitische Post (Steinbach, MB)
Red River (MB) Echo
Steinbach (MB) Post

Secondary Sources

"Agriculture Firm, Village Church Strive Together in Siberia." *Mennonite World Review*, 2 July 2018. http://mennoworld.org/2018/07/02/feature/agriculture-firm-village-church-thrive-together-in-siberia/.

Amstutz, Daniel, and Jan Matthijssen. "Margoredjo Mennonite Mission (Jawa, Indonesia)."

Global Anabaptist Mennonite Encyclopedia Online. 1957. https://gameo.org/index.php?title=Margoredjo_Mennonite_Mission_(Jawa,_Indonesia)&oldid=144334.

Anderson, Cory, and Loren Kenda. "What Kinds of Places Attract and Sustain Amish Populations?" *Rural Sociology* 80:4 (2015): 483–511.

Appadurai, Arjun. "Global Ethnoscapes: Notes and Queries for a Transnational Anthropology." In *Recapturing Anthropology: Working in the Present*, ed. Richard G. Fox, 191–210. Santa Fe, NM: School of American Research Press, 1991.

Barker Devine, Jenny. *On Behalf of the Family Farm: Iowa Farm Women's Activism since 1945*. Iowa City: University of Iowa Press, 2013.

Barnett, Lydia. "The Theology of Climate Change: Sin as Agency in the Enlightenment's Anthropocene." *Environmental History* 20:2 (2015): 217–37.

Bashford, Alison. *Global Population: History, Geopolitics, and Life on Earth*. New York: Columbia University Press, 2014.

Bayly, C. A., Sven Beckert, Matthew Connelly, Isabel Hofmeyr, Wendy Kozol, and Patricia Seed. "AHR Conversation: On Transnational History." *American Historical Review* 111:5 (2006): 1440–64.

Bean, Heather Ann Ackley. "Toward an Anabaptist Mennonite Environmental Ethic." In Redekop, *Creation and the Environment*, 183–205.

Belich, James. *Replenishing the Earth: The Settler Revolution and the Rise of the Anglo-World, 1783–1939*. Oxford: Oxford University Press, 2009.

Bender, Harold S. "A Brief Biography of Menno Simons." In *The Complete Writings of Menno Simons*. Scottdale, PA: Herald Press, 1956.

——. "The Anabaptist Vision." *Mennonite Quarterly Review* 18 (1944): 67–88.

Bender, Harold S., and Michael L. Yoder. "Rural Life." *Global Anabaptist Mennonite Encyclopedia Online*. 1989. https://gameo.org/index.php?title=Rural_life&oldid=143726.

Bergen, Peter. *Sommerfeld Village*. Altona MB: self-published, 1994.

Berninghausen, Jutta, and Birgit Kerstan. *Forging New Paths: Feminist Social Methodology and Rural Women in Java*. London: Zed Books, 1992.

Berry, Wendel. *The Art of the Commonplace: The Agrarian Essays of Wendell Berry*. Ed. Norman Wirzba. Berkeley: Counterpoint, 2002

——. "People, Land and Community." In *The Art of the Commonplace: The Agrarian Essays of Wendell Berry*, ed. Norman Wirzba, 182–94. Berkeley: Counterpoint, 2002.

——. "The Work of Local Culture." *The Contrary Farmer: Gene Logsdon Memorial Blogsite*. Accessed 9 January 2021. https://thecontraryfarmer.wordpress.com/2011/06/10/wendell-berry-the-work-of-local-culture/.

Bieleman, Jan. *Five Centuries of Farming: A Short History of Dutch Agriculture, 1500–2000*. Wageningen, Netherlands: Wageningen Academic, 2010.

Boyd, Brendon. "Working Together on Climate Change: Policy Transfer and Convergence in Four Canadian Provinces." *Journal of Federalism* 47:4 (2017): 546–71.

Braun, Ernest N., and Glen R. Klassen. *Historical Atlas of the East Reserve: Illustrated.* Winnipeg: Manitoba Mennonite Historical Society, 2015.

Brueggemann, Walter. *The Land: Place as Gift, Promise and Challenge to Biblical Faith.* Philadelphia: Fortress, 1977.

Canova, Paola. "Negotiating Environmental Subjectivities: Charcoal Production and Mennonite-Ayoreo Relations in the Paraguayan Chaco." *Journal of Mennonite Studies* 38 (2020): 59–82.

Careless, J. M. S. "Frontierism, Metropolitanism, and Canadian History." *Canadian Historical Review* 35:1 (March 1954): 1–21.

Carlisle, Liz. "Making Heritage: The Case of Black Beluga Agriculture on the Northern Great Plains." *Annals of the American Association of Geographers* 106:1 (2016): 130–44.

Carson, Rachel. *Silent Spring.* Boston: Houghton Mifflin, 1962.

Carter, Sarah. *Imperial Plots: Women, Land and the Spadework of British Colonialism on the Canadian Prairie.* Winnipeg: University of Manitoba Press, 2016.

Charon Cardona, Euridice, and Roger D. Markwick. "The Kitchen Garden Movement on the Soviet Home Front, 1941–1945." *Journal of Historical Geography* 64 (2019): 47–59.

Coen, Deborah R. "Big is a Thing of the Past: Climate Change and Methodology in the History of Ideas." *Journal of the History of Ideas* 77:2 (April 2016): 305–21.

Conquest, Robert. *Harvests of Sorrow: Soviet Collectivisation and the Terror-Famine.* Oxford: Oxford University Press, 1986.

Cronan, William. *Nature's Metropolis: Chicago and the Great West.* New York: Norton, 1992.

Crosby, Alfred W. *Ecological Imperialism: The Biological Expansion of Europe, 900–1980.* Cambridge: Cambridge University Press, 1986.

Cunfer, Geoff, and Fridolin Krausmann. "Adaptation on an Agricultural Frontier: Socio-Ecological Profiles of Great Plains Settlement, 1870–1940." *Journal of Interdisciplinary History* 46:3 (2016): 355–92.

Darian-Smith, Eve, and Philip C. McCarty. "Global Studies as a New Field of Inquiry." In *The Global Turn: Theories, Research Designs, and Methods for Global Studies*, 1–28. Oakland: University of California Press, 2017.

Daschuk, James. *Clearing the Plains: Disease, Politics of Starvation, and the Loss of Aboriginal Life.* Regina, SK: University of Regina Press, 2013.

de Hann, H. "Geert Veenhuizen (1857–1930): The Pioneer of Potato Breeding in the Netherlands." *Euphytica* 7:1 (1958): 31–37.

de Jong, Hielke. "Bread for the Poor: Potatoes and the Church." *Mennonite Historian* 43:3 (September 2017): 2, 4–5.

Deppermann, Klaus. "Melchior Hoffman: Contradictions between Lutheran Loyalty to Government and Apocalyptic Dreams." In *Profiles of Radical Reformers: Biographical Sketches from Thomas Muentzer to Paracelsus*, ed. Hans-Juergen Goertz, 178–90. Kitchener, ON: Herald Press, 1982.

de Stoppelaar, Reinder Jacobus. *Door Zon en Wind.* The Hague: Uitgevers-mij Haga, 1928.

Dharma, Adhi. "The Mennonite Churches of Indonesia." In *Churches Engage Asian Traditions*, Global Mennonite History Series: Asia, ed. C. Arnold Snyder and John A. Lapp, 21–124. Intercourse, PA: Good Books, 2011.

Dilly, Barbara Jane. "A Comparative Study of Religious Resistance to Erosion of the Soil and the Soul among Three Farming Communities in Northeast Iowa." PhD diss., University of California, Irvine, 1994.

Dixon, Ruth B. *Rural Women at Work: Strategies for Development in South Asia.* Baltimore: Johns Hopkins University Press, 1978.

Dop, B. E. "'Terwyl in de Kalkwyk Veele dier Doopsgezinden Woonen': Doopsgezinden in een Groninger Veenkolonie in de Achttiende Eeuw." *Doopsgezinde Bijdragen* 21 (1995): 97–132.

Downes, Rackstraw. *Nature and Art Are Physical: Writings on Art, 1967–2008.* New York: Edgewise, 2014.

Driedger, Leo. "Native Rebellion and Mennonite Invasion: An Examination of Two Canadian River Valleys." *Mennonite Quarterly Review* 46:3 (July 1972): 290–300.

Dube, Bekithemba, Doris Dube, and Barbara Nkala. "Brethren in Christ Churches in Southern Africa." In *Anabaptist Songs in African Hearts*, Global Mennonite History Series: Africa, ed. John A. Lapp and C. Arnold Snyder, 97–190. Intercourse, PA: Good Books, 2006.

Dyck, John. "Edinburg, 1879–1947." In Ens, Peters, and Hamm, *Church, Family and Village*, 271–86.

Eicher, John. "'Every family on their own': Iowa's Mennonite Farm Communities and the 1980s Farm Crises." *Journal of Mennonite Studies* 35 (2017): 75–96.

———. *Exiled among Nations: German and Mennonite Mythologies in a Transnational Age.* Cambridge: Cambridge University Press, 2020.

Enns, F. G. *Gretna: Window on the Northwest.* Gretna, MB: Village of Gretna History Committee, 1987.

Ens, Adolf. *Subjects or Citizens? The Mennonite Experience in Canada, 1870–1925.* Ottawa: University of Ottawa Press, 1994.

Ens, Adolf, Jacob E. Peters, and Otto Hamm, eds. *Church, Family and Village: Essays on Mennonite Life on the West Reserve.* Winnipeg: Manitoba Mennonite Historical Society, 2001.

Ens, Gerhard John. *Volost and Municipality: The Rural Municipality of Rhineland, 1884–1984.* Altona, MB: R. M. of Rhineland, 1984.

Epp, Marlene. "The Semiotics of Zwieback: Feast and Famine in the Narratives of Mennonite Refugee Women." In *Sisters or Strangers? Immigrant, Ethnic, and Racialized Women in Canadian History*, ed. Marlene Epp, Franca Iacovetta, and Frances Swyripa, 314–40. Toronto: University of Toronto Press, 2004.

———. *Women without Men: Mennonite Refugees of the Second World War.* Toronto: University of Toronto Press, 2000.

Epp, Peter. "My Experience Working on the Soviet Collective." Paper presented at "Mennonites, Land and the Environment: A Global History Conference," University of Winnipeg, 29 October 2016.

———. "A Short History of the Omsk Brotherhood." *Journal of Mennonite Studies* 30 (2012): 114–32.

Epp, Piotr. *100 let pod krovom Vsevyshnego: Istoriia Omskikh obshchin EKhB ikh ob'edineniia, 1907–2007.* Omsk: Steinhagen, 2007.

Epp-Tiessen, Esther. *Altona: The Story of a Prairie Town.* Altona, MB: D. W. Friesen & Sons, 1982.

Ericksen, Julia, and Gary Klein. "Women's Roles and Family Production among the Old Order Amish." *Rural Sociology* 46:2 (1981): 282–96.

Eyford, Ryan. *White Settler Reserve: New Iceland and the Colonization of the Canadian West.* Vancouver: UBC Press, 2016.

Fabricant, Nicole. "Ocupar, Resistir, Producir: Reterritorializing Soyscapes in Santa Cruz." In *Remapping Bolivia: Resources, Territory, and Indigeneity in a Plurinational State*, ed. Nicole Fabricant and Bret Gustafson, 145–65. Santa Fe, NM: School for Advanced Research Press, 2011.

Fast, Eduard. "Fast, Johann (1861–1941)." 1956. *Global Anabaptist Mennonite Encyclopedia Online.* https://gameo.org/index.php?title=Fast,_Johann_(1861-1941)&oldid =80746.

Fast, Hans Peter. "Nature and Neighbours in the Netherlands: Talking with *Doopsgezinde* Farmers about the Environment." *Journal of Mennonite Studies* 35 (2017): 61–74.

Fink, Deborah. *Open Country, Iowa: Rural Women, Tradition, and Change.* Albany: SUNY Press, 1986.

Fisher, Susie, ed. *Mennonite Village Photography: Views from Manitoba, 1890–1940.* Altona, MB: Mennonite Historic Arts Committee, 2020.

———. "Seeds from the Steppe: Mennonites, Horticulture, and the Construction of Landscapes on Manitoba's West Reserve, 1870–1950." PhD diss., University of Manitoba, 2017.

———. "(Trans)planting Manitoba's West Reserve: Mennonites, Myth, and Narratives of Place." *Journal of Mennonite Studies* 35 (2017): 127–48.

Flachs, Andrew. *Cultivating Knowledge: Biotechnology, Sustainability, and the Human Cost of Cotton Capitalism in India.* Tucson: University of Arizona Press, 2019.

Foster, Lance M. *The Indians of Iowa.* Iowa City: University of Iowa Press, 2009.

Francis, E. K. *In Search of Utopia: The Mennonites in Manitoba.* Altona, MB: D. W. Friesen & Sons, 1955.

Friesen, Aileen. *Colonizing Russia's Promised Land: Orthodoxy and Community on the Siberian Steppe.* Toronto: University of Toronto Press, 2020.

———. "Sowing Hatred or Producing Prosperity: Agriculture and Believers in Post–World War II Communist Siberia." *Journal of Mennonite Studies* (35) 2017: 287–302.

Friesen, Gerald. *The Canadian Prairies: A History.* Toronto: University of Toronto Press, 1984.

Friesen, John J. *Building Communities: The Changing Face of Manitoba Mennonites.* Winnipeg: CMU Press, 2007.

Friesen, Ted. *Memoirs: A Personal Autobiography of Ted Friesen.* Altona, MB: self-published, 2003.

Fountain, Philip. "Mennonites and Anthropology: An Introduction." *Journal of Mennonite Studies* 38 (2020): 11–22.

Fountain, Philip, and Laura Meitzner Yoder. "Quietist Techno-Politics: Agricultural Development and Mennonite Mission in Indonesia." In *The Mission of Development: Religion and Techno-Politics in Asia*, ed. Catherine Scheer, Philip Fountain, and R. Michael Feener, 214–42. Leiden: Brill, 2018.

Gaard, Greta. "Ecofeminism Revisited: Rejecting Essentialism and Re-Placing Species in a Material Feminist Environmentalism." *Feminist Formations* 23:2 (Summer 2011): 26–53.

Gaard, Greta, and Patrick D. Murphy, eds. *Ecofeminist Literary Criticism: Theory, Interpretation, Pedagogy.* Urbana: University of Illinois Press, 1998.

Geertz, Clifford. *Agricultural Involution: The Processes of Ecological Change in Indonesia.* Berkeley: University of California Press, 1963.

———. *The Interpretation of Cultures: Selected Essays.* New York: Basic Books, 1973.

Geiser, Samuel. "Switzerland." *Global Anabaptist Mennonite Encyclopedia Online.* February 2011. https://gameo.org/index.php?title=Switzerland&oldid=167784.

Gerbrandt, Henry J. *Adventure in Faith: The Background in Europe and the Development in Canada of the Bergthaler Mennonite Church of Manitoba.* Altona, MB: Bergthaler Mennonite Church of Manitoba, 1970.

———. *En Route Hinjawaejis: The Memoirs of Henry J. Gerbrandt.* Winnipeg: CMBC, 1994

Gill, Lesley. *Peasants, Entrepreneurs, and Social Change: Frontier Development in Lowland Bolivia.* Boulder, CO: Westview, 1987.

Gingerich, Melvin. *Mennonites in Iowa.* Kalona: Mennonite Historical Society of Iowa, 1974.

———. *The Mennonites in Iowa: Marking the One Hundredth Anniversary of the Coming of the Mennonites to Iowa.* Iowa City: State Historical Society of Iowa, 1939.

Gjerde, Jon. *The Minds of the West: Ethnocultural Evolution in the Rural Middle West, 1830–1917.* Chapel Hill: University of North Carolina Press, 1997.

Groenveld, S. "Doopsgezinden in Tal en Last: Nieuwe Historische Methoden en de Getalsvermindering der Doopsgezinden, ca. 1700–ca. 1850." *Doopsgezinde Bijdragen* 1 (1975): 81–110.

Gupta, Joyeeta. *The History of Global Climate Governance.* Cambridge: Cambridge University Press, 2014.

Guthrie, Gary. "Fidelity and Fecundity." *Topology Magazine: Artful Dispatches from Places Where We Find Ourselves,* 15 September 2016.

Hall, John W. *Uncommon Defense: Indian Allies in the Black Hawk War.* Cambridge, MA: Harvard University Press, 2009.

Hamm, H. H. *Sixty Years of Progress: The Rural Municipality of Rhineland, 1884–1944.* Altona, MB: Rural Municipality of Rhineland, 1944.

Harker, Ryan D., and Janeen Bertsche Johnson, eds. *Rooted and Grounded: Essays on Land and Christian Discipleship.* New York: Wipf & Stock, 2016.

Hedberg, Anna Sofia. *Outside the World: Cohesion and Deviation among Old Colony Mennonites in Bolivia.* Uppsala: Acta Universitatis Upsaliensis, 2007.

Helms, Douglas. "Conserving the Plains: The Soil Conservation Service in the Great Plains." *Agricultural History* 64:2 (Spring 1990): 58–73.

Hensen, Elizabeth. *Agrarian Revolt in the Sierra of Chihuahua, 1959–1965.* Tucson: University of Arizona Press, 2019.

Heppner, Jack. *Search for Renewal: The Story of the Rudnerweide/Evangelical Mennonite Mission Conference, 1937–1987.* Winnipeg: Evangelical Mennonite Mission Conference, 1987.

Hershatter, Gail. *The Gender of Memory: Rural Women and China's Collective Past.* Berkeley: University of California Press, 2011.

Hettinger, Ned. "Ecospirituality: First Thoughts." *Dialogue and Alliance* 9:2 (1992): 81–98.

Hiebert, Theodore. "Creation, the Fall, and Humanity's Role in the Ecosystem." In Redekop, *Creation and the Environment*, 111–21.

Hoekma, Alle. *Dutch Mennonite Missions in Indonesia: Historical Essays*. Elkhart, IN: Institute of Mennonite Studies, 2001.

Horner Shenton, Rebecca. "Agrarian Tradition vs Modern Farm Business: Monitoring a 'Debate' in the (Old) Mennonite Church." *Journal of Mennonite Studies* 35 (2017): 97–110.

Hughes, J. Donald. "Global Dimensions of Environmental History." *Pacific Historical Review* 70:1 (February 2001): 91–101.

Janzen, Rebecca. *Liminal Sovereignty: Mennonites and Mormons in Mexican Culture*. New York: SUNY Press, 2018.

Janzen, Waldemar. "In Quest of Place." In *Still in the Image: Essays in Biblical Theology and Anthropology*, 137–72. Winnipeg: CMBC, 1992.

Janzen, William. *Limits on Liberty: The Experience of Mennonite, Hutterite, and Doukhobor Communities in Canada*. Toronto: University of Toronto Press, 1990.

Jegathesan, Mythri. *Tea and Solidarity: Tamil Women and Work in Postwar Sri Lanka*. Seattle: University of Washington Press, 2019.

Jellison, Katherine. "Research Note: Amish Women and the Household Economy during the Great Depression." *Mennonite Quarterly Review* 88:1 (January 2014): 97–105.

Kasdorf, Julia. "'Work and Hope': Tradition and Translation of an Anabaptist Adam." *Mennonite Quarterly Review* 69:2 (April 1995): 178–204.

Klaassen, Walter. *Anabaptism: Neither Catholic nor Protestant*. Waterloo, ON: Conrad, 1973.

———. "Pacifism, Nonviolence, and the Peaceful Reign of God." In Redekop, *Creation and the Environment*, 139–53.

Klassen, Judith. "Sounding Spaces: *Lange Wies*, Community, and Environment." *Conrad Grebel Review* 33 (2015):168–75.

Klassen, Shelisa. "Heroes of a Flat Country: Mennonite Life, Agriculture, and Myth-making in Manitoban Newspapers, 1870s–1890s." *Journal of Mennonite Studies* 39 (2021), forthcoming.

———. "'Recruits and Comrades in a War of Ambition': Mennonite Immigrants in Late 19th Century Manitoba Newspapers." MA thesis, University of Manitoba, 2016.

———. "'Working Like Men': Newspaper Examinations of Gender, Respectability, and Mennonite Immigration to Manitoba in the Late Nineteenth Century." *Manitoba History* 84 (Summer 2017): 13–19.

Klein, Herbert S., and Francisco Vidal Luna. *Feeding the World: Brazil's Transformation into a Modern Agricultural Economy*. Cambridge: Cambridge University Press, 2019.

Kline, David. *Great Possessions: An Amish Farmer's Journal*. San Francisco: North Point, 1990.

Klippenstein, Lawrence. "Manitoba Metis and Mennonite Immigrants: First Contacts." *Mennonite Quarterly Review* 48:4 (October 1974): 476–88.

Konersmann, Frank. "Middle-Class Formation in Rural Society: Mennonite Peasant Merchants in the Palatinate, Rhine Hesse and the Northern Rhine Valley, 1740–1880." In *European Mennonites and the Challenge of Modernity over Five Centuries:*

Contributors, Detractors, and Adapters, ed. Mark Jantzen, Mary Sprunger, and John Thiesen, 109–42. North Newton, KS: Bethel College, 2016.

Krahn, Cornelius. *Dutch Anabaptism: Origin, Spread, Life and Thought (1450–1600)*. Scottdale, PA: Herald Press, 1981.

——. "Siberia (Russia)." *Global Anabaptist Mennonite Encyclopedia Online*. 1959. https://gameo.org/index.php?title=Siberia_(Russia)&oldid=162415.

Krahn, Cornelius, and Cornelius J. Dyck. "Menno Simons (1496–1561)." *Global Anabaptist Mennonite Encyclopedia Online*. 1990. https://gameo.org/index.php?title =Menno_Simons_(1496-1561)&oldid=160744.

Kraybill, Donald B., Karen M. Johnson-Weiner, and Steven M. Nolt. *The Amish*. Baltimore: John Hopkins University Press, 2013.

Kristiawan, Danang. "Javanese Wisdom, Mennonite Faith, and the Green Revolution: The Farmers of Margorejo." *Journal of Mennonite Studies* 35 (2017): 173–96.

Kuo, Hui-Ju. "Identifying Sustainability: The Measurement and Typology of Sustainable Agriculture in the United States." *EurAmerica* 48:2 (2018): 195–222.

Ladurie, Emmanuel Le Roy. *Montaillou: The Promised Land of Error and Cathars and Catholics in a French Village*. Trans. Barbara Bray. New York: Victor Books, 1979.

Lanning, James W. "The Old Colony Mennonites of Bolivia: A Case Study." MSc thesis, Texas A&M University, 1971.

Lee, Debbie, and Kathryn Newfont, eds. *The Land Speaks: New Voices at the Intersection of Oral and Environmental History*. New York: Oxford University Press, 2017.

Lefebvre, Henri. *Critique of Everyday Life*. Trans. John Moore. London: Verso, 1947.

Letkemann, Peter. "Mennonite Victims of 'The Great Terror,' 1936–1938." *Journal of Mennonite Studies* 16 (1998): 33–58.

Levine, Philippa. "Is Comparative History Possible?" *History and Theory* 53:3 (October 2014): 331–47.

Limerick, Patricia Nelson, Clyde A. Milner, and Charles E. Rankin, eds. *Trails: Toward a New Western History*. Lawrence: University Press of Kansas, 1991.

Loewen, Royden. "The Anti-Modern and the Medical: Mennonite Settlers and Cultures of Medicine in Latin America." *Histoire Sociale/Social History* 50 (2019): 137–54.

——. "'The Children, the Cows, My Dear Man and My Sister': The Transplanted Lives of Mennonite Farm Women, 1874–1900." *Canadian Historical Review* 73 (1992): 344–73.

——. "Come Watch this Spider: Animals, Mennonites and Indices of Modernity." *Canadian Historical Review* 96 (2015): 61–90.

——. "Competing Cosmologies: Reading Migration and Identity in an Ethno-religious Newspaper." *Histoire Sociale/Social History* 48:96 (May 2015): 87–105.

——. *Diaspora in the Countryside: Two Mennonite Communities in Mid-20th Century North America*. Urbana: University of Illinois Press; Toronto: University of Toronto Press, 2006.

——. *Ethnic Farm Culture in Western Canada*. Ottawa: Canadian Historical Association, 2002.

——. *Horse-and-Buggy Genius: Listening to Mennonites Contest the Modern World*. Winnipeg: University of Manitoba Press, 2016.

——. "The Quiet on the Land: Mennonites and Nature in History." *Journal of Mennonite Studies* 23 (2005): 151–64.

———. "Text, Trains and Time: Mennonite Migrants in South America." In *Place and Replace: Essays in Western Canadian History*, ed. Adelle Perry, Leah Morton, and Essylt Jones, 123–38. Winnipeg: University of Manitoba Press, 2012.

———. *Village Among Nations: "Canadian" Mennonites in a Transnational World, 1916–2006*. Toronto: University of Toronto Press, 2013.

Loewen, Royden, and Ben Nobbs-Thiessen. "The Steel Wheel: From Progress to Protest and Back Again in Canada, Mexico, and Bolivia." *Agricultural History* 92:2 (2018): 172–89.

Loewen, Royden, and Steven M. Nolt. *Seeking Places of Peace: North America; A Global Mennonite History*. Intercourse, PA: Good Books, 2012.

Loo, Tina. "High Modernism, Conflict, and the Nature of Change in Canada: A Look at *Seeing Like a State*." *Canadian Historical Review* 97:1 (2016): 34–58.

Luebke, Frederick C. "Ethnic Group Settlement on the Great Plains." *Western Historical Quarterly* 8:4 (October 1977): 405–30.

MacFayden, Joshua. *Flax Americana: A History of the Fibre and Oil that Covered a Continent*. Montreal: McGill-Queens University Press, 2018.

MacKinnon, Ian. "Javans Fired Up over Reactor next to Volcano." *Guardian* (Mount Muria, Indonesia), 5 April 2007. https://www.theguardian.com/world/2007/apr/05/indonesia.international.

MacMaster, Richard K. *Land, Piety, Peoplehood: The Establishment of Mennonite Communities in America, 1683–1790*. Scottdale, PA: Herald Press, 1985.

Mann, Carolyn, D. Climenhaga, A. Dube, I. Mpofu, K. Ndlovu, and M. Sidible. "Go My Spirit: The Story of the Brethren in Christ in Africa." Paper presented at the 75th Anniversary Conference, Matopo Mission, Rhodesia, 1973.

Massie, Merle. *Forest Prairie Edge: Place History in Saskatchewan*. Winnipeg: University of Manitoba Press, 2014.

Mazingi, Lucy, and Richard Kamidza. "Inequality in Zimbabwe." In *Tearing Us Apart: Inequalities in Southern Africa*, ed. H. Jauch and D. Muchena, 322–83. Johannesburg: OSISA, 2011.

McConnell, David L., and Marilyn D. Loveless. *Nature and the Environment in Amish Life*. Baltimore: Johns Hopkins University Press, 2018.

McCook, Stuart. *Coffee Is Not Forever: A Global History of the Coffee Leaf Rust*. Athens: Ohio University Press, 2019.

McDonald, Shirley. "Settler Life Writing, Georgic Traditions and Models of Environmental Sustainability." *American Review of Canadian Studies* 45:3 (2015): 283–98.

McKay, Ian. "The Liberal Order Framework: A Prospectus for a Reconnaissance of Canadian History." *Canadian Historical Review* 81:4 (2000): 616–45.

McNeill, J. R. *Something New under the Sun: An Environmental History of the Twentieth-Century World*. London: Allen Lane, 2000.

Mills, Gus. "Kalahari Soundscapes: The Functional Significance of Large Carnivore Vocalization." *Conrad Grebel Review* 33 (2015): 212–20.

Moon, David. *The American Steppes: The Unexpected Russian Roots of Great Plains Agriculture, 1870s–1930s*. Cambridge: Cambridge University Press, 2020.

———. "Cultivating the Steppe: The Origins of Mennonite Farming Practices in the Russian Empire." *Journal of Mennonite Studies* 35 (2017): 241–68.

———. *The Plough that Broke the Steppes: Agriculture and Environment on Russia's Grasslands, 1700–1914*. Oxford: Oxford University Press, 2013.

Morton, W. L. *Manitoba: A History*. Toronto: University of Toronto Press, 1957.

Mosley, Stephen. *The Environment in World History*. London: Routledge, 2010.

Multispecies Editing Committee, eds. *Troubling Species: Care and Belonging in a Relational World*. Munich: Rachel Carson Centre, 2017.

Mutel, Cornelia F. *The Emerald Horizon: The History of Nature in Iowa*. Iowa City: University of Iowa Press, 2008.

Nafziger, Tim. "Mennonites and the Conestoga Massacre of 1763." 3 February 2016. Dismantling the Doctrine of Discovery. https://dofdmenno.org/2016/02/03/mennonites-and-the-conestoga-massacre-of-1763/.

Ncube, Belinda. "Mothers, Soil and Substance: Stories of Endurance from Matobo Hills, Zimbabwe." *Journal of Mennonite Studies* 35 (2017): 223–40.

Ndlovu, Callistus P. "Missionaries and Traders in the Ndebele Kingdom: An African Response to Colonialism; A Case Study, 1859–1890." PhD diss., State University of New York, 1973.

Ndlovu, Scotch Malinga. *Brethren in Christ Church among the Ndebele, 1890–1970*. Frederick, ME: America Star Books, 2006.

Neff, Christian. "Brethren of the Common Life." *Global Anabaptist Mennonite Encyclopedia Online*. 1953. https://gameo.org/index.php?title=Brethren_of_the_Common_Life&oldid=146383.

———. "Jansz, Pieter (1820–1904)." *Global Anabaptist Mennonite Encyclopedia Online*. 1957. https://gameo.org/index.php?title=Jansz,_Pieter_(1820-1904)&oldid=145515.

———. "Jansz, Pieter Anton (1853–1943)." *Global Anabaptist Mennonite Encyclopedia Online*. 1957. https://gameo.org/index.php?title=Jansz,_Pieter_Anton_(1853-1943)&oldid=145516.

Neils Conzen, Kathleen. "Making Their Own America: Assimilation Theory and the German Peasant Pioneer." *German Historical Institute Annual Lecture Series* 3 (1990): 1–33.

———. "Peasant Pioneers: Generational Succession among German Farmers in Frontier Minnesota." In *The Countryside in the Age of Capitalist Transformation*, ed. Steven Hahn and Jonathan Prude, 259–92. Chapel Hill: University of North Carolina Press, 1985.

Neufeld, Alfred. *What We Believe Together: Exploring the "Shared Convictions" of Anabaptist-Related Churches*. New York: Good Books, 2015.

Neufeld, Mary. *A Prairie Pilgrim: Wilhelm H. Falk*. Winnipeg: self-published, 2008.

———. *Prairie Pioneers: Schoenthal Revisited*. Winnipeg: Manitoba Mennonite Historical Society, 2015.

Neufeldt, Colin P. "Reforging Mennonite Spetspereselentsky: The Experience of; Mennonite Exiles at Siberian Special Settlements in the Omsk, Tomsk, Novosibirsk and Narym Regions." *Journal of Mennonite Studies* 30 (2012): 269–314.

Nijdam, C., W. F. Golterman, and Lawrence M. Yoder. "Java (Indonesia)." *Global Anabaptist Mennonite Encyclopedia Online*. 1987. https://gameo.org/index.php?title=Java_(Indonesia)&oldid=167536.

Nisbet, Robert. "Community as Typology—Toennies and Weber." In *The Sociological Tradition*, 71–83. New York: Basic Books, 1966.

Nobbs-Thiessen, Ben. "Cheese is Culture and Soy is Commodity: Environmental Changes in a Bolivian Mennonite Colony." *Journal of Mennonite Studies* 35 (2017): 303–28.

——"Cultivating the State: Migrants, Citizenship, and the Transformation of the Bolivian Lowlands, 1952–2000." PhD diss., Emory University, 2016.

——*Landscape of Migration: Transnational Migrants and Agro-Environmental Change on Bolivia's Tropical Frontier, 1952–2000.* Chapel Hill: University of North Carolina Press, 2020.

Obach, Brian K. *Organic Struggle: The Movement for Sustainable Agriculture in the United States.* Cambridge, MA: MIT Press, 2017.

Ong, Walter J. *Orality and Literacy: The Technologizing of the Word.* London: Methuen, 1982.

Packull, Werner O. "The Origins of Swiss Anabaptism in the Context of the Reformation of the Common Man." *Journal of Mennonite Studies* 3 (1985): 36–59.

Page Moch, Leslie. *Moving Europeans: Migration in Western Europe since 1650.* Bloomington: Indiana University Press, 1992.

Palm, Risa, Gregory B. Lewis, and Bo Feng. "What Causes People to Change their Opinion about Climate Change?" *Annals of the American Association of Geographers* 107:4 (2017): 883–96.

Paping, Richard. "General Dutch Population Development, 1400–1850." Unpublished paper, University of Groningen, 2014. https://www.rug.nl/research/portal/files/15865622/articlesardinie21sep2014.pdf.

Patterson, Sean. *Makhno and Memory: Anarchist and Mennonite Narratives of Ukraine's Civil War, 1917–1921.* Winnipeg: University of Manitoba Press, 2020.

Peters, Klaas. *The Bergthaler Mennonites.* Winnipeg: CMBC, 1988.

Phiri, Marko. "El Nino and drought take a toll on Zimbabwe's cattle." *Reuters,* 11 January 2016. https://www.reuters.com/article/zimbabwe-drought-cattle-idUSL8N14U0W820160112.

Plett, Delbert F. *Old Colony Mennonites in Canada, 1875–2000.* Steinbach, MB: Crossway, 2001.

Podolsky, Glenn. "Canada-Manitoba Soil Survey: Soils of the Rural Municipality of Rhineland—Report D76." Winnipeg: Manitoba Department of Agriculture and Department of Soil Science, University of Manitoba, 1991. http://sis.agr.gc.ca/cansis/publications/surveys/mb/mbd76/mbd76_report.pdf.

Polanyi, Michael. "Articulation." In *Personal Knowledge: Towards a Post-Critical Philosophy,* 69–131. New York: Harper & Row, 1958.

Portelli, Alessandro. *The Death of Luigi Trastulli, and Other Stories: Form and Meaning in Oral History.* Albany: SUNY Press, 1991.

Radkau, Joachim. *Nature and Power: A Global History of the Environment.* Trans. Thomas Dunlap. Cambridge: Cambridge University Press, 2008.

——. "Nature and Power: An Intimate and Ambiguous Connection." In "Global Environmental History," special issue, *Social Science History* 37:3 (Fall 2013): 325–45.

Ranger, Terence O. *Voices from the Rocks: Nature, Culture, and History in the Matopos Hills of Zimbabwe.* Bloomington: Indiana University Press, 1999.

Rasmussen, R. Kent. *Mzilikazi.* Harare: Zimbabwe Educational Books, 1985.

Redclift, Michael. "In Our Own Image: The Environment and Society as Global Discourse." *Environment and History* 1:1 (February 1995): 111–23.

Redekop, Calvin W., ed. *Creation and the Environment: An Anabaptist Perspective on a Sustainable World*. Baltimore: Johns Hopkins University Press, 2000.

———. Introduction to Redekop, *Creation and the Environment*, xiii–xix.

Redekop, Magdalene. "Resisting Nostalgia: Little Shtahp on the Prairie." In *Making Believe: Questions About Mennonites and Art*, 124–61. Winnipeg: University of Manitoba Press, 2020.

Regehr, T. D. *Mennonites in Canada, 1939–1970: A People Transformed*. Toronto: University of Toronto Press, 1996.

Reid, Howard. *Henry Ellenberger: Pastor, Poet, Pioneer Organizer of The Zion Mennonite Church 1851*. Bluffton, OH: self-published, 1976.

Reschly, Steven D. *The Amish on the Iowa Prairie, 1840 to 1910*. Baltimore: Johns Hopkins University Press, 2000.

Reschly, Steven D., and Katherine Jellison. "Production Patterns, Consumption Strategies, and Gender Relations in Amish and non-Amish Farm Households in Lancaster County, Pennsylvania, 1935–1936." *Agricultural History* 67:2 (Spring 1993): 134–62.

Richards, John F. *The Unending Frontier: An Environmental History of the Early Modern World*. Berkeley: University of California Press, 2003.

Russell, Peter A. *How Agriculture Made Canada: Farming in the Nineteenth Century*. Montreal: McGill-Queens University Press, 2012.

Rybak, Arkadiusz. "The Significance of the Agricultural Achievements of the Mennonites in the Vistuala-Nogai Delta." Trans. Peter Klassen. *Mennonite Quarterly Review* 66 (1992): 214–20.

Salamon, Sonya. *Prairie Patrimony: Family, Farming, and Community in the Midwest*. Chapel Hill: University of North Carolina Press, 1992.

Saloutos, Theodore. "The Immigrant Contribution to American Agriculture." *Agricultural History* 50:1 (January 1976): 45–67.

Sandwell, Ruth. *Powering Up Canada: A History of Power, Fuel and Energy from 1600*. Montreal: McGill-Queens University Press, 2016.

Sawatzky, Harry Leonard. *They Sought a Country: Mennonite Colonization in Mexico*. Berkeley: University of California Press, 1971.

Schama, Simon. *Landscape and Memory*. New York: Knopf, 1995.

Schroeder, Nina. "'Parks magnificent as paradise': Nature and Visual Art among the Mennonites of the Early Modern Dutch Republic." *Journal of Mennonite Studies* 35 (2017): 11–39.

Scott, James C. *Domination and the Arts of Resistance: Hidden Transcripts*. New Haven, CT: Yale University Press, 1990.

———. *Seeing Like a State: How Certain Schemes to Improve the Human Condition Have Failed*. New Haven, CT: Yale University Press, 1998.

Seven Points on Earth. Director, Paul Plett. Producers, Royden Loewen and Paul Plett. Ode Productions, 2017. Film.

Sherow, James E. *The Grasslands of the United States: An Environmental History*. Santa Barbara, CA: ABC-CLIO, 2007.

Shiva, Vandana. *Making Peace with the Earth: Beyond Resource, Land and Food Wars*. New Delhi: Women Unlimited, 2012.

Shover, John L. *First Majority, Last Minority: The Transforming of Rural Life in America*. Dekalb: Northern Illinois University Press, 1976.

Shumba, Dorcas. "Women and Land: A Study on Zimbabwe." *Journal of Sustainable Development in Africa* 13:7 (2011): 236–44.

Sibanda, Eliakim. "Voices from the Hills vs. Words from the Missionary: Competing Rural Cultures in Southwestern Zimbabwe." *Journal of Mennonite Studies* 35 (2017): 197–222.

Sibanda, Harold. "Sustainable Indigenous Knowledge Systems in Agriculture in Zimbabwe's Rural Areas of Matabeleland North and South Provinces." *Indigenous Knowledge Notes* 2 (1998): 1–2.

Sider, E. Morris. "Davidson, Hannah Frances (1860–1935)." *Global Anabaptist Mennonite Encyclopedia Online.* 1988. https://gameo.org/index.php?title=Davidson, _Hannah_Frances_(1860-1935)&oldid=122476.

Simmons, I. G. "The World Scale." *Environment and History* 10:4 (November 2004): 531–36.

Smith-Howard, Kendra. "Ecology, Economy and Labor: The Midwestern Farm Landscape since 1945." In *The Rural Midwest since World War II*, ed. J. L. Anderson, 44–71. DeKalb: Northern Illinois University Press, 2013.

Snyder, C. Arnold. *Anabaptist History and Theology: An Introduction.* Kitchener, ON: Pandora, 1995.

———. "Margret Hottinger of Zollikon." In *Profiles of Anabaptist Women: Sixteenth-Century Reforming Pioneers*, ed. C. Arnold Snyder and Linda A. Huebert Hecht, 43–53. Waterloo, ON: Wilfrid Laurier University Press, 1997.

Sokolsky, Mark. "Taming Tiger Country: Colonization and Environment in the Russian Far East, 1860–1940." PhD diss., Ohio State University, 2016.

Sprunger, Mary. "A Mennonite Capitalist Ethic in the Dutch Golden Age: Weber Revisited." In *European Mennonites and the Challenge of Modernity over Five Centuries: Contributors, Detractors, and Adapters*, ed. Mark Jantzen, Mary Sprunger, and John Thiesen, 71–90. North Newton, KS: Bethel College, 2016.

Staples, John R. "Afforestation as Performance Art: Johann Cornies' Aesthetics of Civilization." In *Minority Report: Mennonite Identities in Imperial Russia and Soviet Ukraine Reconsidered, 1789–1945*, ed. Leonard G. Friesen, 61–84. Toronto: University of Toronto Press, 2018.

———. *Cross-Cultural Encounters on the Ukrainian Steppe: Settling the Molochna Basin, 1783–1861.* Toronto: University of Toronto Press, 2003.

Stayer, James M., Werner O. Packull, and Klaus Deppermann. "From Monogenesis to Polygenesis: The Historical Discussion of Anabaptist Origins." *Mennonite Quarterly Review* 42:2 (April 1975): 83–121.

Stewart, Ian. "Number Symbolism." *Encyclopedia Britannica Online.* https://www .britannica.com/topic/number-symbolism/7.

Stoesz, Donald. *Canadian Prairie Mennonite Ministers' Use of Scripture, 1874–1977.* Victoria, BC: Friesen Press, 2018.

Stunden Bower, Shannon. *Wet Prairie: People, Land, and Water in Agricultural Manitoba.* Vancouver: UBC Press, 2011.

Sturgeon, Noel. *Ecofeminist Natures: Race, Gender, Feminist Theory and Political Action.* Oxfordshire: Routledge, 1997.

Sukarto, Aristarchus. "Peace and Religious Pluralism as a Worldview." In *Overcoming Violence in Asia: The Role of the Church in Seeking Cultures of Peace*, ed. Donald

Eugene Miller, Gerard Guiton, and Paulus S. Widjaja, 107–36. Telford, PA: Cascadia, 2011.

Sukoco, Sigit Heru, and Lawrence M. Yoder. *Tata Injil di Bumi Muria: Sejarah Gereja Injili di Tanah Jawa.* Semarang, Indonesia: Pustaka Muria, 2010.

——. "Way of the Gospel in the World of Java: A History of the Muria Javanese Mennonite Church, GITJ." Trans. Lawrence M. Yoder. Unpublished manuscript, ca. 2015.

——. *Way of the Gospel in the World of Java: A History of the Muria Javanese Mennonite Church (GITJ).* Goshen, IN: Goshen College, 2020.

Sum, Maisie. "Inspiration, Imitation and Creation in the Music of Bali, Indonesia." *Conrad Grebel Review* 33 (2015): 251–60.

Sutter, Paul S. "The World with Us: The State of American Environmental History." *Journal of American History* 100:1 (June 2013): 94–119.

Swierenga, Robert. "The New Rural History: Defining the Parameters." *Great Plains Quarterly* 1:4 (Fall 1981): 211–23.

Taylor, Jeffrey. *Fashioning Farmers: Ideology, Agricultural Knowledge, and the Manitoba Farm Movement, 1890–1925.* Regina, SK: Canadian Plains Research Center, 1994.

Thiessen, Janis. *Manufacturing Mennonites: Work and Religion in Post-War Manitoba.* Toronto: University of Toronto Press, 2013.

Tickamyer, Ann R., and Siti Kusujiarti. *Power, Change, and Gender Relations in Rural Java: A Tale of Two Villages.* Athens: Ohio University Press, 2012.

Toews, J. A. *A History of the Mennonite Brethren Church: Pilgrims and Pioneers.* Fresno, CA: Mennonite Brethren Churches, 1975.

Toews, John B. *Czars, Soviets and Mennonites.* Newton, KS: Faith and Life Press, 1982.

——. "The Mennonites and the Siberian Frontier (1907–1930): Some Observations." *Mennonite Quarterly Review* 47:2 (April 1973): 83–101.

Trompetter, Cor. *Agriculture, Proto-Industry and Mennonite Entrepreneurship: A History of the Textile Industries in Twente, 1600–1815.* Amsterdam: NEHA, 1997.

——. *Doopsgezinden in Friesland, 1530–1850.* Leeuwarden, Netherlands: Bornmeer, 2016.

——. "Mennonite Capital and Rural Transformation in Friesland and Overijssel, 1700–1850." *Journal of Mennonite Studies* 35 (2017): 41–59.

Turner, Frederick J. "The Significance of the Frontier in American History." *Annual Report of the American Historical Association,* 1893, 197–227.

Urban-Mead, Wendy. *The Gender of Piety: Family, Faith, and Colonial Rule in Matabeleland, Zimbabwe.* Athens: Ohio University Press, 2015.

Urry, James. *Mennonites, Politics, and Peoplehood: Europe—Russia—Canada, 1525–1980.* Winnipeg: University of Manitoba Press, 2006.

van der Zijpp, Nanne. "Annaparochie, Sint (Friesland, Netherlands)." *Global Anabaptist Mennonite Encyclopedia Online.* 1953. https://gameo.org/index.php?title =Annaparochie,_Sint_(Friesland,_Netherlands)&oldid=143836.

——. "Jacobsz, Jan (1542–1612)." *Global Anabaptist Mennonite Encyclopedia Online.* 1957. https://gameo.org/index.php?title=Jacobsz,_Jan_(1542-1612)&oldid=121159.

——. "Klaassen, Johann (1872–1950)." *Global Anabaptist Mennonite Encyclopedia Online.* 1957. https://gameo.org/index.php?title=Klaassen,_Johann_(1872 -1950)&oldid=119319.

van der Zijpp, Nanne, and C. F. Brüsewitz. "Netherlands." *Global Anabaptist Mennonite Encyclopedia Online.* February 2011. https://gameo.org/index.php?title =Netherlands&oldid=167673.

van Engelen, A. F. V., J. Buisman, and F. Ijsen. *A Millennium of Weather, Winds and Water in the Low Countries.* De Bilt: Royal Netherlands Meteorological Institute, 1995. https:// projects.knmi.nl/klimatologie/daggegevens/antieke_wrn/millennium_of_weather.pdf.

Vargas, Carolina, and Martha Garcia Ortega. "Vulnerability and Agriculture among Old Colony Mennonites in Quintana Roo: A Research Note." *Journal of Mennonite Studies* 35 (2017): 329–38.

Vellings, Piet. *75 jaar Fredeshiem.* Steenwijk, Netherlands: Bu'lah Print, 2004.

Verbeek, Annelies, and Alle G. Hoekema. "Mennonites in the Netherlands." In *Testing Faith and Tradition, Global Mennonite History Series: Europe,* ed. Claude Baecher, Neil Blough, James Jakob Fehr, Alle G. Hoekema, Hanspeter Jecker, John N. Klassen, Diether Goetz Lichdi, Ed van Straten, and Annelies Verbeek, 57–96. Intercourse, PA: Good Books, 2006.

Vertovec, Steven. *Transnationalism.* New York: Routledge, 2009.

Viola, Lynne. "Bab'i bunti and Peasant Women's Protest during Collectivization." *Russian Review* 45:1 (January 1986): 23–42.

Visser, Piet, and Grietje Pol-Visse, eds. *Meniste Jistertinzen: Samle Fersen fan Renske Visser-Oosterhof (1915–1984).* Zaandam, Netherlands: Stichting AD&L, 2004.

Vonk, Martine. *Sustainability and Quality of Life: A Study on the Religious Worldviews, Values and Environmental Impact of Amish, Hutterite, Franciscan and Benedictine Communities.* Amsterdam: Buitjen & Schipperheijn Motief, 2011.

Vos, Karel, and Nanne van der Zijpp. "Friesland (Netherlands)." *Global Anabaptist Mennonite Encyclopedia Online.* 1956.

Warkentin, John. *The Mennonite Settlements of Southern Manitoba.* Steinbach, MB: Hanover Steinbach Historical Society, 2000.

Weaver, Dorothy Jean. "The New Testament and the Environment: Toward a Christology for the Cosmos." In Redekop, *Creation and the Environment,* 122–38.

Weaver, J. Denny. *The Nonviolent Atonement.* Grand Rapids, MI: Wm. B. Eerdmans, 2001.

Weaver, John C. *The Great Land Rush and the Making of the Modern World, 1650–1900.* Montreal: McGill-Queen's University Press, 2003.

Weaver-Zercher, David L. *Martyrs Mirror: A Social History.* Baltimore: Johns Hopkins University Press, 2016.

Werner, Hans. "Modelling Mennonites: Farming the Siberian Kulunda Stepp, 1921– 1928." *Journal of Mennonite Studies* 35 (2017): 269–87.

———. "Siberia in the Mennonite Imagination, 1880–1914: Land, Weather, Markets." *Journal of Mennonite Studies* 30 (2012): 159–72.

Wessel, Judith Ann. "The Mennonites in Bolivia: An Historical and Present Socio-Economic Evaluation." Unpublished study, Cornell University, 1967.

White, Lynn. "Christian Myth and Christian History." *Journal of the History of Ideas* 3:2 (1942): 145–58.

Wiebe, Armin. *The Salvation of Yasch Siemens.* Winnipeg: Turnstone, 1984.

Wiebe, Jeremy. "A Different Kind of Station: Radio Southern Manitoba and the Reformulation of Mennonite Identity, 1957–1977." MA thesis, University of Manitoba, 2008.

Wiebe, Joseph R. "On the Mennonite-Metis Borderland: Environment, Colonialism and Settlement in Manitoba." *Journal of Mennonite Studies* 35 (2017): 111–26.

———. *The Place of Imagination: Wendell Berry and the Poetics of Community, Affection, and Identity.* Waco, TX: Baylor University Press, 2017.

Wiebe, Peter. "The Mennonite Colonies of Siberia: From the Late Nineteenth to the Early Twentieth Century." *Journal of Mennonite Studies* 30 (2012): 23–35.

Willems, Yvonne. *Menno Simons Groen, Tentoonstellung: Mennonieten Al 500 Jaar Duursaam* [Menno Simons Groen exhibition: Mennonites over 500 years of sustainability]. Witmarsum, Netherlands, 2018. Publish in conjunction with an exhibition at the Groencentrum, Witmarsum, June–August 2018. http://www .dekoepel.fri/tentoonstelling-menno-simons-groen.

Worster, Donald. "History as Natural History: An Essay on Theory and Method." *Pacific Historical Review* 53:1 (February 1984): 1–19.

———. "World Without Borders: The Internationalizing of Environmental History." *Environmental Review* 6:2 (Autumn 1982): 8–13.

Wright, Wynne, and Alexis Annes. "Farm Women and the Empowerment Potential in Value-Added Agriculture." *Rural Sociology* 81:4 (2016): 545–71.

Yoder, Bill. "Siberian Success: Agriculture Firm, Village Church Strive Together." *Mennonite World Review*, 2 July 2018.

Yoder, Franklin L. *Opening a Window to the World: A History of Iowa Mennonite School.* Kalona: Iowa Mennonite School, 1994.

Yoder, Lawrence M. "Gereja Injili di Tanah Jawa (GITJ)." *Global Anabaptist Mennonite Encyclopedia Online.* 1990. https://gameo.org/index.php?title=Gereja_Injili_di _Tanah_Jawa_(GITJ)&oldid=121094.

———. *The Muria Story: A History of the Chinese Mennonite Churches of Indonesia.* Kitchener, ON: Pandora, 2006.

———. "Tunggul Wulung (Tunggulwulung), Ibrahim (d. 1885)." *Global Anabaptist Mennonite Encyclopedia Online.* 1989. https://gameo.org/index.php?title=Tunggul _Wulung_(Tunggulwulung),_Ibrahim_(d._1885)&oldid=130450.

———. "The Villages of Tunggul Wulung and Pieter Janz: Vision and Reality in the Javanese Countryside." *Journal of Mennonite Studies* 35 (2017): 149–72.

Yoder, Rhonda Lou. "Amish Agriculture in Iowa: A Preliminary Investigation." MA thesis, Iowa State University, 1990.

Yoder Lind, Katie. *From Hazelbrush to Cornfields: The First One Hundred Years of the Amish-Mennonites in Johnson, Washington and Iowa Counties of Iowa, 1846–1946.* Kalona: Mennonite Historical Society of Iowa, 1994.

Zacharias, Committee, 1976.Peter D. *Reinland: An Experience in Community.* Altona, MB: Reinland Centennial

Africans, relations with North Americans, 254

agricultural commodities, 2, 5, 8; alfalfa, 184; barely, 23, 35, 72, 85, 99; beans, 19, 102, 129, 243; casava, 8, 105, 150, 189, 261' clover, 19, 184; coconuts, 131; corn/maize, 5, 8, 22–23, 25, 50, 58, 61–62, 72, 77, 80–84, 86, 98–99, 100, 102–104, 123, 125, 135, 166–67, 172, 190, 217, 223–25, 228, 241, 242–43; flax, 19, 34, 87; hay, 34, 41, 75, 93, 97, 125, 219–22, 287n52; kaffir, 100, 241; lentils, 251; oats, 35 72, 77, 83, 89, 98–99, 102; mustard, 251; napoko, 135, 252; peanuts, 62, 105, 145, 172, 218; peas, 19, 129, 186, 228; rapeseed, 19; rice, 3, 8, 48–49, 53–56, 104–8, 131, 145, 150–51, 188–90, 131, 145, 260–61, 266–67; sorghum, 72, 239, 252; soybeans, 2–3, 58, 80–81, 100, 103–4, 125, 177, 185, 193, 217, 225, 238, 242, 251; sugar beets, 19, 22, 98, 182; sunflowers, 22, 158, 162, 182, 185; sweet potatoes, 8, 14–15, 19, 62, 90, 109, 129, 135, 162, 172, 183, 256, 280n3; viscose, 8, 135; watermelon, 35, 145, 216, 260–61; wheat, 5, 8, 9, 19, 25, 28, 32, 34–36, 40, 42–43, 83, 86, 98, 100, 162, 180, 182, 186–87, 227, 250–52

agricultural extension services, 112, 184, 188, 193–94, 198, 201–2

agricultural history, sustainability focus, 9–10

Agricultural Involution (Geertz), 49

agricultural knowledge: global flow, 11–12, 234, 242–45, 250–51; of Mennonite women, 149–50, 152, 157, 159, 162, 163, 165, 260

agricultural land development, 14

agricultural machinery. *See* agricultural technology

agricultural modernization, 2, 3, 15, 75–114, 266; 1980s-1990s, 82–83; farm specialization, 78; in Global South, 46; Old Colony Mennonites' rejection of, 64, 79, 88–89, 141–42, 143, 171; organic farming, 125; of post-colonial nation-states, 46–47; resistance, 267–68. *See also under specific Mennonite communities*

agricultural out-reach programs, 242–43

agricultural practices, Indigenous people, 46; Bolivia, 72; Iowa, 23–24; Java, 49–50; Manitoba, 86; Mexico, 99–100; Ndebele people (Zimbabwe), 56–59, 61, 62, 63

agricultural practices, Mennonite, 46–47; as climate change adaptations, 219–20, 221–22, 223, 224, 231; crop diversification, 217–18; land quality relationship, 266; rural Anabaptists, 26–27; for specific soil conditions, 266. *See also under specific Mennonite communities*

agricultural technology: combine harvesters, 80, 86, 150, 217, 233, 239; fossil fuel-powered, 75–76, 77–78, 80–81, 91, 96–97; global sourcing of, 232–33, 239, 263; horse-drawn, 79, 97; impact on soil, 216–17; implication for climate change mitigation, 220–21, 223, 224, 226, 229, 231; influence on communitarian agricultural practices, 267; planters and seeders, 29, 88, 173, 229, 232; plows, 3, 28–29, 63, 79, 84, 95, 110, 125, 146, 248; trucks, 108, 142, 199. *See also under specific Mennonite communities;* tractors

agriculture: holistic approach, 13; local culture's influence on, 1, 265–68

à Kempis, Thomas, 18

Allgemeine Doopsgezinde Societaet, "Seven
Points on Earth" team workshops, 11
Alsace Loraine, 26–27, 44
American frontier, 26, 32, 282n62; mythology
of, 27. See also Iowa Mennonite communities; Washington County, IA, Mennonite
communities
Amish, 6; agricultural modernization, 79; in
Iowa, 23–30, 169, 197–200, 267; opposition
to government subsidies, 198–200;
pre-Christian folklore, 123–24; religion/land
management relationship, 117–18, 123,
143–44, 290n13; resistance to modernity,
267; state-farmer relations, 197–200;
women's status, 169
Ammann, Jakob, 26–27
Anabaptists, 4, 6, 17; Dutch, 19–20; global
community, 6–7; history, 25–26; Swiss-South
German, 14, 15; values, 6
animals, draft: cows, 132, 146, 159; donkeys, 3
61, 63, 99, 108, 203; horses, 26, 29, 42, 53, 66,
79, 84, 88, 94, 97, 142, 204, 219, 246; oxen, 42,
97, 108, 110, 153, 238
animals, raising: cattle (beef), 5, 8, 14, 56, 58, 61,
71, 97, chickens, 63, 84–85, 90, 97, 97, 165,
185–86; cows (dairy), 2, 3, 5, 14, 18, 20, 22, 29,
82–84, 89–93, 139, 161–62, 189, 194–96, 198,
203–4, 238, 241, 248, 257; ducks, 124; fish
farming, 107; goats, 135, 201, 254; sheep, 97,
124, 162; turkeys, 78, 80, 186
Animistic religion, 31, 135, 153; cosmology, 45,
59, 257; syncretism, 46, 136, 143, 156
Anishinaabe, 30–31, 180–81, 249
Apollonovka, Siberia, 1–2, 3, 4, 5; agricultural
commodities, 5, 8; agricultural modernization, 96–98; agricultural practices, 95–96,
203, 256; agricultural technology, 203;
climate and environment, 5, 8–9, 163;
colonialism and, 16; ecotone and ecozone, 5,
8–9; foreign investments, 256; global
engagement, 235, 256–57, 263; as joint stock
company, 256–57; state-farmer relations, 37
Apollonovka, Siberia (during Soviet collectivization), 94–98, 178, 267; agricultural
modernization, 93–98, 139, 203; agricultural
practices, 203–6; agricultural technology,
96–97, 204–6; Bolshevik Revolution and, 42,
43, 94, 157, 161, 202; global engagement,

254–56, 257–58, 263; Kolkhoz Kirov
collective, 96–97, 99, 137, 202–4, 206, 287n46;
Mennonites' German identity, 156–57,
158–59, 255, 256–57; post-Communist era,
202–6, 255, 256–57, 258; religion/land
management relationship, 94, 95, 136–40,
143, 255–56, 268–69; religious life, 136–40;
religious persecution, 94–95, 136–37, 156–57;
settlement history, 37–43; soil conditions,
5; Sovkhoz Medvezhinskii, 94, 96–98, 163,
206, 287n52, 288n56; Sovkhoz Novorozhdestvenskii, 94, 98, 138, 206, 254–56, 288n56;
during Stalin's purges, 93–94, 95, 137, 158;
state-farmer relations, 37, 156–73, 176–207,
202–6; women's experiences, 147–48, 156–63,
174; during WW II, 93–94, 158–60
Art of the Commonplace (Berry), 116, 279n2

Banzer, Hugo (president), 192–93
Baptists, Mennonite Brethren-derived, 6
Barker, Jenny (historian), 164
Barnett, Lydia (theologian), 231
Barrientos Ortuño, René (general), 65
Bashford, Alison (historian), 164
Beachy Amish, 224–25, 244–45
Bekker, Elizaveta (memoirist), 157, 160–63
Belich, James (historian), 177
Bender, Harold S., 19, 26, 280n18
Bergthal Colony, Russia, 126
Bergthaler Mennonite Church, 126–29
Bergthaler Mennonites, 32–34, 181, 185, 290n28
Bergthaler-Sommerfelder schism, 126–29
Berry, Wendell, 1, 118, 279n2; Art of the
Commonplace (Berry), 116, 279n2
Bertsche Johnson, Janeen, Rooted and
Grounded, 117
Bieleman, Jan, Five Centuries of Farming, 15,
17–18, 19, 193–94, 280n3
Black Hawk War, 24–25
Bolivia, 1; Easter Revolution, 64, 67, 70, 99;
economic modernization policy, 99; national
food security policy, 64–65, 67–68, 73, 99,
190, 192–93, 235–36, 266; as post-colonial
nation-state, 47. See also Riva Palacio Colony,
Santa Cruz, Bolivia
Bolshevik Revolution, 42, 43, 94, 157, 161, 202
Brandstadt, Terry (governor), 197–98
Brash, Joe, Textbook of Agriculture, 111

Brenneman, Gertrude Mae (memoirist), 165–67
Brethren in Christ Church (BICC), 6
Brethren in Christ Church (BICC) mission
    farms, 109, 112–13; African leadership, 111–13;
    agricultural practices, 109–10; social justice
    ethic, 200; white missionaries, 134–35,
    152–53, 200. *See also* Matopo Mission,
    Matabeleland, Zimbabwe
Brethren of Christ Church, 6
Brethren of the Common Life, 19–20
British Empire, 56, 235
British South African Company, 56–57
Broer, Andries, Luca (pastor), 115–16
Brüsewitz, C. E. (activist), 119–20
Bundy, George (missionary), 112

Canada: as international aid source, 254;
    Mennonite migrations to, 176–77; Mennonite
    settlers' relations with, 263; Old Colony
    Mennonites' migration from, 64, 99, 100; Old
    Colony Mennonites' ongoing relationship,
    239–41, 250; Prairie Farm Rehabilitation Act,
    182; Sommerfelder migration from, 127, 263
Canadian Mennonite communities, 1;
    interaction with Russian Mennonites, 256;
    metropolitan-hinterland thesis, 32, 282n62;
    state-farmer relations, 177–78, 179–87.
    See also Manitoba Mennonite communities;
    Rural Municipality of Rhineland
Canadian Prairies, The : A History (Friesen), 32
Canadian Wheat Board, 186, 187, 250
capitalism: agrarian resistance, 267–68;
    agricultural system, 206; resistance to, 73;
    rural, 120
Carson, Rachel, 147
Carter, Jimmy, 241–42
Catholics: comparison with, 117; as neighbors,
    22, 82, 119, 122, 143
charity, 16, 22, 119, 131, 151, 268
chemical-based agriculture, 239; Iowa, 79–80,
    81, 82; Manitoba, 85–87; Margorejo, Java,
    188; organic farming *vs.*, 82–83; Riva Palacio,
    Bolivia, 103–4, 239; Siberian, 98
China, 239, 242
church buildings, 54, 55, 106, 121, 130, 137–38
climate, 5, 8; Bolivia, 67; Friesland, 17; Iowa, 28;
    Java, 48–49; Manitoba, 34; Matabeleland, 59;
    Siberia, 39

climate change, 12, 208–31, 242, 265;
    deforestation and, 212–13, 214, 215–16;
    denial, 209–10, 218, 222, 224, 226–29;
    Global South vs Global North opinions,
    209–10, 230–31; local contributions to,
    233; positive attitudes toward, 226–28,
    230; religion/land management relation-
    ship and, 210–13, 221, 224–27; religious
    perspective, 221–22
Climenhaga, John A. (missionary), 109, 110,
    133–34
Coen, Deborah R. (historian), 209
collective agriculture. *See* Apollonovka, Siberia
    (during Soviet collectivization)
colonialism, 16; Dutch, in Java, 47–56, 187,
    258–59, 267; in Matopo, Matabeleland, 47,
    56–57, 59–60, 63–64, 152, 153. *See also*
    Indigenous peoples' dispossession
Common Market, 88, 93, 194, 195, 235, 247–48;
    Mansholt Plan, 90, 194
communitarian values, 76, 268, 280n15; Iowa
    Mennonites, 78; Old Colony Mennonites, 66,
    140–41, 190, 237, 263, 267; Sommerfelders,
    126–28
comparative analysis, 3–4, 206
Conestoga people, 26
conscientious objectors, 120. *See also*
    nonviolence; pacifism
Conservative Mennonites, 198, 199–200
Cornies, Johann (reformer), 39, 95
*Creation and the Environment: An Anabaptist
    Perspective on a Sustainable World* (Redkop),
    116
Cronan, William (historian), 25
crop insurance, 185, 186, 197, 198, 219
crop specialization, 78, 79–80
Crosby, Alfred W. (historian), 15, 37, 38, 46
cultural sustainability, 268–69
cultural transfer, 266
culture, influence on perspectives on land
    and nature, 116

Daschuk, James (historian), 30–31
Davidson, Hannah Francis (missionary), 57–58,
    60–64, 108, 152–53, 291n14
deforestation, 64, 70, 71, 76, 101, 267; climate
    change and, 212–13, 214, 215–16
de Stoppelaar, Reinder Jacobus (poet), 90

diaspora, of Mennonite settlers, 6–7, 246–47;
in Global North, 15–16. *See also* migrations
Dijkstra, Sjoukje (Olympian), 92
Dirksen, Jacob Franz (pastor), 137
Dlodlo, Ndeabenduku (pastor), 111
Doerksen, Abram (bishop), 126, 127–28
Doopsgezinde Historisch Kring, 120
Doopsgezinden. *See* Dutch Mennonites
Driedger, Leo (sociologist), 180
Dube, Bekithemba (historian), 134
Dube, Doris (historian), 135
Dutch colonialism, 47–56, 119, 187, 258–59, 267;
Mennonite attitudes toward, 247
Dutch Mennonite Missionary Society, 45
Dutch Mennonites, 14, 52, 115, 116, 119–23, 226;
Anabaptist heritage, 119–20; assistance to
Russian Mennonite refugees, 247; climate
change responses, 218–22; global engage-
ment, 245–49; landowning class, 20–22;
liberalism, 119; migration narratives, 14–15,
44, 141; peace teachings, 119–20, 122;
relationship with non-Dutch Mennonites,
247, 249; relationship with non-Mennonite
neighbors, 119, 121–22; religion/land
management relationship, 119–20; during
WW II, 246, 247. *See also* Friesland
(Netherlands) Mennonite communities;
Old Colony Mennonites
Dutch Missionary Society, 47
Dyck, Abram J, (memoirist), 40–41, 42

East Reserve, Manitoba, 30, 31–33, 34, 272,
283n71; Choritzer Mennonites, 127
Eckhardt, Mrs. Chas. (women's club leader),
167
ecofeminism, 146–47, 148–49, 175
Ecological Imperialism (Crosby), 15
ecospirituality, 116. *See also* nature-based
spirituality; soil and spirituality
Eicher, John (historian), 10, 81, 199, 272
Elias, George G. (barley grower), 85–86
Elias, Peter A. (memoirist), 34–35
Emerald Horizon: The History of Nature in Iowa
(Mutel), 23–24
Engle, Anna A. (missionary), 109–10, 133–34
Engle, Jesse M. (missionary), 52–58, 62–63, 109,
152–53
Enns, Abram (diarist), 1, 2, 3, 42, 101–4

Enns, George (archaeologist), 86
Enns, Jakob (diarist), 101–2
Ens, Adolf (historian), 126–27, 180
Ens, Gerhard (historian), 34, 181
entrepreneurship, 14, 40, 79, 105, 150
environmental adaptation, of Mennonite
agricultural communities, 73; Iowa, 29;
Manitoba, 33–34; Old Colony Mennonites,
Mexico, 99–100; Riva Palacio, Bolivia, 71–72,
100–101; Siberia, 38–40. *See also* agricultural
practices
environmental determinism, 266–67
environmental ethics, 87. *See also* religion/land
management relationship
environmental history, 3–4, 9, 143
environmental management, 182–83, 197;
Bolivian Mennonites, 214–218; Frisian
Mennonites, 195–96; Iowa Mennonites,
197–98; Manitoba Mennonites, 182–83,
186–87; Mennonites' concern for, 269;
Ndebele people, 212–214; Netherlands, 194,
195–96; state-farmer relations and, 182–83,
186–87, 195–96, 197–98. *See also* agricultural
practices; religion/land relationship
environmental movement, 197, 221
Environmental Protection Agency, 197–98
environmental stewardship, 269; state-farmer
relations and, 182–83, 186–87, 195–96,
197–98. *See also* environmental management;
religion/land management relationship
Epp, Johann (memorist), 42–43, 94, 95
Epp, Marlene, *Women without Men*, 157
Epp, Peter (Piotr), *100 let pod krovom
Vseyshnego*, 136–38, 202, 287n46
Evangelical Mennonite Mission Church
(EMMC), 5, 128, 129
everyday faith, 4, 6, 7, 118, 120, 123–26, 129–30,
174, 207, 268
Eyford, Ryan (historian), 32

Falk, Wilhelm (bishop), 128
farmers' markets, 185–86, 202, 277
Fast, Hans Peter (researcher), 10, 271, 272,
274, vii
Fast, Johann (missionary), 54
Fehr, Jacob (memorist), 177
Fehr, Johann (Bolivia), 142–43, 191, 216, 232–34,
236–37, 239

fertilizers, synthetic, 76, 80, 82, 86, 145–46, 183–84, 188, 189–90; climate change and, 212

Fink, Deborah (historian), 154

First Mennonite Church, Iowa City, 223

Fisher, Susie (historian), 10, 34, 180, 227, 272, 274, vii

Fitz family, 84–85

Five Centuries of Farming (Bieleman), 17–18

food consumption, cultural meaning, 12

Fountain, Philip (anthropologist), 7

Friesen, Aileen (historian), 10, 37–38, 272, 273, 274, 302–3, vii

Friesen, Gerald, *The Canadian Prairies: A History*, 32, x

Friesen, Isaac P. (pastor), 128

Friesen, Katherina (diarist), 170–71, 174

Friesland (Netherlands) Mennonite communities, 1–2, 4, 5, 14, 17–23; agrarian, utopian Anabaptist groups, 19–20; agricultural commodities, 5, 8, 15, 19; agricultural landscape, 17–18; agricultural modernization, 88–93, 120–21, 123, 194; agricultural practices, 19, 43, 266; agricultural technology, 267; class divisions, 22; climate and environment, 17, 280n10; climate change, 218–22; communitarian relationships, 121–22; ecotone and ecozone, 5, 8–9; environmental management, 195–96; global engagement, 235, 245–49, 263; Janjacobsgezinden, 20–21, 267; land ownership, 194–95; Mennonite population, 22; migration narratives, 26–27; municipalities, 22, 281n27; peace ethic, 122–23; poetry about, 90, 93, 115–16, 287n38; reclaimed land, 18–19, 91–92, 193, 194–95; relationships with non-Mennonite neighbors, 122–23; religion/land management relationship, 119–23, 143, 221, 268–69; religious affiliations within, 22; resistance to modernity, 267; soil conditions, 17, 22, 266; soil management, 19, 266; state-farmer relations, 193–96, 207; wealthy urban landowners, 20–22

Froese, P. F. (editor), 95

frontier settlements: state's role, 177. *See also* Iowa Mennonite communities; Manitoba Mennonite communities; Riva Palacio, Santa Cruz, Bolivia; Rural Municipality of Rhineland

fungicides, 133, 172, 218, 239

Funk, Johann (bishop), 127

Gaard, Greta (historian), 147

Geertz, Clifford, *Agricultural Involution*, 49, 108

gendered social integration, 147–48, 163–69, 174

gender equality, 150. *See also* Mennonite women

gender relations, 12. *See also* Mennonite women

Gerbrandt, Henry, *Adventure in Faith*, 126

Gereja Injili di Tanah Jawa (GITJ), 49–50, 104, 108, 145, 174, 190, 260; agricultural modernization, 146; agricultural practices, 145–46; history, 129–30; during independence movement, 106; land ownership, 106; membership, 105; religion/land management relationship, 129–31; women members, 145, 148, 152, 174; during WW II, 129–30, 131

German Baptist Union, 136

Germany: invasion of Russia (1941), 158; Mennonites' migration from, 255, 256–57; Mennonites' migration to, 137–38, 250, 255, 256–57, 263; Mennonites' relationship with, 256–57

GITJ. *See* Gereja Injili di Tanah Jawa

global agricultural markets, 259–60

global community, Mennonites' interactions with, 232–65, 266; agricultural aid and out-reach programs, 241, 242–44, 245–46, 253–54; Dutch Mennonites, 245–49; Global North communities, 241–51, 254–58; Global South communities, 232–33, 235–41, 251–54, 258–62; local cultural factors affecting, 262–64; Washington County, IA, Mennonites, 234, 235, 242–43

global labor markets, 240–41

global Mennonite history, 265–66

Global North, Mennonite communities in, 2; climate change concerns, 12; colonialism and, 15–16; comparison with Global South communities, 46–47; global engagement, 263–64; Mennonite diaspora in, 15–16. *See also* Apollonovka, Siberia; Iowa Mennonite communities; Manitoba Mennonite communities; Rural Municipality of Rhineland; Washington County, IA, Mennonite communities

Global South, Mennonite communities in, 9; agricultural modernization resistance, 266–68; agricultural outreach programs, 243–44, 251; agricultural practices, 46–47; climate change concerns, 12; colonialism and, 46–47, 73; global engagement, 263. *See also* Margorejo, Java; Matopo Mission, Matabeleland, Zimbabwe; Riva Palacio, Santa, Cruz, Bolivia

glyphosate (Round-Up), 86, 87, 229

GMO crops, 2–3, 223, 224, 225, 242

government policies, toward Mennonite farmers. *See* state-farmer relations

government subsidies, 185, 186, 189, 197, 198–200

Green Revolution, 64, 73, 104, 107, 187–90

Green Township Farm Bureau Women's Club, 167–68

Groningen, Netherlands, 14–15

Guengerich, Samuel (diarist), 28

Gupta, Joyeeta (environmentalist), 209

Hall, David (missionary), 105

Hamm, H. H. (municipal secretary), 182

Harker, Ryan D., *Rooted and Grounded*, 117

Harwood, W. S., *New Earth*, 96

Heppner, Jack, *Search for Renewal*, 128

herbicides, 78, 85–86, 98, 103–4, 183, 226, 239; chemical drift, 183; resistance to, 87

Hiebert, Theodore (theologian), 116

Hoekema, Alle (historian), 120, 121

Hoogland, Johannes (letter writer), 88–89, 93

horse-and-buggy lifestyle, 141–42, 143

horse-based agriculture, 79, 88–89

Hottinger, Margret (Anabaptist leader), 26

Howe, E. T. (ag rep), 183–84

Hubert, Johann (missionary), 54

Hughes, J. Donald (historian), 3–4, 10

humility, 6, 267, 268; link to Javanese teachings, 122, 123, 132, 133, 267, 268; as negative, 120, 123, 216; rurality and, 20, 23, 27, 114, 117, 124, 126, 144

imperialism, 8, 15

Indigenous people, environmental ethics, 73

Indigenous peoples' dispossession, 16; Bolivia, 67–68, 193; Brazil, 249–50; Iowa, 23–25, 44; Manitoba, Canada, 30–32, 44, 177, 180–81;

249–50; Mennonites' complicity, 249–50; Pennsylvania, 26; Siberia, 37–38, 44; by technologically superior European farmers, 15–16

Indonesia, 1; independence movement, 130–31, 247; national food-supply policy, 263. *See also* Margorejo, Java

inheritance systems, 49, 146, 181

insecticides, 82, 103–4

international aid programs, 241, 242–43, 245–46, 253–54

Iowa Mennonite communities: agricultural modernization, 77–83, 123; agricultural practices, 43, 123, 266; agricultural technology, 29; ethic of the everyday, 123–26; global engagement, 241–45; international agricultural out-reach, 263; land ownership, 26; migration narratives, 23–27; religion/land ownership relationship, 268–69; settlement history, 23–30; soil management, 27, 266; state-farmer relations, 44, 198, 241–42; during WW II, 124. *See also* Washington County, IA, Mennonite communities

Iowa Mennonite School, 124, 269

Isaacksz, Jacob, 20–21

Islam. *See* Muslims

Jacobs, Jan (minister), 20

Janjacobsgezinden, 20, 267

Janz, Pieter (missionary), 45–46, 47, 50–52

Janz, Pieter Anton (missionary), 45–46, 47, 50–56, 105–6, 133, 148, 259

Janzen, Maria (memoirist), 157, 158–60

Javanese Mennonite Church. *See* Gereja Injili di Tanah Jawa (GITJ)

Jellison, Katherine (historian), 169

Jews, 34, 246

Kasdorf, Julia (poet), 23

Kazakhstan, 250, 251, 257–58

Khan, Ahmed H. (ag rep), 87, 184, 186

Khrushhev, Nikita, 137

Kinsinger, Erlis (diarist), 77–78

Kirghiz people, 39

Klaassen, Walter (theologian), 118

Klassen, Johann (missionary), 54, 55–56

Klassen, Menno (memoirist), 84–85

Kletke, Les (columnist), 87
Klippenstein, Johann (village leader), 95
Klippenstein, Lawrence (historian), 30
Krehbiel, John Carl (settler), 27
Kristiawan, Danang (pastor), 10, 151, 187–88, 260, 272, 273, vii
Kumalo, Manhlenhle (pastor), 134

Lancaster County, PA, 26
land, reimagination of, 15–16
land ownership, 26, 280n18; among Mennonite women, 145, 146, 148–49, 152; communitarian values and, 268; *Strassendorf* system, 101. *See also under specific Mennonite communities*
Lanning, James (anthropologist), 71–73
Latin America: religion/land management relationship, 143–44; Sommerfelder Mennonite migration to, 183
Lenin, Vladimir, 94
Levine, Philippa (historian), 3
local culture, 1; agriculture effects, 1, 265–68; 279n2; global engagement effects, 262–64
Loewen, Isaac D. (Kolkhoz leader), 96
Loewen, Royden, 260, 272
Logengula, King, 56, 134
London Mission Society, 57
Loo, Tina (historian), 178
Loveless, Marilyn D., *Nature and the Environment in Amish Life*, 117–18, 290n13
Low German-speaking Mennonites, 5, 105, 138, 204, 240, 255; diaspora, 100. *See also* Old Colony Mennonites; Riva Palacio, Santa Cruz, Bolivia, Mennonite community
Lutherans, 84

MacFayden, Joshua (historian), 34
Making Peace with the Earth (Shiva), 9
Manitoba Mennonite communities, 30–37; agricultural practices, 32, 34–37, 43, 266; bicameral governance system, 181; comparison with Washington County, IA Mennonites, 32; East Reserve, 30, 31–33, 34, 127, 272, 283n71; global engagement, 235, 263; religion/land management relationship, 268–69; settlement history, 30–37; soil management, 229, 266; state-farmer relations, 32–33, 44, 176–77, 179–84, 206, 207; West Reserve, 30, 31, 32–37, 83, 126–29,

180–81. *See also* East Reserve, Manitoba; West Reserve
Mansholt Plan, 90, 194
Margorejo, Java, 1–2, 3, 4, 5, 8, 73; agricultural modernization, 104–8, 187–90; agricultural practices, 105, 259, 266; agricultural technology, 108; as Christian farm-village, 45–46, 50–56, 261–62; climate and environment, 47–48, 49–50; Dutch colonialism and, 47–56, 187, 258–59, 267; ecozone and ecotone, 5, 8–9, 47–48; global engagement, 235, 249–51, 258–62; Green Revolution, 107, 187–90; independence movement and, 130–31, 247; Japanese WWII occupation, 106; land justice movement, 131; land ownership/land-lease agreement, 49, 52–53, 105–7, 146, 187–89; literacy and education, 50, 51, 52, 53, 104; Mennonite missionaries' leadership, 45–46, 47, 49–56, 54–55; Muslim converts and residents, 45–46, 47, 53, 132, 133, 260–61, 260–62, 289n69; national anti-Communist campaign and, 129–30, 131; nature-based spirituality, 45–46, 47, 51–52, 130, 131–33, 267; post-colonial history, 106–8, 187–90; religion/land management relationship, 129–33, 143, 267, 268–69; residents' occupations, 105, 288n68; settlement history, 47–56; soil conditions, 47–48, 266; state-farmer relations, 50–51, 105, 107, 187–90, 207; village charter, 51–53; women's experiences, 145–46, 147–52, 174. *See also* Gereja Injili di Tanah Jawa (GITJ)
Margorejo Farmers Union, 148
Martyrs Mirror, 22–23, 268
Matopo Mission, Matabeleland, Zimbabwe, 56–64, 73; African leadership, 111–13, 133–34; agricultural modernization, 110–13, 201; agricultural practices, 108–13, 266, 267; climate and environment, 58–59, 61–62, 63–64, 109; climate change, 208–9, 210–14, 266; colonialism, 47, 56–57, 59–60, 63–64, 152, 153; economic crises, 252, 263; global engagement, 252–54; during independence movement and civil wars, 111–12, 113, 263; patriarchy (male privilege), 152–53, 154–56; post-colonialism, 152, 154, 155; post-Zimbabwe independence, 252; racialized

Matopo Mission, Matabeleland, Zimbabwe
(continued)
    segregation, 108, 113, 263, 266; religion/land
        management relationship, 133–36, 143,
        210–13, 267; resistance to modernity, 110–11;
        soil conditions, 58, 59; state-farmer relations,
        200–202, 207; white missionaries'
        leadership, 109–13, 134–35, 152–53, 200,
        291n14; women as farmers, 110, 147–48,
        152–56. See also Ndebele people
McConnell, David L., 117–18, 290n13
McNeill, J. R., 3, 76
Menno, Simons, 6, 17, 18, 20, 120, 121, 140,
    247
Mennonite agricultural communities:
    agricultural commodities, 5, 8; agricultural
    transformation, 2; commonalities, 7–8;
    diversity, 7; ecozones and ecotone differ-
    ences, 5, 8–9; political and national
    influences on, 8, 12. See also Apollonovka,
    Siberia; Friesland (Netherlands) Mennonite
    communities; Iowa Mennonite communities;
    Manitoba Mennonite communities;
    Margorejo, Java; Matopo Mission,
    Matabeleland; Riva Palacio, Bolivia; Rural
    Municipality of Rhineland; Washington
    County, IA, Mennonite communities
Mennonite Brethren Church (Apollonian,
    Siberia), 5, 6, 37, 41, 136–40, 160–61, 203
Mennonite Central Committee, 125, 131, 243,
    244, 245–46, 253–54; Pax program, 243
Mennonite Central Committee (MCC), 100
Mennonite Church USA, 197–98, 222, 242,
    245
Mennonite Committee for Foreign Needs, 14
Mennonite Cooperative Economic Commission,
    107–8
Mennonite Disaster Service (MDS), 245
Mennonite District Service, 125
Mennonite Economic Development Associates
    (MEDA), 243, 244
Mennonites: agricultural background, 20;
    assimilated vs. traditional, 243–44, 245;
    identities, 1, 279n1; origin, 4, 6; values, 4, 6
Mennonite women, 145–75; class and wealth,
    147–52, 174; gendered social integration, 29,
    147–48, 163–69, 174; gender/ethnicity and,

147–48, 156–63, 174; generational succession,
    169–74; land ownership, 145, 146, 148–49, 152;
    as pastors, 119; patriarchal social structure
    and, 12, 110, 146, 148, 149, 150, 169, 170; race
    and, 152–56, 174; religion/land management
    relationship, 160–61, 162, 174; within
    technologically progressive communities,
    147–48, 163–69, 174. See also agricultural
    knowledge, of Mennonite women
Mennonite World Conferences: Third (1936),
    120, 246–47; 2003, 254
Meskwaki, 25, 28
Metis, 30–32, 180
Mexico. See also Old Colony Mennonites
    (Mexico); Old Colony Mennonites'
    migration from, 99; Sommerfelder migration
    to, 127
migrant workers, Mennonites as, 238–41, 250,
    261
migrations: as diaspora, 6; Dutch Mennonites,
    14–15, 44, 88, 141; Friesland (Netherlands)
    Mennonites, 26–27; Iowa Mennonites,
    23–27; migrations from Germany, 255,
    256–57; migrations to Germany, 137–38, 250,
    255, 256–57, 263; Old Colony Mennonites, 63,
    65–70, 99, 100, 141, 143, 232, 236–37; to
    Paraguay, 127, 191, 238, 241, 247, 249–50;
    from Russia, 54–56, 100, 128–29; Swiss
    Mennonites, 15, 25–26, 44
military exemption, 30, 66, 141, 176
Miller, Chester A. (columnist), 82
Miller, Katie (land owner), 75
missionaries, 6; Beachy Amish, 245; Brethren
    in Christ Church (BICC), 134–35, 152–53,
    200; Matabeleland, 109
Moellinger, David (reformer), 27
Moffat, Robert (missionary), 57
monocrop agriculture, 15, 76; conversion
    to monoculture and specialty crops,
    83–87, 182
Moon, David, The Plough that Broke the Steppe,
    39, 197
Morelles, Evo, 193
Morton, W. L. (historian), 181
Mugabe, Robert, 152
Muslims: converts from, 53, 281; as neighbors,
    106, 133, 184, 186, 258

Mutel, Cornelia F., *Emerald Horizon*, 23–24
Mzilikazi, King, 6, 61, 113, 133–34, 252

Nachfolge Christi, 6, 267, 280n15
national food security policies, 7; Bolivia, 64–65, 67–68, 73, 99, 190, 192–93; Indonesia, 263; USA, 197
nature, Anabaptist/Mennonite view, 115–17
Nature and Power (Radkau), 9, 178, 207, 268
Nature and the Environment in Amish Life (McConnell and Loveless), 117–18, 290n13
nature-based spirituality, 116, 267, 289n3; Javanese, 45–46, 47, 55–56, 130, 131–33, 207; Ndebele people, 47, 59–60, 61, 62, 63–64. *See also* religion/land management relationship; soil and spirituality
Ncube, Belinda (researcher), 10, 58, 112, 154, 211, 272, 273, vii
Ncube, Mazilbopela (deacon), 133–34
Ndebele people, 56–59; agricultural practices, 56–59, 61, 62, 63; BICC white missionaries impact, 134–35; forced relocation, 111–12; foreign aid to, 253–54; Gukurahundi massacres, 152; land ownership, 200, 201; nature-based spirituality, 113, 153–54, 210; poverty, 252, 253, 254; pre-Christian history, 252; religion/land management relationship, 133–36; soil management, 202; state-farmer relations, 200–202; women as farmers, 152–56, 174
Ndlovu, Scotch Malinga (historian), 112–13, 200
Netherlands, 1; agricultural expertise, 248–49; colonialism in Indonesia, 47–56, 50, 51, 119, 187, 258–59, 267; environmental management, 194, 196; environmental movement, 221; Golden Age, 21. *See also* Dutch Mennonites
Neufeld, Mary (biographer), 128
New Earth (Harwood), 96
New Russia, Mennonite communities, 37, 38–40, 41–43, 44
NGOs (nongovernmental organizations), 107, 131, 241, 253–54
Nkomo, Joshua (politician), 200
Nobbs-Thiessen, Ben, 10, 237, 272, 67–68, 285n60

nonviolence/nonresistance, 4, 6, 119–20, 122–23, 140–41, 267, 268
no-till agriculture, 223

Old Colony Mennonites: rejection of modernity, 64, 141–42, 143, 171. *See also* Riva Palacio, Santa Cruz, Bolivia
Old Colony Mennonites (Mexico), 101, 127, 215; agricultural practices, 102; environmental adaptation, 99–100; migrations from, 99, 141, 143, 232, 236–37; migration to, 99, 100, 141
Old Flemish Mennonite Church, 14
Old Frisians, 119
Old Order Amish, 6, 29, 222, 225–26
100 let pod krovom Vseyshnego (Epp), 136–38, 287n46
oral history, 11; ethics protocol, 11; Manitoba Mennonites, 184–87
organic farming, 125, 129, 220, 225
Organization of Rural Associations for Progress (ORAP), 253

pacifism, 27, 72, 130–31, 140–41, 190, 280n15; rejection by Dutch Mennonites, 119; rejection by Javanese Mennonites, 106; rejection by Ndebele people, 112–13
Palatinate, Mennonite migrations from, 26–27, 44
Pan-Russian Mennonite Agricultural Society, 95–96, 98
Paraguay: Mennonite migrations from, 250; Mennonite migrations to, 127, 191, 238, 241, 247, 249–50; Old Colony Mennonites in, 238
Paz Estenssoro, Victor (president), 64, 99
peace ethic, 6, 122–23, 279n13, 279n14
Peasants' War, 4, 6
Peenstra, Siobe (memoirist), 91–93
Penn, William, 26
Pennsylvania, Amish and Mennonite communities, 26
Peter, Samuel (farmer), 14–15
Peters, Bernhard F. (bishop), 70, 140–42
Peters, Johann K. (diarist), 68
Peterson, Duane (diarist) 78
Plough that Broke the Steppe, The (Moon), 39, 197
Polanyi, Michael, 265

Radkau, Joachim, *Nature and Power,* 9, 178, 207, 268

Rahn, Catarina Dueck (memoirist), 41–42

railroads, transcontinental, 37, 41, 44, 96

Ranger, Terence, *Voices from The Rocks,* 59–60

Reber, John (estate), 28

Redekop, Calvin, *Creation and the Environment,* 116

Reformed Church, 22, 119, 121, 122

religion/land management relationship, 7–8, 16, 17, 44, 115–44, 280n18; climate change and, 221, 224–27; as cultural sustainability, 268–69; ethic of the everyday, 123–26; Global North, 73–74; Global South, 73–74. *See also under specific Mennonite communities*

Reschly, Steven (historian), 169

Reyndertz, Tjaert (martyr), 20

Rhineland Agricultural Society, 182, 185, 267

Rhodes, Cecil, 56–57

Rhodesia, 46, 56, 57, 60–61, 109, 111–12. *See also* Matopo Mission, Matabeleland; Zimbabwe

Richards, John F., *Unending Frontier,* 9, 16

Riel, Louis (Metis leader), 30

Riva Palacio, Bolivia, 1–2, 3, 4, 5, 8, 12, 64–73; agrarian simplicity, 236–37; agricultural modernization, 99–104; agricultural practices, 101, 102–3, 104, 170, 171, 172–73, 216–18, 236, 237–39, 244; agricultural technology, 103, 232–33, 236–37, 239–40; chemical-based agriculture, 239; climate and environment, 66–67, 100, 102–3, 170, 171–72, 239, 266; climate change response, 214–18; communitarian values, 237, 263, 267; ecotone and ecozone, 5, 8–9, 215–16; environmental adaptation, 100–101; environmental management, 191–92; global engagement, 232–33, 234–41, 235–41, 263; horse-and-buggy lifestyle, 141–42, 143; Iowan Mennonites' attitudes toward, 243–44; land ownership, 236; migration from Russia, 100; ongoing relationship with Canada, 239–41; parent community, 101; postcolonialism and, 67–68; rejection of modernity, 64, 141–42, 143, 171; religion/land management relationship, 140–43, 144, 268–69; resistance to modernity, 266, 267–68; settlement history, 64–73; soil conditions, 67, 266; soil

management, 216–18; state-farmer relations, 47, 64–66, 64–68, 67–69, 70, 71, 99, 140, 141, 143, 144, 190–93, 235–36; women's experiences, 147–48, 169–74

Rooted and Grounded (Harker and Johnson), 117

Ropp, Ernie (equipment dealer), 81

rural-development programs, 241, 242–44

rural electrification, 85, 91, 93, 122

Rural Municipality of Rhineland, Manitoba, 1–2, 3, 4, 5, 34, 181, 267; agricultural commodities, 5, 8; agricultural moderniza-tion, 83–87, 181–84, 185, 267; agricultural practices, 267; Bergthaler-Sommerfelder schism, 126–29; chemical-based agriculture, 183–84; climate change response, 226–30; colonialism and, 16; ecotone and ecozone, 5, 8–9; environmental management, 182–83; global engagement, 249–50; land ownership, 181–82; monocrop conversion to monocul-ture, 83–87; Neubergthal community, 35, 128–29, 272; religion/land management relationship, 127–29, 143–44, 269; settlement history, 30–37, 249; soil management, 182–83, 267; state-farmer relations, 179–87

Russia, 1; Mennonite migrations from, 54–56, 100, 128–29, 250; Mennonite missionaries from, 54–56. *See also* Apollonovka, Siberia; Siberian Mennonite communities

Russian Mennonites, as refugees from Soviet Union, 247

Salamon, Sonya (sociologist), 164

Saritruno, Samuel, 148

Sauk nation, 24–25

Schroeder, David (theologian), 127

Schroeder, Nina (historian), 20

Scott, James C., 267

Search for Renewal (Heppner), 128

settlement histories, of Mennonite communi-ties, 249. *See also* settlement history *under specific Mennonite communities*

"Seven Points on Earth" team workshops, 11

Shantz, Jacob Y., 31–32

Shiva, Vandana, *Making Peace with the Earth,* 9

Sibanda, Eliakim (historian), 60

Siberian Mennonite communities, 37–43; agricultural modernization, 93–98;

agricultural practices, 37, 38–40, 43, 93–98, 266, 267; climate and environment, 37, 39, 40, 41, 266; colonialism and, 37–38; environment, 42; migration narratives, 37; resistance to collectivization, 267; settlement history, 37–43; soil conditions, 39, 40, 266; state-farmer relations, 37–38, 39–40, 44
Siemens, J. J. (reformer), 185, 251
Simmons, L. G. (historian), 4, 233
Simons, Menno. *See* Menno, Simons
simplicity, relationship to agrarian life, 6, 15–16, 26–28, 30, 44, 47, 168, 268, 279n14; among Dutch and Friesland Mennonites, 14, 15, 17, 19–20, 23, 119, 121; among Old Colony Mennonites, 47, 140–41, 143, 169, 190, 215, 216, 232, 235, 236; among Sommerfelders, 127; among Swiss-descendant Mennonites, 26–27; among the Amish, 117–18, 123, 197; as environmental value, 168, 197
Smith, Ian, 200, 201
Smith-Howard, Kendra (historian), 197
social justice, 121–22, 200
Soemareh, Wirjo (village leader), 106
soil, human interaction with, 12–13
soil and spirituality, 23, 60, 110, 134–136, 156
soil management: faith-based approach, 125–26; Midwest US, 197. *See also under specific Mennonite communities*
soil types: alluvial clays and sand, 9, 67; black lacustrine clay, 3, 8, 33, 36, 84, 183, 229; loess soils, 8, 22, 123, 266; peat, 18, 58; savannah-based soils, 58, 109; semi-arid chernozem soils, 39, 205, 266; subterranean soils, 8, 17, 33, 266; tropical (i.e. Grumosols, Regasols, Latosols), 48, 266
Something New Under the Sun (McNeill), 3
Sommerfelder Mennonites, 127–29, 182, 183
Stalin, Joseph, 94, 95, 158, 205
Staples, John (historian), 39
state biopower. *See* state-farmer relations
state-farmer relations, 12, 176–207, 265; complementarity in, 179–84, 206; environmental management and, 182–83, 186–87, 195–96; government subsidies, 185, 186, 189, 197, 198–200. *See also under specific Mennonite communities*
Stunden Bower, Shannon (historian), 32
Suharto (president), 188

Sukarno (president), 187, 188, 247
Sukoco, Sigit Heru. *See* "Way of the Gospel in the World of Java"
Susquehannocks, 26
sustainability, 9–10, 268; Anabaptist theology and, 117; cultural, 268–69
Sutter, Paul S. (historian), 9
Swiss Anabaptists, 4, 6, 15, 25–26
Swiss Church Aid of Bulawayo, 273
Swiss Mennonites: in Friesland, 14–15; in Iowa, 23, 25; migrations, 15, 25–26, 44; in Palatinate and Alsace Loraine, 26–27, 44
Swiss-South German Anabaptists, 14–15
Swiss-South German Mennonites, 26–27, 79, 272

technological change, 12. *See also* agricultural technology
Thiessen, Nickolai (missionary), 54–55, 105–6
Toews, Isaac (village mayor), 95
Toews, John B. (historian), 39–40
tractors: controversial, 64, 72, 102, 123, 142, 216, 237; expansion, 90, 97; hand tractors, 108, 146, 190; importation, 232, 239, 248; innovation, 76, 91, 93, 244; operation, 92, 138, 160, 204, 206
transnationalism. *See* global community
Trompetter, Cor (historian), 21, 22
Tunggul Wulung, Ibrahim (mystic), 45–46, 50, 51, 53, 133

Unending Frontier (Richards), 9, 16
United States, 1; agricultural exports, 241–42; soil, 242
University of Winnipeg, Mennonite Studies, 10
Urban-Mead, Wendy (historian), 219n12

Van der Zijpp, Nanne (historian), 119
Veenhuizen, Geert (potato breeder), 15
Verbeek, Annelise (pastor), 120, 121
Visser, Piet (historian), 22, 23, 271
Visser-Oosterhof, Renske (poet), 90, 93
Voices from The Rocks (Ranger), 59–60

Waldheim. *See* Apollonovka, Siberia
Warkentin, John, 31, 33, 286n20

Washington County, IA, Mennonite communities, 1–3, 4, 5, 23–30; agricultural commodities, 5, 8; agricultural modernization, 29, 77–83, 169; Anabaptist heritage, 25–27; climate and environment, 164; climate change response, 222–26; colonialism and, 16; comparison with Manitoba Mennonites, 32; ecotone and ecozone, 5, 8–9; global engagement, 234, 235, 242–43; migration narratives, 23, 25–30; religion/land relationship, 124–26, 143–44; settlement history, 23–30; state-farmer relations, 196–200, 207; urban markets, 25; women's gendered social integration, 147–48, 163–69, 174

Waterlanders, 20

"Way of the Gospel in the World of Java" (Sukoco and Yoder), 50, 53, 54, 104–5, 107, 130, 131, 284n2

weather: cyclones, 214; deluges, 59, 102, 156, 215, 220, 223, 228; drought, 59, 60, 131, 134, 143, 156, 171, 182, 188, 208, 212, 214ff, 223ff, 258; dust and sand storms, 39, 142, 216; hail, 101, 228; hurricanes, 93; snow falls and snow storms, 40, 163, 165, 206, 211, 229; storms, 172, 219, 221, 275, 277; tornadoes, 223, 228; tsunamis, 261

Weaver, John (historian), 15–16

Weaver-Zercher, David (historian), 23

Wendell, Berry, 1, 116, 118

Werner, Hans (historian), 40

West Reserve, Manitoba, 30, 31, 32–37, 83, 126–29, 180–81

white cultural privilege, 153–54, 291n15

Wiebe, Johann (memoirist), 68–71, 72, 73

Wiens, Peter (entrepreneur), 40

Wittrig, Nettie (diarist), 164–65, 292n42

Women. *See* Mennonite women

Worster, Donald (historian), 9, 233

Yayasan Kerkasama Ekonomi Muria (YAKEM), 108

Yoder, Dan W. [Bill Jr.], 82, 286n15

Yoder, Franklin (historian), 124

Yoder, Lawrence M. *See* "Way of the Gospel in the World of Java"

Yoder Lind, Kate (historian), 29

Zernike, Anne (pastor), 119

Zimbabwe, 1; economy, 156; independence movement, 200, 252; women's role, 152. *See also* Matopo Mission, Matabeleland, Zimbabwe

Zimbabwe Africa National Union (ZAPU), 200–201

Zimbabwe Africa People's Union (ZAPU), 200–201

Zulu Empire, 56–57, 252